ENGINEERING STRATEGIES AND PRACTICE I & II

2009/2010

A Wiley Canada Custom Publication
for the

University of Toronto

WILEY

Wiley Canada Custom Services

JOHN WILEY & SONS CANADA, LTD.

Cover Photo:
Beer—Cover Photo: © IT Stock

Marketing Manager: Anne-Marie Seymour
Custom Coordinator: Rachel Coffey

Printed and bound in the United States of America
10 9 8 7 6 5 4 3 2 1

John Wiley & Sons Canada, Ltd
6045 Freemont Blvd.
Mississauga, Ontario
L5R 4J3
Visit our website at: www.wiley.ca

Dym and Little, Engineering Design:
A Project-Based Introduction, 3rd Edition
&
Selected chapters from
Beer and McMurrey, A Guide to
Writing as an Engineer, 3rd Edition

THIRD EDITION

ENGINEERING DESIGN: A PROJECT-BASED INTRODUCTION

CLIVE L. DYM and PATRICK LITTLE
with
ELIZABETH J. ORWIN and R. ERIK SPJUT
Harvey Mudd College

JOHN WILEY & SONS, INC.

EXECUTIVE PUBLISHER	Don Fowley
ASSOCIATE PUBLISHER	Dan Sayre
ACQUISITIONS EDITOR	Michael McDonald
EDITORIAL ASSISTANT	Rachael Leblond
EXECUTIVE MARKETING MANAGER	Chris Ruel
SENIOR PRODUCTION EDITOR	Ken Santor
COVER DESIGNER	Miriam Dym

This book was set by Thomson Digital Inc. and printed and bound by Malloy, Inc. The cover was printed by Malloy Inc.

This book is printed on acid free paper. ∞

ISBN 978-0-470-22596-7

Printed in the United States of America

10 9 8 7 6 5 4 3 2 1

To

Joan Dym
whose love and support are distinctly nonquantifiable

cld

Charlie Hatch
a teacher's teacher

pl

Carl Baumgaertner
who inspired me to teach

ejo

Karen Spjut
who has been there through sickness and health

res

CONTENTS

FOREWORD*

To design is to imagine and specify things that don't exist, usually with the aim of bringing them into the world. The "things" may be tangible—machines and buildings and bridges; they may be procedures—the plans for a marketing scheme or an organization or a manufacturing process, or for solving a scientific research problem by experiment; they may be works of art—paintings or music or sculpture. Virtually every professional activity has a large component of design, although usually combined with the tasks of bringing the designed things into the real world.

Design has been regarded as an art, rather than a science. A science proceeds by laws, which can sometimes even be written in mathematical form. It tells you how things must be, what constraints they must satisfy. An art proceeds by heuristic, rules of thumb, and "intuition" to search for new things that meet certain goals, and at the same time meet the constraints of reality, the laws of the relevant underlying sciences. No gravity shields; no perpetual motion machines.

For many years after World War II, science was steadily replacing design in the engineering college curricula, for we knew how to teach science in an academically respectable, that is, rigorous and formal, way. We did not think we knew how to teach an art. Consequently, the drawing board disappeared from the engineering laboratory—if, indeed, a laboratory remained. Now we have the beginnings—more than the beginnings, a solid core—of a science of design.

One of the great gifts of the modern computer has been to illuminate for us the nature of design, to strip away the mystery from heuristics and intuition. The computer is a machine that is capable of doing design work, but in order to learn how to use it for design, an undertaking still under way, we have to understand what the design process is.

We know a good deal, in a quite systematic way, about the rules of thumb that enable very selective searches through enormous spaces. We know that "intuition" is our old friend "recognition," enabled by training and experience through which we acquire a great collection of familiar patterns that can be recognized when they appear in our problem situations. Once recognized, these patterns lead us to the knowledge stored in our memories. With this understanding of the design process in hand, we have been able to reintroduce design into the curriculum in a way that satisfies our need for rigor, for understanding what we are doing and why.

* Herb Simon graciously contributed this foreword to our first edition. Unfortunately, the passage of time since was marked by the loss of one of our great heroes, a true Renaissance mind, because Herb passed away on 4 March 2002. We still feel the loss.

One of the authors of this book is among the leaders in creating this science of design and showing both how it can be taught to students of engineering and how it can be implemented in computers that can share with human designers the tasks of carrying out the design process. The other is leading the charge to integrate the management sciences into both engineering education and the successful conduct of engineering design projects. This book thus represents a marriage of the sciences of design and of management. The science of design continues to move rapidly forward, deepening our understanding and enlarging our opportunities for human-machine collaboration. The study of design has joined the study of the other sciences as one of the exciting intellectual adventures of the present and coming decades.

Herbert A. Simon
Carnegie Mellon University
Pittsburgh, Pennsylvania
August 6, 1998

PREFACE

Why should you read this book on engineering design?

Writing a later edition of a text presents both an opportunity and a challenge. Put simply, things change. The world certainly has changed since the first edition of this book appeared in 1999. Events such as the 9/11 attack changed our sense of security and independence, and made us rethink how designed artifacts like passenger planes can be misused and about the unexpected conditions buildings might encounter. Wars in Iraq and Afghanistan have shown how biomedical technologies can save even catastrophically wounded individuals, calling forth technical skills that can help disabled veterans return to productive and satisfying lives. The rise of the global economy has shifted production around the world and made communication and teamwork skills even more important elements of professional engineering practice. Our increased awareness of global warming has led designers to think much more broadly about the entire life cycle of product design, from inception, through use, and even unto disposal.

The engineering profession has responded and adapted to these larger forces, and engineering education has been both a driver and passenger in that adaptation. At the time of our first edition, project-based, first-year design courses were considered controversial, if not impossible or meaningless. Now such courses are offered by many engineering programs, and we are proud to have helped bring that curricular adaptation to life. Correspondingly, capstone courses were often undertaken more in response to accreditation needs than a desire for real world projects. Today capstone courses, such as Harvey Mudd College's Engineering Clinic, not only give students authentic design experience, but they introduce them to working with students from around the world. The students in the classroom or design studio have also changed: Many more women and underrepresented minority students now major in engineering — and the faculty teaching engineering has also begun to change similarly.

Finally, as authors we have changed, becoming not only older, but also, we hope, wiser. The broader changes in the world and the engineering community have given us the chance to see which of the design ideas we taught worked well, which needed refinement, and which worked not at all. We have tried to adapt this edition to these circumstances and to our increased knowledge of the world, the engineering profession, and the educational mission.

Of course, many things have not changed at all. It has always been the case that engineering design requires a careful ear to the wishes of the client, users and the larger public. It will always be the case that engineers must organize their processes into ways that communicate the geography of design spaces to their design partners. That design practice should mean well managed design activities conducted by effective teams whose members respect one another is not a new idea, and it is unlikely ever to grow out of fashion. Perhaps most of all, a commitment to ethical design by and on behalf of a diverse community must remain at the forefront of what it is we do as engineers.

This third edition offered an opportunity to change and update the book in a way that reflects a changed world without merely reacting to current events, fashion, or styles. The challenge was much the same — to guide young engineers in their development to understand the context, content, and skills of design in order that they can be responsible members of this changing world.

Today there are a lot of books on design, engineering design, project management, team dynamics, project-based learning, and the other topics we cover in this volume. This book grew out of our desire to combine these topics in a single, introductory work that focused particularly on conceptual design. That original desire arose from our teaching at Harvey Mudd College, where our students do team-based design projects in a first-year design course, *E4: Introduction to Engineering Design*, and in the Engineering Clinic. Clinic is an unusual capstone course taken during junior (for one semester) and senior (for both semesters) years in which students work on externally sponsored design and development projects. In both E4 and Clinic, Mudd students work in multidisciplinary teams, under specified time deadlines, and within specified budget constraints. These conditions replicate to a significant degree the environments within which most practicing engineers will do most of their professional design work. In looking for books that could serve our audience, we found excellent texts covering detailed design, usually targeted toward senior capstone design courses, or "introductions to engineering" that focused on describing the branches of engineering. We could not find a book that introduced the processes and tools of conceptual design in a project or team setting that we found suitable for first- and second-year students. Since our original inspiration to write this book, other more "skills-oriented" texts and series have come onto the market. While these are no doubt valuable to many educators, we think there remains a place for a book that addresses our original concerns, allowing the teacher to guide students in the design process while giving them freedom to add their own focus.

In writing all three editions of this book we confronted many of the same issues that we discuss in the pages that follow. It was important for us to be very clear about our overall objectives, which we outline below, and about the particular objectives we had for each chapter. We asked about the pedagogic function served by the various examples, and whether some other example or tool might provide a better means for achieving that pedagogical function. The resulting organization and writing represents our implementation of our best design. Thus, this and all books are definitely designed artifacts: They require the same concern with objectives, choices, functions, means, budget, and schedule, as do other engineering or design projects.

This book is directed to three related audiences: students, teachers, and practitioners. While each group has its own special concerns, those interests are intimately tied together by the topics of this book.

The book is intended to enable *students* to learn the nature of design, the central activity of engineering, either directly or as an ultimate aim. (Even the most focused materials engineers, for example, hope that any new materials they develop will be used in the design of *something*.) We also hope the book will help students learn the tools and techniques of formal design that will be useful in framing the design problems they will face during their education and during their careers. The issues and tools of project management will also be faced by graduates when they go on to careers in industry, government, and academe. Since design and research are increasingly done in teams, the insights and tips on team dynamics will be valuable as young engineers reflect on their occupational life as well. We have included examples of work done by our students on actual projects in E4, both to show how the tools are used and to highlight some frequently made mistakes. We hope student readers will take heart from these illustrations and use them in learning to be effective engineers (or at least effective engineering students).

Given that we are professors at an undergraduate college of engineering and science, we wrote this book with *teachers* very much in mind. We considered both the delivery of this material to students and the ways in which professors might teach introductory design courses. Thus, this book is structured to allow a teacher to use ongoing examples for illustration and as homework (or in-class) exercises. The material is ordered so that professors can decide for themselves whether to cover the ideas in the text prior to or concurrently with particular stages in design projects. We have tried both approaches in our courses and find that each has its benefits. An accompanying *Instructor's Manual* outlines sample syllabi and organizations for teaching the material in the book, as well as additional examples. We also discuss the structure of the material in the book later in the preface to help teachers decide how to use the book.

Finally, we hope the book will be useful to *practitioners*, either as a refresher of things learned or as an introduction to some essential elements of conceptual design that were not formally introduced in engineering curricula in years past. We do not assume that the examples given here substitute for an engineer's experience, but do believe that case studies show the relevance of these tools to practical engineering settings. Some of our friends and colleagues in the profession like to point out that the tools we teach would be unnecessary if only we all had more common sense. Notwithstanding that, the number and scale of failed projects suggests that common sense may not, after all, be so commonly distributed. In any case, this book offers both practicing engineers (and engineering managers) a view of the design tools that even the greenest of engineers will have in their toolbox in the coming years.

SOME REMARKS ON VOCABULARY AND WORD USAGE

There is no engineering design community that transcends all engineering disciplines or all types of engineering practice. For that very reason, words are used differently in different domains, and so differing technical jargons have developed. Since our intent was to provide a unified shared understanding that would be a useful foundation for all of our students' future design work, both in their remaining formal studies and in their chosen fields of endeavor, we chose to begin our discussions of the major concepts and terms of art with formal dictionary definitions (largely drawn from (Wolf 1977)). We did this for two reasons. First, to remind both readers (and practitioners) that word usage does have its roots in a shared understanding of vocabulary, and in our case, the English vocabulary. Even technical jargon has — or should have — a visible and traceable path back toward common usage. Thus, in this third edition we have worked very hard to be as crisp and consistent as possible with the words we have chosen to use.

Second, it is clear that word usage has flourished differently across the many domains of engineering practice. For example, different authors (in both the research literature and in textbooks) define different phases of the design process with varying activities occurring within them. Thus, we have worked very hard to provide as crisp an articulation as possible of our model of the design process (see Chapter 2). Still further, the notions of *design requirements* and *design specifications* seem to be evolving somewhat rapidly. Thus, we choose to speak in terms of *design requirements* that specify in engineering terms how a design is supposed to demonstrate the behavior of its *functions* and, where appropriate, of any *attributes* the design is supposed to display.

SOME SPECIFICS ABOUT WHAT'S COVERED

Design is an *open-ended* and *ill-structured* process, by which we mean there is no unique solution, and that the candidate solutions cannot be generated with an algorithm. As we discuss in the early chapters, designers have to provide an orderly process for organizing an ill structured design activity in order to support making decisions and tradeoffs among competing possible solutions. As such, algorithms and mathematical formulations cannot replace the imperative to understand the needs of various stakeholders (clients, users, the public, and so on), even if those mathematical tools are used later in the design process. This lack of structure and available formal mathematical tools make the introduction of conceptual design early in the curriculum possible and, we think, desirable. It provides a framework in which engineering science and analysis can be used, while not demanding skills that most first- and second-year students have not yet acquired. We have, therefore, included in this book a number of specific tools for conceptual design, for acquiring and organizing design knowledge, and for managing the team environment in which design takes place.

The following *formal conceptual design methods* are delineated:

- objectives trees
- establishment of metrics for objectives
- pairwise comparison charts (PCCs)
- functional analysis
- morphological ("morph") charts
- function-means trees
- development of requirements

Since both the framing or defining of a design problem and conceptual design thinking require and produce a lot of information, we introduce a variety of means of acquiring and processing information, including literature reviews, brainstorming, synectics and analogies, user surveys and questionnaires, reverse engineering (or dissection), simulation and computer analysis, and formal design reviews.

The successful completion of any design project by a team requires that team members estimate a project's scope of work, schedule, and resources early in the life of the project. To this end, we introduce several *design management tools*:

- work breakdown structures (WBSs)
- linear responsibility charts (LRCs)
- schedules
- budgets

We also discuss several topics, some new and some with extended content, that we feel are increasingly important in a first exposure to design. In a new Chapter 6, we have for the first time introduced some ideas about the *modeling and analysis of designs*, placed in the context of doing preliminary and detailed design. The material will introduce the reader to the basic principles of mathematical modeling, thus reinforcing concepts behind applying mathematics and physics to engineering. Then we go on to illustrate a few of the kinds of calculations that might be done in the preliminary and detailed phases of design. Our illustration vehicle is the design of a basic rung or step for a ladder, and we apply some results from elementary beam theory and some basic aspects of materials selection. Needless to say, in one chapter and in the kinds of courses that this book is aimed toward, we could not delve into preliminary and detailed design in all engineering disciplines. What we do present is, we feel, representative of the ''good habits of thought'' needed for modeling and analyzing designs in all disciplines.

We discuss the endgame and completion of a design project, with a strong emphasis on the ways and means of reporting design results, in Chapters 8 and 9. These chapters allow instructors to focus on engineering communication as an integral part of the design process, including engineering drawings, reports, and presentations. Thus, we have included a *new discussion about building physical models and prototypes* in Chapter 7 and a *new discussion of drawing* in Chapter 8. This was done in part because of our desire to bring together some basic skills need in design, such as communicating through drawings by adhering to appropriate standards and conventions (e.g., geometric dimensioning and tolerances). In addition, we have also observed in our own students that they no longer start college — if they ever did — with much hands-on experience even in basic woodcraft. Thus, since we expect our students to build elementary (physical) models and prototypes, it seemed useful to include some cautionary notes about ways to work in a shop or laboratory, as well as some basic tips on how to actually make (and fasten!) some basic wooden parts.

In Chapter 11 we discuss ''design for *X*'' issues, including manufacturing and assembly, affordability (engineering economics), reliability and maintainability, sustainability, and quality. This chapter provides a vehicle for faculty who want to expand on these topics and lead students into issues such as concurrent design, DFM, or emerging areas such as sustainability and carbon footprints.

We top off our exploration of engineering design with our capstone Chapter 12, where we discuss important ethics issues in design. This chapter reflects a wider notion of engineering ethics than in the past, as we invite faculty to address traditional notions of liability and responsibility and also newer ideas of the social and political dimensions of engineering design.

INTEGRATIVE DESIGN EXAMPLES

As a rare, if not unique, feature, we use one case study and two integrative examples to follow the design process through to completion, thus showing each of the tools and techniques as they are used on the same design project. In addition to numerous one-time examples, we detail the following case study and integrative examples:

1. Design of a *microlaryngeal surgical stabilizer*, a device used to stabilize the instruments during throat surgery. This case study derives from a Harvey Mudd College first-year design project sponsored by the Beckman Laser Institute of the University of California Irvine.

2. Design of a *beverage container*. The designer, having a fruit juice company as a client, is asked to develop a means of delivering a new beverage to a market predominantly composed of children and their parents. There are clearly a number of possibilities (e.g., mylar bags, molded plastics), and issues such as environmental effects, safety, and the costs of manufacturing are considered.

3. Design of a *support arm* to be used by very young students diagnosed with cerebral palsy (CP). The support arm was designed by teams of Harvey Mudd College students in our E4 design course. Prototypes were subsequently built by the students and delivered to the Danbury School, a special education elementary school within the Claremont Unified School District of Claremont, California.

Finally, an accompanying *Instructor's Manual* includes a case study of the design of a *transportation network* to improve enable automobile commuter traffic between Boston and its northern suburbs, through Charlestown, Massachusetts. This conceptual design problem clearly illustrates the many factors that go into large-scale engineering projects in their early stages, when choices are being made among highways, tunnels, and bridges. Among the design concerns are cost, implications for future expansion, and preservation of the character, environment, and even the view of the affected neighborhoods. This project is also an example of how conceptual design thinking can significantly influence some very "real-world" events.

As noted at the outset, this edition has presented both an opportunity and a challenge for us as authors. We now share those with our readers.

Clive L. Dym
Patrick Little
Elizabeth J. Orwin
R. Erik Spjut
Claremont, California
September 16, 2007

ACKNOWLEDGMENTS

A book like this does not get written without faith, support, advice, criticism, and help from many people. We continue to be grateful to the many colleagues and friends who helped us bring the first two editions to fruition, and our thanks are detailed in those prior edition.

For this third edition, we want to add the following thanks to:

The *HMC E4 student design teams* that developed the design work products that we used as illustrations. Those teams and their projects are listed in our bibliography as: (Attarian et al. 2007), (Best et al. 2007), (Both et al. 2000), (Chan et al. 2000), (Feagan et al. 2000), and (Saravanos et al. 2000).

Jamal Ahmad, Petroleum Institute (UAE), for candid, incisive comments resulting from classroom use of the second edition.

David C. Brown, Worcester Polytechnic Institute, for trenchant comments on our model of the design process.

Miriam Dym for designing the cover for this third edition.

Joe Hayton, formerly of John Wiley & Sons and now at Elsevier Scientific Publishers, for his continuing support of this project from its inception through the first two editions, and for initiating and fostering the third.

Mike McDonald of John Wiley & Sons who has become our new editor and champion.

Greg S. Moore, Vice President of Engineering of DeWalt Tools, for providing useful insights about and graphics of the DeWalt DW21008K corded power drill.

Katherine Elizabeth ("Katie") Near, HMC'10, for an assiduous reading of the typescript.

Carl A. Reidsema, University of New South Wales, for a careful review of several early chapters.

Ken Santor and his production staff at Wiley for once again making a manuscript into a lovely book.

The late *Herbert A. Simon* of Carnegie Mellon University, whose foreword and generous encouragement of CLD continue to provide inspiration.

Darin Barney and others at McGill University provided a collaborative environment for PL to explore the social and political dimensions of engineering design on a recent sabbatical, which is reflected in the new Chapter 12.

Finally, *to each and all of our spouses and families,* for tolerating us during the absences that such projects entail, and for listening to each of us as we worked through differences to find a common voice.

CHAPTER 1

ENGINEERING DESIGN

What does it mean to design something? How does engineering design differ from other kinds of design?

PEOPLE HAVE been designing things for as long as we can "remember" or archaeologically uncover. Our earliest ancestors designed flint knives and other basic tools to help meet their most basic needs. Their wall paintings were designed to tell stories and to make their primitive caves visually more attractive. Given the long history of people designing things, it is useful to wonder what an engineer who designs a building's structure does differently than a decorator who designs the building's interior decor. We will use this chapter to set some contexts for engineering design and to start developing both a vocabulary and a shared understanding of what we mean by engineering design.

1.1 WHERE AND WHEN DO ENGINEERS DESIGN?

There are a lot of questions we could ask about engineers doing design, and there are probably more answers than there are questions: What does it mean for an *engineer* to design something? When do engineers design things? Where? Why? For whom?

An engineer might work for a large company that processes and distributes various food products, where she could be asked to design a new container for a new juice product. He could work for a design-and-construction company, for which he is designing part of a highway bridge embedded in a larger transportation project. An engineer might work for an automobile company that wants to develop a new concept for the instrumentation cluster in its cars, perhaps to enable drivers to check various parameters without taking their eyes off the road. Or an engineer could work for a school system that wants to design specialized facilities to better serve students with orthopedic disabilities.

Clearly, this list could easily be lengthened, so it is worth asking: Are there common elements in the engineers' situations, or in the ways that they do their designs? In fact, there are common features both in their situations and their design work, and these commonalities make it possible to describe a design process and the context in which it occurs.

We can start by identifying three "roles" being played as the design of a product unfolds. Obviously there is the *designer*. Then there will be a *client*, the person or group or company that wants a design conceived, and there is the *user*, the person (or the set of people) who will actually use whatever is being designed. For the working engineer, the client could be internal (e.g., the person who decides the food company should start selling a new juice product) or external (e.g., the government agency that contracts for the new highway system). And while the designer may relate differently to internal and external clients, in either case it is the client who presents a *problem* or *project statement* from which all else flows. Design project statements may be oral, and often they are quite short. These two qualities suggest that the designer's first task is to clarify what the client really wants and translate it into a form that is useful to her as an engineering designer. We'll say more about this in Chapter 3 and beyond, but we want to emphasize that *a design is motivated by a client* who wants some sort of device, system, or process.

Design is motivated by a client.

The user is the third player or stakeholder in the design effort. In the contexts mentioned above, the users are, respectively, consumers who buy the new juice drink, drivers on the interstate highway system, drivers of the new line of autos, and students with orthopedic disabilities (and their teachers). The users hold a stake in the design process because a product won't sell if its design doesn't meet their needs. Thus, the designer, the client, and the user form a triangle, as shown in Figure 1.1. The designer has to understand what the client wants, but the client also has to understand what his users need or what the markets want and communicate that to the designer. In Chapter 2 we will describe design processes that model how the designer can interact and communicate with both the client and potential users to help inform her own design thinking, and we will identify some tools (discussed in detail in Chapters 3–5) that she can use to organize and refine her thinking.

In addition, there is a stakeholder we've not yet named, that is, the *public*. That's in part because the public is implicitly embodied in the notion of the user. Including the notion of a public that is affected by a design as much as (or more than) the users we've already identified is interesting because it suggests that we may well confront ethical issues in such design projects. We will explore this further in Chapter 12. It's also worth noting that often the client speaks on behalf of the intended users, although anyone who has sat in cramped coach seating on a commercial flight would have to ask both airlines and airplane manufacturers who their customers really are!

The designer and the client have to understand what the users want in a design.

As described above, engineering designers work in many different kinds of environments, including: small and large companies, start-up ventures, government, not-for-profit organizations, and engineering services firms (one breed of which is the industrial design consultancy). Apart from the salaries and perks of working in these various places to do design, designers will most likely see differences in the size of a project, the number of colleagues on

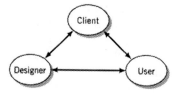

FIGURE 1.1 The designer–client–user triangle. There are three parties involved in a design effort: the client, who has objectives that the designer must clarify; the user of the designed device, who has his own requirements; and the designer, who must develop specifications such that something can be built to satisfy everybody!

the design team, and their access to relevant information about what users want. On large projects, many of the designers will be working on pieces of a project that are so detailed and so confined that much of what we describe in this book may not seem immediately useful. Thus, the designers of a bridge abutment, an airplane's fuel tank, or components on a computer motherboard are not likely to be as concerned with the larger picture of what clients and users want because the system-level design context has already been established when this level of design is reached. Indeed, as we will explain in Chapter 2, these kinds of design problems are the part of the process called *detailed design*, in which the choices and procedures are well understood because more general design issues have already been resolved. However, even for such large projects, the response to a client's project statement is initiated with *conceptual design*. Some thinking about the size and mission of the airplane will have been done to identify constraints surrounding fuel tank design, while the performance parameters that the computer motherboard must display will be determined by some assessment of the market for and the price of the computer in question.

Large, complex projects often lead to very different interpretations of client project statements and of user needs. One has only to look at the many different kinds of skyscrapers that decorate our major cities to see how architects and structural engineers envisage different ways of housing people in offices and apartments. Visible differences also emerge in airplane design (Figure 1.2) and wheelchair design (Figure 1.3). Each of these sets of devices could result from a simple, common design statement: The airplanes

FIGURE 1.2 A few "devices that safely transport people and goods through the air," i.e., airplanes. No surprises here, right? We've all seen a lot of airplanes (or, at least, in pictures or movies). But even these planes, albeit of different eras and origins, show that they clearly were designed to achieve very different missions.

FIGURE 1.3 A collection of "personal mobile devices that transport people who are unable to use their legs," i.e., wheelchairs. Here too, as with the airplane, we see some sharp differences in the configurations and components of these wheelchairs. Why are the wheels so different? Why are the wheelchairs so different?

are "devices to transport people and goods through the air," and the wheelchairs are "personal mobile devices to transport people who are unable to use their legs." However, the different products that have emerged represent different concepts of what clients and users wanted (and what designers perceived they wanted!) from these devices. Designers have to clarify what a client wants and translate those wants into an engineered product.

The designer-client-user triangle also prompts us to (a) recognize that the interests of the three players might diverge, and (b) consider that the consequences of such divergence could mean more than financial problems resulting from a failure to meet users' needs. This is because the interaction of multiple interests creates an interaction of multiple obligations, and these obligations may well conflict. For example, the designer of a juice container might consider metal cans, but easily "squashed" cans are a hazard if sharp edges emerge during the squashing. There could be tradeoffs among design variables, including the material of which a container is to be made and the container's thickness. The choices made in the final design could easily reflect different assessments of the possible safety hazards, which in turn could lay a foundation for potential ethics problems. Ethics problems, which we discuss in Chapter 12, occur because designers have obligations not only to clients and users, but also to their profession and, as detailed in the codes of ethics of engineering societies, to the public at large. Thus, ethics issues are always part of the design process.

Designers have obligations to the profession and to the public.

Another aspect of engineering design practice that is increasingly common in projects and firms of all sizes is the use of *teams* to do design. Many engineering problems are inherently multidisciplinary (e.g., the design of medical instrumentation), so there is a need to understand the requirements of clients, users, and technologies in very different environments. This, in turn, requires that teams be assembled to address such different sets of environmental needs. The widespread use of teams clearly affects how design projects are managed, another recurring theme of this book.

Engineering design is a multifaceted subject, and in no way do we think that the reader will truly understand this wonderfully complex activity by reading this one short book (or by doing a single design project). However, we do think that we can provide some frameworks within which the reader can think productively about some of the conceptual issues and the resulting choices that are made very early in the design of many different kinds of engineered products.

1.2   A VOCABULARY PRIMER FOR ENGINEERING DESIGN

It is already clear that the word *design* is used both as a noun (*n*) and a verb (*vb*). *Webster's New Collegiate Dictionary* defines the two usages as:

- **design** *n*: a mental project or scheme in which means to an end are laid down; the arrangement of elements that go into human productions (as of art or machinery).
- **design** *vb*: to conceive and plan out in the mind; to devise for a specific function or end.

The points behind these two definitions are clear: Designing is about people planning and creating ways to produce things that achieve some known goals.

There are many, many definitions of *engineering design* in the literature, and there is a fair bit of variation in the ways in which design actions and attributes are described by engineers. So we will now define what we mean by engineering design and then will go on to define some of the related terms that are commonly used by engineers and designers.

1.2.1 Our definition of engineering design

Engineering design is a thoughtful process for generating designs for devices, systems, or processes that attain given objectives while adhering to specified constraints.

The following formal definition of engineering design is the most useful one for our purposes:

- **Engineering design** is a systematic, intelligent process in which designers generate, evaluate and specify designs for devices, systems or processes whose form(s) and function(s) achieve clients' objectives and users' needs while satisfying a specified set of constraints.

It is important to recognize that when we are designing devices, systems and processes we are designing artifacts: artificial, human-made objects, the "things" or devices that are being designed. They are most often physical objects like airplanes, wheelchairs, ladders, cell phones and carburetors. But "paper" products such as drawings, plans, computer software, articles, and books are also artifacts in this sense, as are the "soft" electronic files that become "real" when displayed on a computer screen. In this text we will use device or system rather interchangeably as the objects of our design.

With further recourse to our dictionary, we note (and then comment on) the following definitions:

- **form** *n*: the shape and structure of something as distinguished from its material

  So what we mean by *form* is pretty straightforward, and its meaning in the engineering context is consistent with its more common use.

- **function** *n*: the action for which a person or thing is specially fitted or used or for which a thing exists; one of a group of related actions contributing to a larger action

  Simply put, *functions* are those things the designed device or system is supposed to do. As we describe in Sections 2.3 and 4.1, *engineering functions* involve the transfer or flow of energy, information and materials. Note, too, that we view energy transfer quite broadly: It includes supporting and transmitting forces, the flow of current, the flow of charge, and so on.

- **means** *n*: an agency, instrument, or method used to attain an end

  Although not explicitly recognized in our definition of engineering design, *means* is nonetheless important as in this context it refers to a way of making a function happen.

- **objective** *n*: something toward which effort is directed; an aim or end of action

  An *objective* in our context is consistent with its common usage.

- **constraint** *n*: the state of being checked, restricted, or compelled to avoid or perform some action

  This definition, too, is what we would expect from standard use. It is worth stressing that *constraints* are extremely important in engineering design because they impose absolute limits which, if violated, mean that a proposed design is simply not acceptable.

Anticipating another point we will stress again (in Chapter 3), note that objectives for a design are entirely separate from the constraints placed on a design. Objectives may be completely achieved, may be achieved in some measure, or may not be achieved at all. Constraints, on the other hand, are binary: They are satisfied or they are not satisfied; they are black or white, and there are no intermediate states. Thus, if we were designing a corn degrainer for Nicaraguan farmers to be cheaply built of indigenous (local) materials, an objective might be that it should be as cheap as possible, while a constraint might be that it could not cost more than $20.00 (US). Making the degrainer of indigenous materials could either be an objective, if it is a *desired* feature, or a constraint, if it's absolutely required.

Our definition of engineering design states that designs emerge from a *systematic, intelligent process*. This is not to deny that design is a creative process. However, at the same time, there are techniques and tools we can use to support our creativity, to help us think more clearly, and to make better decisions along the way. These tools and techniques, which form much of the subject of this book, are not formulas or algorithms. Rather, they are ways of asking questions, and presenting and reviewing the answers to those questions as the design process unfolds. We will also present some tools and techniques for managing a design project. Thus, while demonstrating ways of thinking about a design as it unfolds in our heads, we will also talk about ways to deploy the resources needed to complete a design project on time and within budget.

1.2.2   The assumptions behind our definition of engineering design

Design is a thoughtful process that can be understood.

There are some implicit assumptions behind our definition of engineering design and the terms in which it is expressed. It is useful to make them explicit.

First, design is a *thoughtful* process that can be *understood*. Without meaning to spoil the magic of creativity or the importance of innovation in design, people do *think* while designing. So it is important to have tools to support that thinking, to support design decision making and design project management. (One piece of supporting evidence for this obvious hypothesis is that computer programs have been written to emulate design processes. We couldn't write such programs if we couldn't articulate and describe what goes on in our heads when we design things.)

The idea that there are *formal methods* to use when generating design alternatives is strongly related to our inclination to think about design. This might seem pretty obvious because there's not much point in considering new ways of looking at design problems or talking about them — unless we can exploit them to do design more effectively.

Form cannot be deduced from function.

Form and *function* are two related yet independent entities. This is important. We often think of the design process as beginning when we sit down to draw or sketch something, which suggests that form is a typical starting point. However, we should keep in mind that function is an altogether different aspect of a design that may not have an obvious relationship to shape or form. In particular, while we can often infer the purpose of an object or device from its form or structure, we can't do the reverse, that is, we cannot automatically deduce what form a device must have *from the function alone*. For example, we can look at a pair of connected boards and deduce that the devices that connect them (e.g., nails, nuts and bolts, rivets, screws, etc.) are fastening devices whose function is to attach the individual members of each pair. However, if we were to start with a statement of purpose that we wish to attach one board to another, there is no obvious link or inference that we can use to create a form or shape for a fastening device. That is, knowing that we want to achieve the *function* of attaching two boards does not lead us to (or even suggest) any of the *forms* of welds, screws, rivets, or glues.

The relationship of form and function is important in understanding the creative aspects of design. If we can systematically articulate all of the functions that a device is expected to perform, then we can be creative in developing forms within which these functions can be realized. In this sense, the use of organized, thoughtful processes *adds* to the creative side of design.

There are benchmarks available to assess how we expect a design to perform and to implicitly measure the progress made toward a successful design. These benchmarks derive from a questioning process (see Chapter 2) that begins with the designer:

- translating the client's desires into *objectives* for the device or system being designed;
- establishing a set of *metrics* that can be used to ascertain or measure the extent to which a proposed design will meet the client's objectives;
- establishing the *functions* that a successful design will perform; and
- establishing the *requirements* that express in engineering terms both the design's attributes and its behavior, that is, the design's functions.

Let us formally define the two new terms we have just introduced, that is, metrics and requirements. Our definitions, while presented in standard dictionary format, represent a blend of actual dictionary definitions with our understanding of the "best practices" of engineering design as it is currently done in industry. Thus:

- **metric** *n*: a standard of measurement; in the context of engineering design, a scale on which the achievement of a design's objectives can be measured and assessed

  Metrics provide scales or rulers on which we can measure the degree to which objectives are achieved. To offer a truly simple example, let us suppose an objective of being able to jump 10 meters. A metric for a jump would award 1 point for each meter jumped, so that a jump of 2 meters earns 2 points, while a jump of 8 meters earns 8 points, and so on. As we discuss at length in Chapter 3, not all objectives are so easily quantified, and not all measurements are so easily made. Thus, there are interesting issues that must be addressed when we talk about metrics in depth.

- **requirement(s)** *n*: thing(s) wanted or needed; thing(s) essential to the existence or occurrence of something else; in the context of engineering design, engineering statements of the functions that must be exhibited and the attributes that must be displayed by a design

  The design requirements, which are often called *design specifications*, are stated in a number of ways, depending on the nature of the requirements the designer chooses to articulate. As we explain in Chapter 5, design requirements may specify: *values* for particular design attributes; the *procedures* used to calculate attributes or behavior of the design; or *performance levels* of the functional behavior that must be attained by the design. We shall explore the nature of design requirements (or specifications) extensively in Chapter 4.

Fabrication specifications enable implementation independent of the designer's involvement.

The endpoint of a successful design is a set of plans for making the designed device. This set of plans, often called *fabrication specifications*, may include drawings, assembly instructions, and lists of parts and materials, as well as a host of text, graphs, and tables that explain: what the artifact is; why it is what it is; and how it can be realized or brought to life. This will be the case whether the artifact is a physical object, a process description, or some soft representation.

Further, fabrication specifications must be clear, unambiguous, complete, and transparent. This is because fabrication specifications must, *by themselves*, enable someone other than the designer (or others involved in the design process) to make what the designer intended so that it performs as the designer intended. This is a facet of modern engineering practice that represents a departure from a (long-ago) time when designers were often craftsmen who made what they designed. These designer-fabricators could allow themselves latitude or shorthand in their design plans because as fabricators they knew exactly what they intended as designers. Nowadays engineers rarely make what they design. Sometimes designs are "thrown over the wall" to a manufacturing department or to a fabricator who acts entirely on "what's in the specs." But increasingly manufacturing issues are addressed during the design process, which means that manufacturing engineers and even suppliers become part of the design team, which also means there are further needs for designers to be good communicators!

It often happens that the manufacture or use of a device highlights deficiencies that were not anticipated in the original design. Designs often produce *unanticipated consequences* that may become *ex post facto* evaluation criteria. For example, the automobile does provide the intended personal transportation. On the other hand, some regard the automobile as a failure because of its contribution to air pollution and traffic congestion. In addition, changing societal expectations have dictated serious redesign of many of the automobile's attributes and behaviors.

Communication is a key issue in design.

Finally, our definition of engineering design and the related assumptions we have identified clearly rely heavily on the fact that communication is central to the design process. Some set of languages or representations is inherently and unavoidably involved in every part of the design process. From the original communication of a design problem through the specification of requirements and of fabrication specifications, the device or system being designed must be described and "talked about" in many, many ways. Thus, *communication is a key issue*. It is not that problem solving and evaluation are less important; they are extremely important. But problem solving and evaluation are done at levels and in styles — whether spoken or written languages, numbers, equations, rules, charts, or pictures — that are appropriate to the immediate task at hand. Successful work in design is inextricably bound up with the ability to communicate.

1.3 LEARNING AND DOING ENGINEERING DESIGN

Design is rewarding, exciting, fun, even exhilarating. But good design doesn't come easily. In fact, achieving excellence in design is hard. That is why learning and doing (and teaching!) design is hard.

1.3.1 Engineering design addresses hard problems

Engineering design is an ill structured, open-ended activity.

Engineering design problems are generally difficult because they are usually *ill structured* and *open-ended*:

- Design problems are *ill structured* because their solutions cannot normally be found by applying mathematical formulas or algorithms in a routine or structured way. Mathematics is both useful and essential in engineering design, but much less so in the early stages when "formulas" are both unavailable and inapplicable. In fact, some engineers find design difficult simply because they can't fall back on structured, formulaic knowledge — but that's also what makes design a fascinating experience.

- Design problems are *open-ended* because they typically have several acceptable solutions. Uniqueness, so important in many mathematics and analysis problems, simply does not apply to design solutions. In fact, more often than not designers work to reduce or bound the number of design options they consider lest they be overwhelmed by the possibilities.

Evidence for these two characterizations can be seen in the familiar ladder. Several ladders are shown in Figure 1.4, including a stepladder, a portable stepladder, and a rope ladder. If we want to design a ladder, we can't identify a particular ladder type to target unless and until we determine a specific set of uses for that ladder. Even if we decide that a particular form is appropriate, say a stepladder for the household handyman, other questions arise: Should the ladder be made of wood, aluminum, plastic, or a composite material? How much should it cost? And, which ladder design would be the *best*? Can we identify the *best* ladder design, or the *optimal design*? The answer is, "No," we can't stipulate a ladder design that would be universally regarded as the best or that would be mathematically optimal in every dimension.

How do we talk about some of the design issues, for example, purpose, intended use, materials, cost, and possibly other concerns? In other words, how do we articulate the choices and the constraints for the ladder's form and function? There are different ways of representing these differing characteristics by using various "languages" or representations. But even the simple ladder design problem becomes a complex study that shows how the two characteristics of poorly defined endpoints (e.g., what kind of ladder?) and ill defined structure (e.g., is there a formula for ladders?) make design a tantalizing yet difficult subject. How much more complicated and interesting are projects to design a new automobile, a skyscraper, or a way to land a person on the moon?

1.3.2 Learning design by doing design

For someone who wants to learn *how to do it*, design is not all that easy to grasp. Like riding a bicycle or throwing a ball, like drawing and painting and dancing, it often seems easier to say to a student, "Watch what I'm doing and then try to do the same thing yourself." There is a *studio* aspect to trying to teach any of these activities, an element of *learning by doing*.

One of the reasons that it is hard to teach someone how to do design — or to ride a bike or throw a ball or draw or dance — is that people are often better at *demonstrating* a

FIGURE 1.4 A collection of "devices that enable people to reach heights they would be otherwise unable to reach," i.e., ladders. Note the variety of ladders, from which we can infer that the design objectives involved a lot more than the simple idea of getting people up to some height. Why are these ladders so different?

skill than they are at *articulating* what they know about applying their individual skills. Some of the skill sets just mentioned clearly involve some physical capabilities, but the difference of most interest to us is not simply that some people are more gifted physically than others. What is really interesting is that a softball pitcher cannot tell you just how much pressure she exerts when holding the ball, nor exactly how fast her hand ought to be going, or in what direction, when she releases it. Yet, somehow, almost by magic, the softball goes where it's supposed to go and winds up in the hands of a catcher. The real point is that the thrower's nervous system has the knowledge that allows her to assess distances and choose muscle contractions to produce a desired trajectory. While we can model that trajectory, given initial position and velocity, we do not have the ability to model the knowledge in the nervous system that generates that data.

Design is best learned by both doing and studying.

Note also that designers, like dancers and athletes, *use drills and exercises* to perfect their skills, *rely on coaches* to help them improve both the mechanical and interpretive aspects of their work, and *pay close attention* to other skilled practitioners of their art. Indeed, one of the highest compliments paid to an athlete is to say that he or she is "a student of the game."

1.4 ON THE EVOLUTION OF DESIGN AND ENGINEERING DESIGN

People have been doing design from time immemorial. People have also been talking and writing about design for a long time, but for much less time than they have actually been designing things. So, let's have a brief look at the evolution of design and engineering design over time.

1.4.1 Remarks on the evolution of design practice and thinking

Thinking back to early elementary artifacts, it is almost certainly true that the "designing" was inextricably linked with the "making" of these primitive implements. We have no record of a separate, discernible modeling process, so we can't know that for sure. Who can say that small flint knives were not consciously used as models for larger, more elaborate cutting instruments? The inadequacy of small knives for cutting into the hides and innards of larger animals could have been a logical driver for enlarging a small flint knife. People must have *thought* about what they were making, recognized shortcomings or failures of devices already in use before they made more sophisticated versions.

But we really have no idea of *how* these early designers thought about their work, what kinds of languages or images they used to process their thoughts about design, or what mental models they used to assess function or judge form. If we can be sure of anything, it is that much of what they did was done by trial and error. (Nowadays, when trial solutions are generated by unspecified means and tested to eliminate error, we call it *generate and test*.)

We find examples of ancient works that must have been designed, such as the Great Pyramids of Egypt, the cities and temples of Mayan civilization, and the Great Wall of China. Unfortunately, the designers of these wonderfully complex structures did not leave a paper trail of recorded thoughts about their designs. However, there are some discussions of design that go pretty far back, one of the most famous being the collection of works by the Venetian architect Andrea Palladio (1508–1580). His works were apparently first translated into English in the eighteenth century. Since then, discussions of design have been developed in fields as diverse as architecture, organizational decision making, and various styles of professional consultation, including the practice of engineering. That's one reason there are a lot of definitions of engineering design.

We see that even in ancient times, designers evolved from pragmatists who likely designed artifacts as they made them, to more sophisticated practitioners who sometimes designed immense artifacts that others constructed. It has been said that the former approach to design, wherein the designer actually produces the designed object directly, is a distinguishing feature of a *craft*, and is found in such modern and sophisticated endeavors as graphics and type design.

1.4.2 A systems-oriented definition of design

We identified several key words when we defined *engineering design*, including form, function, requirements, and specifications. Could we (and how would we) define *design*? Here, too, it is as hard to define design generally as it is to define the particular endeavor of engineering design. For example, design could be defined as a goal-directed activity, performed by humans, and subject to constraints. The product of this design activity is a *plan* to realize those goals.

Herbert A. Simon, late Nobel laureate in economics and founding father of several fields, including design theory, offered a broad definition of design that is closely related to our engineering concerns:

- **Design** is an activity that intends to produce a "description of an artifice in terms of its organization and functioning — its interface between inner and outer environments."

Designers are thus expected to describe the shape and configuration of a device (its "organization"), how that device does what it was intended to do (its "function"), and how the device (its "inner environment") works ("interfaces") within its operating ("outer") environment. Simon's definition is interesting for engineers because it places designed objects in a *systems* context that recognizes that any artifact operates as part of a system that includes the world around it. In this sense, all design is systems design because devices, systems, and processes must each operate within and interact with their surrounding environments.

1.4.3 On the evolution of engineering design

We noted earlier that engineering designers do not typically produce their artifacts. Rather, they produce fabrication specifications for making the artifacts. The designer in an engineering context produces a detailed description of the designed device so that it can be assembled or manufactured, thus separating the "designing" from the "making." This specification must be both complete and quite specific; there should be no ambiguity and nothing can be left out.

Traditionally, fabrication specifications were presented in a combination of drawings (e.g., blueprints, circuit diagrams, flow charts, etc.) and text (e.g., parts lists, materials specifications, assembly instructions, etc.). We can achieve completeness and specificity with such traditional specifications, but we may not capture the designer's intent — and this can lead to catastrophe. In 1981, a suspended walkway in the Hyatt

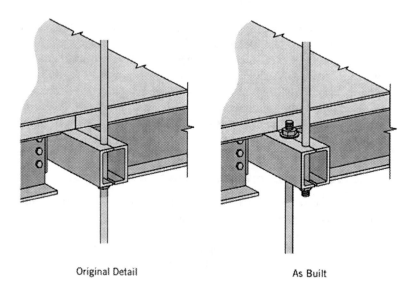

Original Detail   As Built

FIGURE 1.5 The walkway suspension connection — as originally designed and as built — in the Regency Hyatt House in Kansas City. We see that the change made during construction resulted in the second-floor walkway being hung from the fourth-floor walkway, rather than being connected directly to the roof truss.

Regency Hotel in Kansas City collapsed because a contractor fabricated the connections for the walkways in a manner different than intended by the original designer.

In that design, walkways at the second and fourth floors were hung from the same set of threaded rods that would carry their weights and loads to a roof truss (see Figure 1.5). The fabricator was unable to procure threaded rods sufficiently long (i.e., 24 ft) to suspend the second-floor walkway from the roof truss, so instead, he hung it from the fourth-floor walkway with shorter rods. (It also would have been hard to screw on the bolts over such lengths and attach walkway support beams.) The fabricator's redesign was akin to requiring that the lower of two people hanging independently from the same rope change his position so that he was grasping the feet of the person above. That (upper) person would then be carrying both people's weights with respect to the rope. In the hotel, the supports of the fourth-floor walkway were not designed to carry the second-floor walkway in addition to its own dead and live loads, so a collapse occurred, 114 people died, and millions of dollars of damage was sustained. If the fabricator had understood the designer's intention to hang the second-floor walkway directly from the roof truss, this accident might never have happened. Had there been a way for the designer to explicitly communicate his intentions to the fabricator, a great tragedy might have been avoided.

The design may be the only connection between the designer and the fabricator.

There's another lesson to be learned from the separation of the "making" from the "designing." If the designer had worked with a fabricator or a supplier of threaded rods while he was still designing, he would have learned that no one made threaded rod in the lengths needed to hang the second-floor walkway directly from the roof truss. Then the designer could have sought another solution in an early design stage. In much of the manufacturing and construction business, it was the case for many years that there was a "brick wall" between the design engineers on one side and the manufacturing engineers and fabricators on the other. Only recently has this wall been penetrated. Manufacturing and assembly considerations are increasingly addressed *during* the design process, rather than afterward. One element in this new practice is *design for manufacturing*, in which the ability to make or fabricate an artifact is specifically incorporated into the design requirements, perhaps as a set of manufacturing constraints. Clearly, the designer must be aware of parts that are difficult to make or of limitations on manufacturing processes as her design unfolds.

Concurrent engineering, another recent idea, refers to the process in which designers, manufacturing specialists, and those concerned with the product's life cycle (e.g., purchase, support, use, and maintenance) work together, along with other design stakeholders, so that they can collectively and *concurrently* design the artifact together. Concurrent engineering thus works to capture the designer's intent by integrating the design and fabrication activities. Clearly, then, concurrent engineering demands teamwork of a high order. Research in this area focuses on ways to enable teams to work together on complex design tasks when team members are dispersed not only by engineering discipline, but also geographically, culturally, and by time zones.

The designer's intent must be clearly communicated to the fabricator.

The Hyatt Regency tale and the lessons drawn from it suggest that fabrication specifications are really important. Unless a design's fabrication specifications are complete and unambiguous, and unless they clearly convey a designer's intentions, the device or system won't be built in accord with the requirements set out by the designer. It is also the case that fabrication specifications provide a basis for evaluating how well a design meets its original design goals because those specifications emerge from the design's requirements, which in turn result from our ability to translate the original objectives (and constraints) of the client into those requirements.

But while worrying about design requirements and fabrication specifications, as well as all of the other issues we have raised, we should also keep in mind that design is a human activity, a social process. This means that communication among stakeholders remains a pre-eminent, consistent concern.

1.5   MANAGING ENGINEERING DESIGN

Good design doesn't just happen.

Good design doesn't just happen. Rather, it results from careful thought about what clients and users want, and about how to articulate and realize design requirements. That is why this book focuses on tools and techniques to assist the designer in this process. One particularly important element of doing good design is *managing* the design project. Just as thinking about design in a rigorous way does not imply a loss of creativity, using tools to manage the design process doesn't mean we sacrifice technical competency or inventiveness. On the contrary, there are many organizations that foster imaginative engineering design as an integral part of their management style. At 3M, for example, each of the more than 90 product divisions is expected to generate 25 percent of its annual revenues from products that didn't even exist five years earlier. We will also introduce a few tools and techniques of management that are applicable to design projects.

Just as we began by defining terms and developing a common vocabulary for design, we will do the same for management, project management, and the management of design projects. We will go from the general to the specific in these definitions. Later we will look less at definitions and more at "hands-on" matters. First we define management as follows:

- **management** *n*: the process of achieving organizational goals by engaging in the four major functions of planning, organizing, leading, and controlling

  This definition emphasizes that organizational goals are not achieved without certain processes. In this sense, management has something in common with design insofar as both are goal-directed and can be considered in terms of steps or processes. We will carry this analogy a little further in Chapter 3 when we consider the phases or stages of design.

The four functions of management can also be defined and discussed in ways that help us see how management might relate to design.

- **planning** *n*: the process of setting goals and deciding how best to achieve them

  Planning involves considering the mission or purpose of the organization, and translating that aim into appropriate strategic and tactical goals and objectives for the organization.

- **organizing** *n*: the process of allocating and arranging human and nonhuman resources so that plans can be carried out successfully

  Put another way, the *organizing* function of management is concerned with "creating a framework for developing and assigning tasks, obtaining and allocating resources, and coordinating work activities to achieve goals."

- **leading** *n*: the ongoing activity of exerting influence and using power to motivate others to work toward reaching organizational goals

  That *leadership* results from influence is very important in design settings, where a number of different types of influence can come into play. For example, one member of the design team may have influence because of her position (e.g., the team leader), while another may have influence because the team recognizes his expertise in a particular domain.

- **controlling** *n*: the process of monitoring and regulating the progress of an organization toward achieving its goals

  Many people confuse leading and *controlling*, perhaps because we use the term "controlling" as a less-than-flattering synonym for using power in some settings. For engineers it is more complex because we use the term control to refer to directing system performance by monitoring and regulating. In this book, control means ensuring that actual performance conforms to expected standards and goals.

Project management is the application of these four functions to accomplish the goals and objectives of a project. A project is "a one-time activity with a well defined set of desired end results." Examples of engineering design projects abound, ranging from designing new highways (civil engineering) to designing new computer memories (electrical engineering) to designing the flow of materials and manufacturing on a factory floor (industrial engineering). The common thread in these three projects is that each can be well defined in terms of its goals, has finite resources, and is to be accomplished in a fixed time frame (sometimes simply as soon as possible). To help project managers in carrying out the four functions (i.e., planning, organizing, leading, and controlling), a number of tools and techniques have been developed. These include tools for understanding and listing the work to be done; scheduling the tasks to be done logically and efficiently; assigning tasks to individuals; and monitoring progress. We will explore some of these tools and techniques as they are most applicable to design projects later in the book.

Note that the precision in goals and objectives that is spoken of regarding projects is somewhat at odds with some of the previous discussion of the open-ended nature of design activities. This is certainly the case when we try to predict the final form or outcome of a design project. Unlike a construction project, where the desired and expected results are clear and generally well articulated, a design project, and especially a conceptual design project, may have a number of possible successful outcomes, or none! This makes the task and tools of project management only partially useful in design settings. As a result, we will only present project management tools that we have found to be useful in managing design projects conducted by small teams.

In addition to a more restricted set of tools for managing the design project, we will introduce formal tools for guiding the design process itself. These tools are also a form of project management as they help the team to understand and agree upon goals, organize their activities, organize resources to realize the goals, and monitor whether the alternatives they generate and ultimately select are consistent with their objectives.

1.6 NOTES

Section 1.2: Our definition of engineering design draws heavily on (Dym and Levitt 1991, Dym 1994, Dym et al. 2005).

Section 1.4: Simon's definition of design was given in a set of lectures that were published as *The Sciences of the Artificial* (1981).

Section 1.5: Definitions of management and the four major functions of planning, organizing, leading, and controlling are found in (Bartol and Martin 1994) and (Bovee et al. 1993). The project is defined in (Meredith and Mantel 1995).

1.7 EXERCISES

1.1 Draft a definition of engineering design in terms more colloquial than used in the definition presented in Section 1.2.1.

1.2 List at least three questions you would ask if you were, respectively, a user (purchaser), a client (manufacturer), or a designer who was about to undertake the design of a portable electric guitar.

1.3 List at least three questions you would ask if you were, respectively, a user (purchaser), a client (manufacturer), or a designer who was about to undertake the design of a greenhouse for a tropical climate.

1.4 All aspects of management may be said to be goal directed. Explain how this description is exemplified for each of the four functions of management identified in Section 1.5.

CHAPTER 2

THE DESIGN PROCESS

Is there a way to do engineering design? And can you give me a roadmap of where is this book is going?

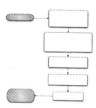

HAVING DEFINED engineering design and its attendant vocabulary, we go on to explore the activity of doing a design, that is, the *process* of design. Some of this may seem abstract because we are trying to describe a very complex process by breaking it down into smaller, more detailed *design tasks*. Further, as we define those design tasks, we will identify places in the design process where we can use the design tools and methods we present in Chapters 3–5. Please keep in mind, however, that *we are **not** presenting a recipe* for doing design. Instead, we are outlining a framework within which we can *articulate and **think*** about what we are doing as we design something.

2.1 THE DESIGN PROCESS AS A PROCESS OF QUESTIONING

Imagine you are working in a company that makes diverse consumer projects, and your boss calls you into her office and says, "Design a safe ladder." You wonder to yourself, Why does anyone need still another ladder? Aren't there a lot of safe ladders already on the market (including those we saw in Figure 1.5)? And what does she mean by a "safe ladder"?

This scenario is clearly contrived, so its not a big surprise that a whole bunch of questions immediately come to mind. Typically, however, design projects or problems begin with a verbal statement that talks about a client's intentions or goals, the design's form or shape, its purpose or function, and perhaps some things about legal requirements. The designer's first task is to *clarify* what the client wants so as to be able to translate those wishes into meaningful *objectives* (goals) and *constraints* (limits). This clarification task proceeds as the designer asks the client to be more precise about what she really wants.

In fact, asking questions is an integral part of the entire design process. To paraphrase an observation made long ago by Aristotle, *knowledge resides in the questions that can be asked and the answers that can be provided*. Thus, it is important to think about the questions that should be asked and to identify others in the designer-client-user triangle who may have answers (and may also have other useful questions). Further, by looking at the kinds of questions that can be asked throughout the design process we can develop and articulate that process as a sequence of *design tasks*. In fact, the following array of questions we might ask about designing a ladder suggests a sequence of design tasks (in italics) that we will consider later.

Questions such as:

- Why do you want another ladder?
- How will the ladder be used?
- How much can it cost?

help *clarify and establish the client's objectives* for the design.

Questions such as:

- What does "safe" mean?
- What's the most you're willing to spend?

help *identify the constraints* that govern the design.

Questions such as:

- Can the ladder lean against a supporting surface?
- Must the ladder support someone carrying something?

help *establish functions* that the design must perform and suggest *means* by which those functions can be performed.

Questions such as:

- How much weight should a safe ladder support?
- How high should someone on the ladder be able to reach?

help *establish requirements* for the design.

Questions such as:

- Could the ladder be a stepladder or an extension ladder?
- Could the ladder be made of wood, aluminum, or fiberglass?

help *generate design alternatives.*

Questions such as:

- What is the maximum stress in a step supporting the "design load"?
- How does the bending deflection of a loaded step vary with the material of which the step is made?

help *model* and *analyze* the design.

Questions such as:

- Can someone on the ladder reach the specified height?
- Does the ladder meet OSHA's safety specification?

help *test* and *evaluate* the design against its objectives and its constraints.

Questions such as:

- Are there other ways to connect the steps?
- Can the design be made with less material?

help *refine* and *optimize* the design.

Finally, questions such as:

- What is the justification for the design decisions that were made?
- What information does the client need to fabricate the design?

help *document* the design process and *communicate* the completed design.

Thus, the questions we asked about the ladder design establish steps in a process that move us from an abstract statement of a client's desires through increasing levels of detail toward an engineering solution. The early design tasks move toward translating the client's wishes into a set of *requirements* that state in engineering terms how the design is to function or perform. These requirements, also called *design specifications*, serve as benchmarks against which the design's performance is measured. Design specifications are typically stated in one of three forms, so the requirements might: *prescribe* values for attributes of the design; *specify procedures* for calculating attributes or behavior; or they might *specify the performance* of the design's behavior. (We discuss this further in Chapter 4.) Continuing on, as we generate different *concepts* of how the design might work or function, we also create *design alternatives*. Then we choose one concept (say, here, a stepladder) and *build and analyze a model* of that ladder's design; *test and evaluate* the design; *refine and optimize* some of its details; and then *document* the justification for the stepladder's final design and its fabrication specifications. In Section 2.2 we will present *all* of the tasks of the engineering design process in greater detail.

Some of the early questions asked (to clarify the client's wishes) clearly connect to later tasks in the process where we make choices, analyze how competing choices interact, assess tradeoffs in these choices, and evaluate the effect of these choices on our top-level goal of designing a safe ladder. For example, the ladder's *form* or shape and layout are strongly related to its *function*: We are more likely to use an extension ladder to rescue a cat from a tree and a stepladder to paint the walls of a room. Similarly, the weight of the ladder has an impact on the efficiency with which it can be used: Aluminum extension ladders have replaced wooden ones largely because they weigh less. The material of which a ladder is made affects not only its weight, but also its cost and its feel: Wooden extension ladders are much stiffer than their aluminum counterparts, so users of aluminum ladders feel a certain amount of "give" or flex in the ladder, especially when it is significantly extended.

There are no equations for safety, color, marketability....

Some of the questions in the later design tasks can be answered by applying the mathematical models such as those used in physics. For example, Newton's equilibrium law and elementary statics can be used to analyze the stability of the ladder under given loads on a specified surface. We can use beam equations to calculate deflections and stresses in the steps as they bend under the given foot loads. But there are no equations that define the meaning of "safe," or of the ladder's marketability, or that help us choose its color. Since there are no equations for safety, marketability, color, or for most of the other issues in the ladder questions, we must find other ways to think about this design problem.

It also seems clear that we will face a vast array of choices as our design evolves. At some point in our ladder design, for example, we have to choose a *type* of ladder, say a stepladder (for painting) or an extension ladder (for rescuing cats). We then have to decide how to fasten the steps to the ladder frame. The choices will be influenced by the desired behavior (e.g., although the ladder itself may flex, we don't want individual steps to have much give with respect to the ladder frame), as well as by manufacturing or assembly considerations (e.g., would it be better to nail in the steps of a wooden ladder, use dowels and glue, or nuts and bolts?). Note that we are now decomposing the complete ladder into its components and selecting particular types of components.

We should also note that as we work through these questions (and design tasks), we are constantly communicating with others about the ladder and its various features. When we question our client about its desired properties, for example, or the laboratory director about evaluation tests, or the manufacturing engineer about the feasibility of making certain parts, we are interpreting aspects of the ladder design in terms of *languages* and parameters that these experts use in their own work: We draw pictures in graphical languages; we write and apply formulas in the language of mathematics; we ask verbal questions and provide verbal descriptions; and we use numbers all of the time to fix limits, describe test results, and so on. Thus, the design process can't proceed without recognizing different design languages and their corresponding interpretations.

This simple design problem illustrates how we might *formalize* the design process to make explicit the design tasks that we are doing. We are also *externalizing* aspects of the process, moving these aspects from our heads into a variety of recognizable languages to be able to communicate with others. Thus, we learn two important lessons from our ladder design project:

Clarifying objectives and translating them into the right "languages" are essential elements of design.

- *Clarifying* the client's objectives is a crucial part of an engineering design project. The designer must fully understand what the client wants (and the users need) from the resulting design. Good communication among all three parties in the designer-client-user triangle is essential, and we must be very careful when eliciting the details of what the client really wants.

- Performing the design tasks in the design process requires *translating* the client's objectives into the kinds of words, pictures, numbers, rules, properties, etc., that are needed to characterize and describe both the object being designed and its behavior. The tasks of analyzing and modeling, testing and evaluating, and refining and optimizing cannot be done just with words. And the documentation of the final design cannot be done with words alone. We need pictures and numbers, and likely other ways of representing the desired design. Thus, the designer translates the client's verbal statement into whatever languages are appropriate to completing the various design tasks at hand.

2.2 DESCRIBING AND PRESCRIBING THE DESIGN PROCESS

We have just seen that asking increasingly detailed questions exposed several design tasks. We will now formalize such design tasks into a design process. Many design process models are *descriptive*: they *describe* the elements of the design process. Other models are *prescriptive*: they *prescribe* what must be done during the design process. After introducing some descriptive models, we will introduce an extended set of our design tasks into one of them and so convert that (chosen and revised) descriptive model into a prescriptive model.

2.2.1 Describing the design process

The simplest descriptive model of the design process defines three phases:

1. *Generation*: the designer *generates* or creates various design concepts.
2. *Evaluation*: the designer *tests* the chosen design against metrics that reflect the client's objectives and against requirements that stipulate how the design must function.
3. *Communication*: the designer *communicates* the final design to the client and to manufacturers or fabricators.

Another three-stage model splits up the design process differently: *Doing research, creating*, and *implementing* a final design, with the contexts providing meanings for these three steps. While these two models have the virtue of simplicity, they are so abstract that they provide little useful advice on how to do a design. They also assume that the designer understands the client's objectives and the users' needs, and they both accept that the identification of the design problem has already occurred (and is, implicitly, not a part of the design process). Further, and perhaps most importantly, these models tell us nothing about *how* we might generate or create designs.

We show another, widely accepted descriptive model of the design process in Figure 2.1(a), with three "active" stages shown in boxes with rounded corners. It also shows the client's problem statement, sometimes identified as the *need* for a design, as the starting point. The final design (or its fabrication specifications) is the endpoint. The model's first phase is *conceptual design*, in which different *concepts* (also called *schemes*) are generated to achieve the client's objectives. Thus, the major functions and the means for achieving them are identified, as are the spatial and structural relationships of the principal components. Enough details have been worked out so that we can estimate costs, weights, and overall dimensions. For the ladder project, for example, conceptual designs might be an extension ladder, a stepladder, and a rope. The evaluation of these concepts will depend on the client's objectives, such as its intended use, its allowable cost, and even the client's aesthetic values.

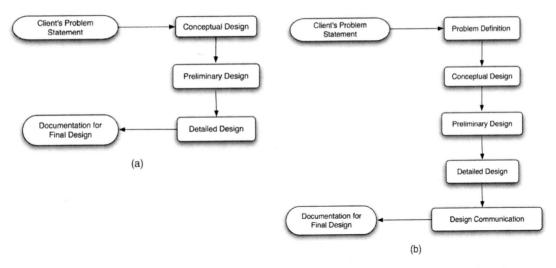

FIGURE 2.1 Two descriptive, "linear" models of the design process: (a) three stages; (b) five stages. Both models show the design process as simple, linear sequences of objects (*need* and *final design*) connected by three (*conceptual, preliminary, and detailed design*) or five (*problem definition, conceptual design, preliminary design, detailed design*, and *design communication*) design phases. As descriptive models, they provide very little guidance on how to do the various stages.

With its focus on tradeoffs between high-level objectives, conceptual design is clearly the most abstract and open-ended part of the design process. The output of the conceptual stage may include several competing concepts. Some argue that conceptual design *should* produce two or more schemes since early commitment to or fixation on a single design choice may be a mistake. This tendency is so well known among designers that it has produced a saying: "Don't marry your first design idea."

The second phase in this model of the design process is *preliminary design* or the *embodiment of schemes*. Here proposed concepts are "fleshed out," that is, we hang the meat of some preliminary choices upon the abstract bones of the conceptual design. We embody or endow design schemes with their most important attributes. We select and size the major subsystems, based on lower-level concerns that take into account the performance and operating requirements. For a stepladder, for example, we size the side rails and the steps, and perhaps decide on how the steps are to be fastened to the side rails. Preliminary design is more technical in nature, so we might use various back-of-the-envelope calculations. We make extensive use of rules of thumb about size, efficiency, and so on, that reflect the designer's experience. And, in this phase of the design process, we solidify our final choice of design concept.

The final stage of this model is *detailed design*. We now refine the choices we made in preliminary design, articulating the final choice in far greater detail, down to specific part types and dimensions. This phase typically follows design procedures that are quite well understood by experienced engineers. Relevant knowledge is found in design codes (e.g., the ASME Pressure Vessel and Piping Code, the Universal Building Code), handbooks, databases, and catalogs. Design knowledge is often expressed in specific rules, formulas and algorithms. This stage of design is typically done by component specialists who use libraries of standard pieces.

The classic model just outlined can be extended to a five-stage model that delineates two additional sets of activities that precede and follow the three-stage model sequence:

- *Problem definition*: a pre-processing stage that *frames the problem* by clarifying the client's original problem statement *before* conceptual design begins.

- *Design communication*: a post-processing phase that identifies the work done *after* detailed design to collect, organize, and present the final design and its fabrication specifications.

Note that in practice much of the documentation will have been developed along the way (e.g., a designer might write out the design rationale behind her choice for some details when the choice is actually made, rather than at the end), so the communication phase is as much about tracking and organizing work products as it is about writing a "new" report from scratch.

This five-stage model of the process, displayed in Figure 2.1(b), is more detailed than the three-stage models discussed above, but it does not bring us much closer to knowing *how* to do a design because it too is *descriptive*. In the next section we present a process model that prescribes what ought to done.

2.2.2 Prescribing the design process

In Figure 2.2 we show the descriptive model of Figure 2.1(b) converted into a five-phase *prescriptive* model that prescribes or specifies what is done in each phase in terms of fifteen design tasks, as delineated in the five charts below. It begins with the client's problem statement (chart 1) and ends with design documentation (chart 5). Each phase requires *input(s)*, has *design tasks* that must be performed, and produces *output(s)* or product(s). Thus, charts 1–5 detail the five phases of design with their inputs, tasks and output. Each chart is followed by a brief listing of the *sources of information, design methods*, and *means* relevant to the design tasks in that phase. (The sources of information and the corresponding means are discussed in Section 2.3; the design methods identified here will be outlined in Section 2 and discussed in detail in Chapters 3–5.) Note, too, that the output for each phase serves as the input to the following phase.

1. During *problem definition* we *frame the problem* by clarifying the client's objectives and gather the information needed to develop an unambiguous statement of the client's wishes, needs and limits.

      Input: *client's problem statement*
      Tasks: *clarify design objectives (1)*
            establish metrics for those objectives (2)
            identify constraints (3)
            revise client's problem statement (4)
    Outputs: *revised problem statement*
            list of final objectives
            metrics for final objectives
            list of final constraints

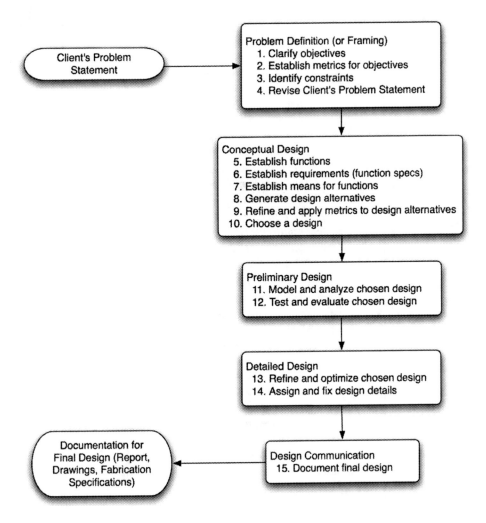

FIGURE 2.2   A five-stage *prescriptive* model of the design process. Like the descriptive model of Figure 2.1 (b), this model also styles the process as a linear sequence of objects (*need* and *final design*) and design phases, within which are situated the design tasks.

The *sources of information* for problem definition include current technical literature, codes, regulations and experts. *Design methods* include objectives trees and pairwise comparison charts. The *means* include literature reviews, brainstorming, user surveys and questionnaires, and structured interviews.

2. In the *conceptual design* stage of the design process we generate *concepts* or *schemes* of *design alternatives* or possible acceptable designs.

Inputs: *revised problem statement*
list of final objectives
metrics for final objectives
list of final constraints

Tasks: *establish functions (5)*
establish requirements (function specs) (6)
establish means for performing functions (7)
generate design alternatives (8)
refine and apply metrics to design
alternatives (9)
choose a design alternative (10)

Outputs: *requirements (specifications for functions)*
a chosen design

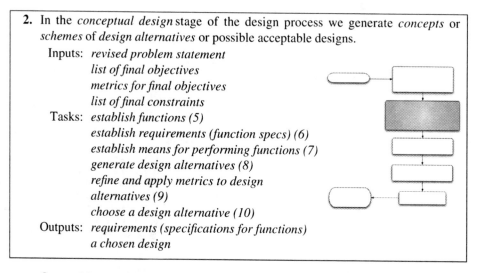

Competitive products are the additional principal *sources of information* for conceptual design. *Design methods* include functional analysis, function-means trees, morphological charts, requirements matrices, the performance specification method, and quality function deployment (QFD). *Means* include brainstorming, synectics and analogies, benchmarking, and reverse engineering (dissection).

3. In the *preliminary design* phase we identify the principal attributes of the chosen design concept or scheme.

Inputs: *requirements or specifications for functions*
the chosen design

Tasks: *model, analyze chosen design (11)*
test, evaluate chosen design (12)

Output: *an analyzed, tested, evaluated design*

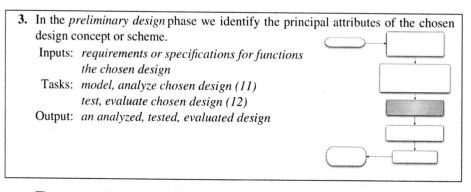

The *sources of information* during preliminary design include heuristics (rules of thumb), simple models and known physical relationships. *Design methods* include appropriate physical modeling and verifying that design requirements are met. *Means* include computer modeling and simulation, prototype development, laboratory and field tests, and proof-of-concept testing.

4. During *detailed design* we refine and optimize the final design and assign and fix the design details.

Input: *the analyzed, tested, evaluated design*

Tasks: *refine, optimize the chosen design (13)*
assign and fix the design details (14)

Output: *proposed design and design details*

The *sources of information* for detailed design include design codes, handbooks, local laws and regulations, and suppliers' component specifications. *Design methods* include discipline-specific CADD (computer-aided design and drafting). *Means* include formal design reviews, public hearings (if applicable), and beta testing.

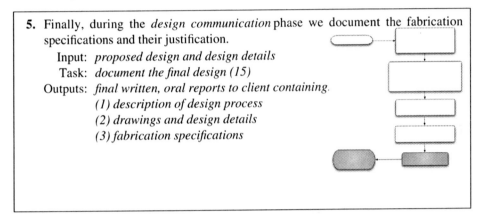

5. Finally, during the *design communication* phase we document the fabrication specifications and their justification.

    Input: *proposed design and design details*
    Task: *document the final design (15)*
  Outputs: *final written, oral reports to client containing*
         (1) description of design process
         (2) drawings and design details
         (3) fabrication specifications

The *sources of information* for the design communication phase are feedback from clients and users, and itemized lists of required deliverables.

We now have a checklist we can use to ensure that we have done all of the required steps. Lists like this are often used by design organizations to specify and propagate approaches to design within their firms. However, we should keep in mind that this and other detailed elaborations add to our understanding of the design process only in a limited way. At the heart of the matter is our ability to model the tasks done within each phase of the design process. With this in mind, we will present some means and formal methods for doing these 15 design tasks in Section 2.3.

2.2.3 Feedback and iteration in the design process

All of the models presented so far have been "linear" or sequential. The design process is not linear or sequential, however, and two very important elements must be added. The first of these is *feedback*, that is, the process of feeding information about the output of a process back into the process so it can be used to obtain better results. Feedback occurs in two notable ways in the design process, as illustrated in Figure 2.3:

- First, there are *internal feedback loops* that come *during the design process* and in which the results of performing the test and evaluation task are fed back from the preliminary design phase to *verify* that the design performs as intended. As noted in Section 1.3, feedback comes from internal customers, such as manufacturing (e.g., can it be made?) and maintenance (e.g., can it be fixed?).
- Second, there is an *external feedback loop* that comes *after the design reaches its intended market* and in which user feedback then *validates* the (presumably successful) design.

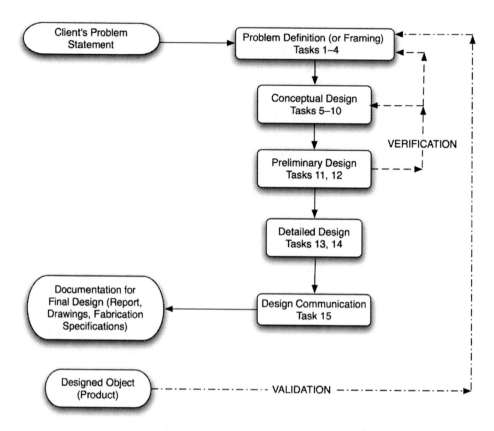

FIGURE 2.3 A five-stage prescriptive model of the design process that shows (in dashed lines) feedback loops for: (a) *verification*, internal feedback *during* the design process; and (b) *validation*, external feedback obtained after the design process when the designed object or device is in actual use.

The second element that we have thus far left out of our process models is *iteration*. We iterate when we repeatedly apply a common method or technique at different points in a design process (or in analysis). Sometimes the repeated applications occur at a different *levels of abstraction* wherein we know different degrees of detail, so we might use different scales. Thus, as we fix more detail, we become less abstract. Such iterations or repetitions typically occur at more refined, less abstract points in the design process (and on a finer, more detailed scale in analysis). In terms of the linear five-stage model depicted in Figure 2.2, we should anticipate repeating tasks 1–3 and 5 in some form in conceptual, preliminary, and detailed design. That is, we always want to keep the original objectives in mind to ensure that we have not strayed from them as we get deeper and deeper into the details of our final design. Of course, this may also mean that we may have reason to do some redesign, in which case we will certainly repeat tasks 11 (analyzing the design) and 12 (testing and evaluating the design) as well.

Given that there are feedback loops and that we will repeat or reiterate some tasks, why did we present our process models as linear sequences? The answer is simple. We noted in Chapter 1 that "design is a goal-directed activity, performed by humans." As

Iteration and feedback are integral parts of the design process.

important as the feedback and iterative elements of design are, it is equally important not to be overly distracted by these adaptive characteristics when learning about — and trying to do — design for the first time. It is also true that, in some sense, feedback loops and the need to repeat some design tasks will occur naturally as a design project unfolds. When we are doing a design for a client, it is only natural to go back and ask the client if the original project statement was properly revised. And it is just as natural to show off emerging design concepts, and to refine these schemes by responding to feedback and by reiterating objectives, constraints, and requirements.

2.2.4   On opportunities and limits

The primary focus of this book is conceptual design, the first phase of the design process. As a result, we will often be dealing with some very broad themes and approaches in ways that are logical, but not as neat and tidy as a set of formulas or algorithms. In fact, conceptual design tools lend themselves to answering questions that are not easily posed in formal mathematical terms. It is ironic that this seeming lack of rigor of the tools we will present for use in conceptual design also makes them very useful more generally for *problem solving*.

2.3   STRATEGIES, METHODS, AND MEANS IN THE DESIGN PROCESS

Even prescriptive descriptions of design processes can fail us because they don't tell us *how* to generate or create designs. Here we will briefly introduce some of the formal design methods and some of the means of acquiring design-related information, as a prelude to the more detailed descriptions given in Chapters 3–5. Remember that we are introducing these decision-support techniques and tools to explain *how* we go about designing systems or devices, that is, we are describing *thought processes* or *cognitive tasks* that will be done during the design process. We begin with ideas for strategic approaches to design thinking.

2.3.1   Strategic thinking in the design process

It is generally unwise to commit to a particular concept or configuration until forced to by the exhaustion of additional information or of alternate choices or of time. This general strategy for thinking about design is called *least commitment*. (Remember that saying, "Never marry your first design.") Least commitment is less a method than it is a good strategy or habit of thought. It militates against making decisions before there is a reason to make them. Premature commitments can be dangerous because we might become attached to a bad concept or we might limit ourselves to a suboptimal range of design choices. Least commitment is of particular importance in conceptual design because the consequences of any early design decision are likely to be propagated far down the line.

Another important strategy of design thinking is to apply the power of *decomposition*, that is, of breaking down, subdividing or decomposing larger problems (or entities or ideas) into smaller subproblems (or sub-entities or sub-ideas). These smaller subproblems are usually easier to solve or otherwise handle. That is why decomposition is sometimes identified as *divide and conquer*. We have to keep in mind that subproblems can interact, so we must ensure that the solutions to particular subproblems do not violate the assumptions or constraints of complementary subproblems.

2.3.2   Some formal methods for the design process

We now present brief introductions to the formal design methods listed in the five charts representing the five stages of the design process (Section 2.2.2).

We build *objectives trees* in order to clarify and better understand a client's project statement. Objectives trees are hierarchical lists of the client's objectives or goals for the design that branch out into tree-like structures. The objectives that designs must attain are clustered by sub-objectives and then ordered by degrees of further detail. The highest level of abstraction of an objectives tree is the top-level design goal, derived from the client's project statement. In Section 3.2 we explain how to construct objectives trees and explore the kinds of information we learn from them.

We rank design objectives using *pairwise comparison charts* (PCCs), a relatively simple device in which we list the objectives as both rows and columns in a matrix or chart and then compare them on a pair-by-pair basis, proceeding in a row-by-row fashion. Pairwise comparison charts are useful for rank ordering objectives early in the design process, and they are described in Section 3.4.

In Section 3.5 we discuss the process of *establishing metrics* for assessing how well a design's objectives are achieved. That is, we explain how we can measure the attributes and behavior of proposed *design alternatives*, so that we can choose a final design that best reflects the client's objectives.

Functional analysis is used to identify what a design must do. One starting point for analyzing the functionality of a proposed device is a "black box" with a clearly delineated boundary between the device and its surroundings. Inputs to and outputs from the device occur across that boundary, and we often assess the inputs and outputs by (1) tracking the flow of energy, information and materials through the device and (2) detailing how energy is used or converted and how information and/or materials are processed to produce the desired functions. We present functional analysis and the related tools in Section 4.1.

The *performance specification method* provides support for the elaboration of the *requirements* that reflect, in engineering terms, how a design will function. The aim is to list solution-independent attributes and performance specifications (i.e., "hard numbers") that specify the requirements of a design concept. We describe performance specifications (requirements) and their role in Section 4.2.

The *morphological chart* is used to identify the ways or means that can be used to make the required function(s) happen. The *functions* are expressed as *verb-noun action pairs*, and the *means* are specific ways or devices for using or converting energy and for processing information and/or materials. The morph chart, a matrix, provides a visual framework of the *design space*, that is, of an imaginary "plane," "room," or "space" that we can use to generate, collect, identify, store, and explore all of the potential design alternatives that might solve our design problem. We describe design spaces and morph charts in Section 5.1.

2.3.3   Some means for acquiring and processing design knowledge

Here we describe the means by which information can be gathered and analyzed for use in the formal design methods. These means are tools that have been developed in a number of disciplines. We organize them into three categories: means for acquiring information; means for analyzing the information obtained and testing outcomes against desired results; and means for obtaining feedback from clients, users, and stakeholders (i.e., other interested parties). In many cases the means are so widely applied that we view their detailed description as beyond our scope. In other cases, however, the means are so important that we will provide expanded discussions later in the book.

2.3.3.1   Means for acquiring information   The classic (and familiar) method of determining the state of the art and prior work in the field is the *literature review*. In the preprocessing and conceptual stages we need literature reviews to enhance our understanding of the nature of potential users, the client, and the design problem itself. We can also consider prior or existing solutions, including product advertising and vendor literature. In preliminary design we are more likely to look to the technical literature regarding the physical properties of possible solutions. In the detailed design stage we want to look at handbooks, material properties tables, and design and legal codes.

User surveys and questionnaires are used in market research. They focus on identifying user understanding of the problem space and user response to possible solutions. Market research can help a designer to clarify and better understand the design problem in its early stages, so that questions are necessarily open-ended. Later surveys may be used together with pairwise comparison and morphological charts to assist in selection and choice.

Focus groups are an expensive way of allowing a design team to observe the response of appropriately selected users and others to potential designs. Since the intelligent use of focus groups demands considerable sophistication in matters psychological, they are not generally used by student design teams.

On the other hand, *informal interviews* are often undertaken very early in a design project, when the team is still trying to define the problem sufficiently to plan an approach. While informal interviews are relatively easy to conduct, it is important to be sensitive to the time and other constraints of the interviewee. All too often a design team will simply show up and ask seemingly random questions, with the dual effect of highlighting their ignorance and doing little or nothing to dispel it. There are ways to reduce this problem, including sending the interviewee copies of the topics and questions in advance and undertaking extensive literature research prior to conducting interviews.

The *structured interview* is another means for eliciting information that combines the consistency of a survey with the flexibility of informal interviews. Here the interviewer uses a previously defined set of questions that may or may not be made available to the interviewees. In addition to using the question set to get direct answers, the interviewer can follow up a particular response and open up new areas. A structured set of questions also assures the interviewee that the interview has both purpose and focus, and it ensures that interesting side issues do not prevent the key matters from being covered.

Design teams can also acquire further insight by doing *brainstorming,* an activity that allows the participants to generate related (or even unrelated) ideas that are listed but not evaluated until a later time. The free-wheeling nature of brainstorming can be very helpful in opening up new avenues for research and analysis. However, as we note in more detailed discussions of brainstorming in Sections 5.2.3 and 9.1, it is quite important that team members maintain a high level of respect for the ideas of others and that *all* ideas are captured and listed as they are offered.

Design teams can further draw upon their own abilities to discover and explore relationships and similarities between ideas and solutions that initially seem unrelated. In an environment free of criticism and evaluation, similar to that in brainstorming, the team conducts a *synectic* activity in which it tries to uncover or develop *analogies* between one type of problem and other types of problems or phenomena. For example, a team might seek to redefine a problem into terms of some other matter solved by an earlier team. Or, the team might initially seek to find the most outrageous solutions to a problem and then look for ways that such solutions could be made useful. We will say more about synectics in Section 3.6, but we note that it is time consuming and demands serious commitment on the part of the design team. Interestingly, synectics is generally more widely used in industrial design settings than in academia.

Finally, design teams often use two (unrelated) activities to look at "what's out there" when developing new products:

- *Competitive products* are *benchmarked,* that is, designers look at similar products that are already available and try to evaluate how well those products perform certain functions or exhibit certain features. These competitive products serve as standards for "building a better mousetrap" since proposed new designs can be compared to existing mousetraps.

- *Reverse engineering* or *dissection* consists of dissecting or taking competitive or similar products apart. The idea is to determine why a given product or device was designed the way it was, with the intention of finding better ways of performing the same or similar subfunctions.

2.3.3.2 *Means for analyzing information and testing outcomes*   An important first step in determining whether or not design concepts might work is to establish ways to measure outcomes. Of particular interest are the development of *metrics* that can be used to measure whether a design's objectives are achieved, and the statement of *requirements* or *design specifications* that set forth in engineering terms the functional performance of a design. This is an extremely important topic that we will discuss in depth in Sections 4.4 and 5.3.

In some cases we find it possible to gather and assess data about a potential design solution by undertaking *experiments* in the field or in a laboratory. For example, if the solution involves a structure, it may be possible to measure the stress or strength relationships of critical parts of the design in a laboratory test.

A crucial step along the path from conceptual design to detailed design is *proof-of-concept testing.* Such testing, which is more likely physical but may be computational, involves establishing a formal means of determining whether or not the concept under consideration can reasonably be expected to meet the design requirements. The idea of a proof-of-concept test is to show that a design concept will perform as anticipated under certain prespecified conditions. For example, suppose we are given a batch of pins made of different materials: acrylic, aluminum, nylon, stainless steel, steel, or wood. We propose to separate the pins by material, using each pin's electrical properties. A proof-of-concept test might be to show that each pin has an electrical property (e.g., its resistance) that can be measured to determine that pin's material. As with any scientific test, we must define outcomes that are sufficient for accepting or rejecting a concept.

Prototype development is a very important means of determining whether or not a design can perform its required functions. Here a *prototype* or test unit embodies the principal functional characteristics of the final design, even though it may not look at all like the expected end product. In fact, early prototypes typically have only a subset of the required functionality, both to shorten development time and to reduce costs. Prototypes can be instrumented to support laboratory or other tests. In addition, rapid prototyping, wherein a simplified version of the designed object is rapidly created by some type of electrodepositing, has become very common in product development and design.

In many cases, however, we can't develop or test a prototype, perhaps because of cost, size, or hazards. In such cases we often resort to *simulation* in which we exercise an analytical, computer or physical model of a proposed design to simulate its performance under a stated set of conditions. This presumes that the device being modeled, the conditions under which it operates, and the effects of the operations are all quite well understood. Such deep understanding enables modeling to produce appropriate and useful data that in turn enables evaluation against constraints, requirements and applicable standards. One outstanding example of such simulation is the use of wind tunnels and related computer analyses to assess the effects of wind loading on tall buildings and on long, slender suspension bridges.

Closely related to simulation, *computer analysis* involves the development of a computer-based model, which may consist simply of analytic, discipline-based techniques relevant to describing the design. These include finite element analysis, integrated circuit modeling, failure mode analysis, criticality analysis, etc. Computer-based models are used widely in all engineering disciplines and they become even more important as a design project moves into detailed design.

2.3.3.3 *Means for obtaining feedback*

Among the most important means for obtaining feedback from both clients and users are *regularly scheduled meetings* at which the progress of the design project, including articulation of the various phases of the design process, is tracked and discussed. We assume throughout our discussions that the design team is always communicating with both clients and users. We suggest that various formal design results be reviewed with them often.

It is a standard practice to hold a *formal design review* at specified intervals of the design process in which the current design is presented to the client(s), selected users, and/or other stakeholders. Design review presentations typically include sufficient technical detail that the implications of the design can be fairly explored and assessed. It is particularly important that young designers become comfortable with the "give and take" that often accompanies such reviews. While it may seem harsh to be asked to justify various technical details to clients and outside experts, it is usually beneficial because implicit unwarranted assumptions and errors or oversights are often uncovered.

In some design environments, relevant civil laws or public policies require that *public hearings* be held for the purpose of exposing the design to public review and comment. While it is beyond our scope to consider such hearings in detail, it is useful for designers to understand that public hearings and meetings are increasingly the norm for major design projects, even when the client is a private corporation.

We have already noted that *focus groups* are important sources of user input on problem definition. Such groups are also widely used to assess user reaction to designs as they near adoption and marketing.

In some industries, most notably software design, an almost-but-not-quite-finished version of a product is released to a small number of users for *beta testing*. Beta tests allow designers to expose design or implementation errors and to get feedback about their product before it reaches a larger market.

2.4 GETTING STARTED ON MANAGING THE DESIGN PROCESS

Just as there are many models describing the design process, there are also many models of how projects are managed. We show one such road map of the project management process for design projects in Figure 2.4, in direct analogy with Figure 2.2 for the design process. This model presents project management as a four-phase process:

- *project definition* or *scoping*, developing an initial understanding of the design problem and its associated project;
- *project framing*, developing and applying a plan to do the design project;
- *project scheduling*, organizing that plan in light of time and other resource constraints; and as the project unfolds,
- *project tracking*, *evaluation* and *control*, keeping track of time, work, and cost.

In Chapter 9 we will detail a number of tools to help us proceed along this path. We defer discussion of project management tools to a later chapter not because they are unimportant, but only so as to maintain a continuity of focus on the design process. Since design is generally done by teams and it consumes serious resources, project management is critical to project success. However, one design team activity that cannot be entirely deferred is the organization and development of the project team. We briefly turn to that here.

Design is a social activity. Design is a social activity: Design is increasingly done by teams, rather than by individuals acting alone. For example, new products are often developed by teams that include designers, manufacturing engineers, and marketing experts. These teams are assembled to gather the diverse skills, experiences, and viewpoints needed to successfully design, manufacture, and sell new products. This dependence on teams is not surprising if we reflect on the information, methods, and means for design that we have been discussing. Many of the activities and methods are devoted to applying different talents and

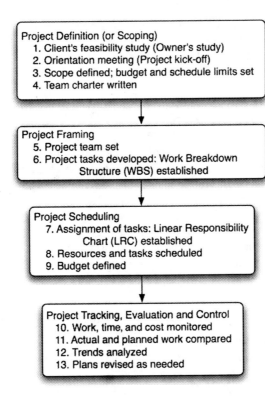

Project Definition (or Scoping)
1. Client's feasibility study (Owner's study)
2. Orientation meeting (Project kick-off)
3. Scope defined; budget and schedule limits set
4. Team charter written

Project Framing
5. Project team set
6. Project tasks developed: Work Breakdown
Structure (WBS) established

Project Scheduling
7. Assignment of tasks: Linear Responsibility
Chart (LRC) established
8. Resources and tasks scheduled
9. Budget defined

Project Tracking, Evaluation and Control
10. Work, time, and cost monitored
11. Actual and planned work compared
12. Trends analyzed
13. Plans revised as needed

FIGURE 2.4 Managing a design project follows an orderly process, beginning with the client's understanding of the problem. At the early stages, the design team is concerned with understanding the problem and making plans to solve it. Later, the focus shifts toward project control and staying on plan. Adapted from (Orberlander 1993).

skills to achieving a common understanding of a problem. Consider, for example, the difference between laboratory testing and computer-based analysis of a structure. Both require common knowledge of structural mechanics, yet years of investment are required to master the specific testing and laboratory skills or the analysis and computer skills. Thus, there may be considerable value in constructing teams whose membership has, collectively, all of the needed skills and can work together successfully.

Groups and teams are such an important element of human enterprise that we should not be surprised to learn that they have been extensively studied and modeled. One of the most useful models of group formation suggests that almost all groups typically undergo five stages of development as they transition from groups into successfully functioning teams. Those five stages have been rather memorably named as:

- *forming,*
- *storming,*
- *norming,*
- *performing,* and
- *adjourning.*

Suffice it to say here that the names attached to this five-stage model of group dynamics suggest that starting a new design team (or even entering an existing design team for the first time) is not simple or straightforward. Thus, it is worth both thinking about how we relate to others on teams (and in this context, groups are not teams!) and reading more about team dynamics in Section 9.1.

2.5 CASE STUDY AND ILLUSTRATIVE EXAMPLES

John Maynard Keynes, a famous economist, once said that "Nothing is required, and nothing will avail, except a little, a very little, clear thinking." Design, however, is best learned by combining clear thinking with doing.* It is also fair to say that *design is best experienced by doing*. To that end, we strongly encourage fledgling designers and engineers to participate actively in teams doing design projects. Some of the formal techniques can be learned by doing exercises and by observing how others have applied the techniques. To this end, we will elaborate a design case study and begin two design examples that illustrate the kinds of problems that engineers face when doing conceptual design, that is, when they have to devise concepts or ideas to solve a design problem.

2.5.1 Case study: Design of a microlaryngeal surgical stabilizer

We now present a case study of the conceptual design of a device to help stabilize the instruments used during *laryngeal* or vocal cord surgery. This design project was undertaken by four teams of students in Harvey Mudd College's first-year design class on behalf of Brian Wong, M.D., of the Beckman Laser Institute of the University of California Irvine. The case study is an edited combination of results obtained by the four teams that shows *how a design team thought through the design process while they were designing a device for a client.*

Laryngeal, or vocal cord, surgery is often required to remove growths such as polyps or cancerous tumors. The "lead" cells of such growths must be removed accurately and completely. Patients also incur the risk of damage to their vocal cords — and so their speech — during these surgeries. In spite of many other surgical advances over recent decades, laryngeal surgery has not changed much. One change that has occurred is that surgeons now access the vocal cords through the mouth, rather than by cutting open the throat. This has made it harder to insert and stabilize both optical devices and surgical instruments that cut, suck, grasp, move, and suture. Surgeons must be able to control their own tremors in order make accurate and precise cuts during the procedure.

Tremor is the natural, small-scale shaking of the hand. (Watch the movement of your own fingertips as you hold your hands straight out in front of you.) In the context of laryngeal surgery, such tremors tend to be amplified as the surgeons insert and control foot-long instruments in the patient's throat.

---

* One of the Harvey Mudd founding fathers, the late Jack Alford, said that learning to design was like to learning to dance: "You have to get out on the dance floor and get your toes stepped on." We are simply making explicit the idea that you would do well to take some lessons as you get out on the floor!

The project began when Dr. Wong presented the following *initial problem statement* to the four teams of three or four students who chose to work on this project:

> *Surgeons who perform vocal cord surgery currently use microlaryngeal instruments, which must be used at a distance of some 12–14 in to operate on surfaces with very small structure (1–2 mm). The tremor in the surgeon's hand can become quite problematic at this scale. A mechanical system to stabilize the surgical instruments is required. The stabilization system must not compromise the visualization of the vocal cords.*

The teams began by *defining* or *framing* this design problem. They talked with Dr. Wong and other physicians and did some basic library research to gain further information about laryngeal surgery. They learned that the abnormalities that were operated on were typically 1–2 mm wide, while the vocal cords themselves are approximately 0.15 mm thick. This meant that the physiological surgical tremors of the surgeon's hands had to be reduced from 0.5–3.0 mm to an acceptable tremor amplitude of 0.1 mm. They also learned that the surgeons needed to control the instruments at distances far from the patient's mouth (and vocal cords). One of the teams wrote the following *revised problem statement*:

> *Microlaryngeal surgery seeks to correct abnormalities in the vocal cords. The abnormalities, such as tumors and cysts, are often 1–2 mm in size and are typically removed from the vocal cords, which are only 0.15 mm in size. During the operation, the surgeon must control his or her surgical instruments from a distance of 300–360 mm (12–14 in) due to the difficulties in accessing the vocal cords. At this small scale, the physiological tremor in the surgeon's hand can be problematic. Design a solution that minimizes the effects of hand tremors in order to reduce unintentional movements at the distal end of the instrument to an amplitude of no more than 1/10 of a millimeter. The solution must not compromise visualization of the vocal cords.*

Note that this revised problem statement contains more detail and also excludes an implied "mechanical" solution referenced in the original problem statement.

Continuing with problem definition, the teams also developed a list of the client's *objectives* for the designed stabilizing device that would summarize the attributes the client hoped the device would have. The objectives and sub-objectives are routinely displayed in an *objectives tree*, and one team's objectives tree for this project is displayed in Figure 2.5. Two of the objectives are that the device should minimize obstruction of the surgeon's vision and that the cost of manufacture should be minimized. At the same time, the teams also developed lists of *constraints*, that is, the strict limits within which the designed device must remain. A constraint list for the device includes:

- it must be made of nontoxic materials;
- it must be made of materials that do not corrode;
- it must be sterilizable;
- its cost must not exceed $5,000;
- it must not have sharp edges;
- it must not pinch or gouge the patient; and
- it must be unbreakable during normal surgical procedures.

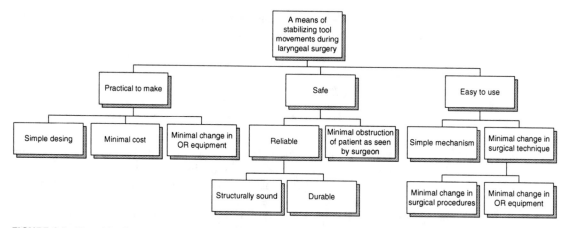

FIGURE 2.5   The *objectives tree* displaying the client's objectives and sub–objectives for the microlaryngeal stabilization device whose design is presented as a case study in Section 2.5.1. This tree is developed largely from the work of one of the three teams who worked on this project. After (Chan et al. 2000).

We see that there is an upper limit on the cost and on the device having sharp edges, among others.

Another facet of framing the problem involved rank ordering the objectives for the design in terms of their perceived relative importance. This ranking was done using a *pairwise comparison chart* (PCC), which is an extension of what people normally do when comparing two objects one against the other. The PCC, which will be explained in detail in Section 3.3, enables each objective to be compared to every one of the other objectives, The PCC produced by one of the teams is shown in Table 2.1, and it shows that the most important objective is to reduce the surgeon's tremor, while the least important is the cost of the instrument. This ranking helped focus the team's attention, as well as seeming to accord with our intuitions.

TABLE 2.1 A *pairwise comparison chart* created by one of the student teams to compare objectives for the microlaryngeal stabilization device. An entry '1' indicates the objective in that row is more important than that of the column in which it is entered. It shows that the reduction of the surgeon's tremor is the most important objective for this project. After (Both et al. 2000)

GOALS	Reduce tremor	Sturdy	Safe	Inexpensive	Easily used	SCORE
Reduce tremor	••••	1	1	1	1	4
Sturdy	0	••••	0	1	0	1
Safe	0	1	••••	1	1	3
Inexpensive	0	0	0	••••	0	0
Easily used	0	1	0	1	••••	2

The teams then set about establishing *metrics* that would enable them (later in the design process) to measure whether various designs would achieve the objectives set out for the project. The metrics for two of the objectives of Figure 2.5, along with their units and scales, are:

> Objective: *Minimize viewing obstruction.*
>
> Units: Rating percentage of view blocked on a scale from 1 (worst) to 10 (best).
>
> Metric: Measure the percentage of view blocked by the instrument. On a linear scale from 1 (100%) to 10 (0%), assign ratings to the percentage of view blocked.

> Objective: *Minimize the cost.*
>
> Units: Rating cost on a scale of 1 (worst) to 5 (best).
>
> Metric: Determine a bill of materials. Estimate labor, overhead, and indirect costs. Calculate the total cost. On a scale from 1 (worst) to 5 (best), assign ratings to the calculated cost as follows:

Cost ($)	*Points*
4,000–5,000	1
3,000–4,000	2
2,000–3,000	3
1,000–2,000	4
1–1,000	5

Having thus evolved to a deeper understanding of what the client wanted from this design, and of what is desired of the design's attributes, the design teams then turned to *conceptual design*, that is, to determining what a successful design will actually do. That is, the teams set about determining the *functions* that their proposed devices will perform, and writing the *requirements*, which are the engineering statements or specifications of the performance of the functions. The teams identified the required functions by applying some of the tools that will be discussed in detail in Section 4.1, including the *black box*, the *glass box*, and the *functions-means tree*. One such list of functions states that the microlaryngeal stabilizer must:

- stabilize the instrument;
- move the instrument;
- stabilize the distal end of the instrument;
- reduce surgeon's muscle tension (shaking tremors) during surgery; and
- stabilize itself.

The requirements for the first of these functions were written as:

> Function: *Stabilize the instrument.*
>
> Requirements: This function is not achieved if the design cannot reduce the amplitude of a trembling hand to less than 0.5 mm; it is optimally achieved if it

controls the amplitude of a trembling hard to make it less than 0.05 mm; and it is overly restrictive if it inhibits or disallows any instrument or hand use.

With the functions and the specifications now largely determined, the design process turns to *creating* or *generating alternative designs*. One excellent way to begin creating designs is to list each of the required functions in the left-hand column of a matrix, and then list across each functional row the various *means* by which each function can be implemented. As we explain in Section 5.1, the resulting matrix or chart is called a *morphological chart* or morph chart. We show a morph chart for the microlaryngeal stabilizer in Figure 2.6. Such morph charts effectively tell us how large a *design space* we are working in because each design candidate must achieve every function, no matter which of its implementations or means are used. Thus, for each of the five functions displayed in Figure 2.6, we select one of its following means to produce a possible design. As we will detail in Section 5.1, a given function's means does not necessarily connect with all of the means of all of the other functions, so there will inevitably be combinations that are excluded. Still, the combinatorial effects can be daunting for designs that have many functions since each function can be implemented by several different means.

For the present design project, one design alternative positions the surgical instrument at the end of a *lever*. A second design alternative has the instrument supported on a *stand*, moved by a system of *pulleys*, and supported by the *stand* itself. The instrument stand removes the need for surgeon to operate in a fixed position, which thus reduces tremor-inducing muscle tension. In the third design alternative, the surgeon's *hands* hold and move the instrument. Distal support is provided by *crosswires* attached directly to the laryngoscope. A *forearm rest* reduces the surgeon's tremor-inducing muscle tension. Figure 2.7 shows concept drawings for these three design alternatives.

We complete the conceptual design phase of the design process by narrowing the field of possible designs and, eventually, *selecting a final design*. We do this by exercising a *decision* or *selection matrix* in which we rate each possible design based on how well it achieves each of the design objectives as measured by the metrics described just above. The scores earned by each objective are then added for each design to get a cumulative total for each design, as shown in Figure 2.8. As we will point out in Chapters 3 and 5, this decision matrix should be used with care because sometimes the numbers are subjective (e.g., some of the metrics may be qualitative rather than quantitative measurements) and we cannot establish any weighting of the objectives beyond ranking them in perceived order of importance. In the present instance, for example, the design that was finally chosen by the client, the *crosswire* support, came in second in the selection process, although its total (770) was only marginally lower than the total points earned by the first-place design (790).

Design concepts or ideas must also be tested in meaningful ways in order to ensure that they work. One of the student design teams tested its concept by attaching a pencil lead to the end of a surgical tool and traced a pre-drawn square with the instrument both with and without their design being attached. As we can see from their test results, displayed in Figure 2.9, their designed device removed almost all of the tremor.

Finally, after all of the work and the selections, the design process is complete and a design is offered to the client. In the present case, one of the selected designs is being used in laryngeal surgeries and being prepared for manufacture as a medical product.

FUNCTION	POSSIBLE MEANS						
Stabilize instrument	Hand	Stand	Clamp	Magnet	Edge of Laryngoscope	Wire	
Move instrument	Hand	Gears	Pneumatics	Ball bearing	Lever	Pulley	
Stabilize distal end of instrument	Magnet	Crosswires	Track system	Spring	Gyroscopes	Ball Bearings	Stand
Reduce muscle tension of surgeon during surgery	Instrument Stand	Hand Platform	Pillow	Elbow Platform	Forearm Rest	Shoulder Sling	
Stabilize itself	Gyroscope	Springs System	Stand	Magnet	Suspension System	Rest against stable surface	Attach to laryngoscope

FIGURE 2.6 A *morphological chart* for the microlaryngeal stabilization device showing the *functions* and the corresponding *means* or *implementations* for each function. Possible designs are assembled in a "Chinese menu" fashion, that is, one from row *A*, one from row *B*, etc. After (Chan et al. 2000).

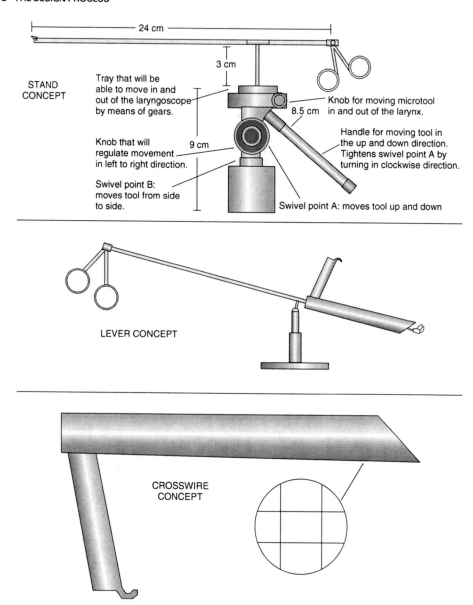

STAND CONCEPT

24 cm

3 cm

Tray that will be able to move in and out of the laryngoscope by means of gears.

Knob for moving microtool in and out of the larynx.

8.5 cm

9 cm

Knob that will regulate movement in left to right direction.

Handle for moving tool in the up and down direction. Tightens swivel point A by turning in clockwise direction.

Swivel point B: moves tool from side to side.

Swivel point A: moves tool up and down

LEVER CONCEPT

CROSSWIRE CONCEPT

FIGURE 2.7 Three *design alternatives* produced by the student teams that worked on the microlaryngeal stabilization device for surgeons at the University of California at Irvine. Dr. Wong and his colleagues have adopted the crosswire concept for clinical trials. After (Chan et al. 2000, Saravonas et al. 2000).

This case study has presented many of the design tools that are the main focus of this book. We have not shown the management tools, although the design teams on this project did use them, and we have said nothing about the dynamics of each of the three design teams. These as yet unsung elements are also very important for achieving effective design results.

DESIGN CONSTRAINTS	LEVER	INSTRUMENT STAND	LARYNGOSCOPE CROSSWIRES
C: Must not fall apart during surgery	y	y	y
C: Noncorrosive materials	y	y	y
C: Must withstand medical sterilization procedures (autoclave, enzymatic, bleach, etc.)	y	y	y
C: Cannot get in the way of surgical instruments	y	y	y
C: Cannot block the view of vocal cords	y	y	y
C: Materials must be compatible with human body	y	y	y
C: Must be easy to clean with conventional means (scrub brush, water jet, soaking, etc.)	y	y	y
C: Cannot cost more than $5,000	y	y	y
DESIGN OBJECTIVES	**Score**	**Score**	**Score**
O: Structurally sound	75	85	80
O: Strong materials	85	90	85
O: Minimum obstruction of vocal cords	100	100	65
O: Minimum obstruction between patient and surgeon/nurse	65	70	100
O: Simple design	60	70	90
O: Minimum cost	50	70	55
O: Compatible with existing instruments	30	80	50
O: Minimal alteration of existing surgical procedures	45	80	70
O: Compatible with existing instruments	30	85	80
O: Simple mechanism	70	60	95
TOTAL	610	790	770

FIGURE 2.8 A *decision* or *selection matrix* used by one of the student teams that worked on the microlaryngeal stabilization device to select a final design. The decision matrix, whose numbers should be taken with caution, suggests which designs are preferred. After Chan et al. 2000).

2.5.2 Illustrative examples: Descriptions and project statements

We now describe the two illustrative examples that will be carried through the next four chapters to illustrate the various design techniques in developing yet familiar contexts. The first illustrative example is the design of a container for a new juice product. It will

FIGURE 2.9 An example of the *testing* done by one student team to show that their concept successfully stabilized the surgeon's hand and reduced tremor, as demonstrated by the successful tracing of a pre-drawn square. After (Chan et al. 2000).

be the design project for which we introduce and explain formal design methods. A summary of the beverage container project is:

> *Design of a container to deliver a children's beverage.* This is a stylized industrial design project that highlights some of the early questions that must be addressed before a designer can apply more conventional engineering science knowledge to the problem.
>
> Designers: Dym, Little, Orwin and Spjut LLC.
>
> Clients: American Beverage Company (ABC) and National Beverage Company (NBC).
>
> Users: Children living both in the United States and abroad.
>
> Problem statement: Design a bottle for a new children's juice product.

The second illustrative design example is based on the work done by students in the first-year design course at Harvey Mudd College. We use their results, with both their permission and some post-project critiques of our own, to illustrate and further explain how formal design methods are used. The design teams' results and some commentary on those results appear at the ends of the chapters as particular formal design methods are introduced. Note, too, that the design teams' results will be displayed *exactly as they presented them in their final reports* at the end of the semester, without any further edits. This second illustrative example project is:

> *Design of an arm restraint device for children with cerebral palsy.* The arm restraint was designed by a team of students in Harvey Mudd College's first-year design class.
>
> Designers: Teams of students in HMC's first-year design class.
>
> Client: Danbury School, Claremont, California
>
> Users: Students at Danbury School with diagnosis of cerebral palsy (CP).
>
> Abbreviated project statement: Design a device to stabilize the arm of a student and counteract her involuntary, CP-induced tremors as she writes or draws.

2.6 NOTES

Section 2.1: The stepladder example derives from the first-year design course taught at Harvey Mudd College and is briefly described in (Dym 1994b). The paraphrasing of Aristotle's observation is from (Dym 2005).

Section 2.3: As with definitions of design, there many descriptions of the design process and many of them can be found in (Cross 1989), (Dym 1994a), (French 1985, 1992), (Pahl and Beitz 1984), and (VDI 1987). Further descriptions of the tasks of design can be found in (Asimow 1962), (Dym and Levitt 1991a), and (Jones 1981). Examples of the application of conceptual design tools as problem-solving tools can be found in (Schroeder 1998) for automobile evaluation and in (Kaminski 1996) for college selection.

Section 2.3: Further elaboration of strategic thinking in design appears in (Dym and Levitt 1991a). More detailed descriptions of the formal design methods can be found in (Cross 1989), (Dym 1994a), (French 1985, 1992), (Pahl and Beitz 1984), and (VDI 1987). A strongly related discussion of concurrent engineering can be found in (Carlson-Skalak, Kemser and Ter-Minassian 1997). More detailed descriptions of the means for acquiring and processing knowledge are (also) given in (Bovee, Houston, and Thill 1995), (Ulrich and Eppinger 1995), and (Jones 1992).

Section 2.4: Leifer propounded the notion that engineering is a social activity (1991). The basic stages of group formation are discussed in most recent management texts.

Section 2.5: The microlaryngeal stabilization device case study is detailed in (Both et al. 200, Chan et al. 2000, Feagan et al. 2000, and Saravanos et al. 2000).

2.7 EXERCISES

2.1 Describe in your own words the similarities and differences among the four models of the design process shown in Figures 2.1–2.4.

2.2 When would you be likely to use a descriptive model of the design process? When would you use a prescriptive model?

2.3 Map the management process shown in Figure 2.5 onto the design process shown in Figure 2.4.

2.4 Explain the differences among tasks, methods, and means.

2.5 You work for HMCI, a small engineering design company. You have been named team leader for a four-person design project that will be described in more detail in Exercises 3.2 and 3.5. You have not previously worked with any of these team members. Describe several strategies for moving the team quickly to the performing stage of group formation.

2.6 As Director of Engineering at HMCI, you notice that one of your team leaders and the team's client are unable to agree on a schedule. How might you advise the team leader to resolve this matter constructively?

CHAPTER 3

DEFINING THE CLIENT'S DESIGN PROBLEM

What does this client really want? Are there any limits?

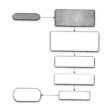

IN THE PRECEDING chapters we have defined engineering design, explored and described the process of design, and talked briefly about managing a design project. Now we turn to *problem definition*, the pre-processing phase of design, during which we frame the problem to cast it into engineering terms. Thus, we focus here on the first four design tasks identified in Figure 2.3.

3.1 IDENTIFYING AND REPRESENTING THE CLIENT'S OBJECTIVES

The starting point of most design projects is the identification by a *client* of a *problem* to be solved or a *need* to be met (recall Figures 2.1–2.4). The design team then works to solve the client's problem. The client's need is typically presented as a verbal statement in which the client identifies a gadget that will appeal to certain markets (e.g., a container for a new beverage), a widget that will perform some specific functions (e.g., a chicken coop), or a problem to be fixed through a new design (e.g., a new transportation network and hub).

3.1.1 The client's original problem statement

Sometimes clients' project statements are quite brief. For example, imagine that you are on a design team either at the American Beverage Company (ABC) or the for National Beverage Company (NBC), the clients identified in the beverage container problem given in Section 2.5. You might simply get an order from upper management that says: "Design a bottle for our new children's fruit juice product." Your design team could respond to this directive by choosing an existing bottle, designing on a clever label, and calling its work done. However, is this a *good* design? Is it the *right* design? There is no way to answer these questions because the problem statement was so brief that it gave no hint of other considerations that might enter into thinking about or assessing the design, e.g., the intended market, the shape or materials choice of the container, and so on.

Another project statement might take the this form: "The Claremont Colleges need to reconfigure the intersection of Foothill Avenue and Dartmouth Avenue so students can cross the road." While communicating someone's idea of what the problem is, statements like this one have limitations because they often contain errors, show biases, or imply solutions. *Errors* may include incorrect information, faulty or incomplete data, or simple mistakes regarding the nature of the problem. Thus, the problem statement just given should refer to the correct Foothill *Boulevard*, not Foothill Avenue. *Biases* are presumptions about the situation that may also prove inaccurate because the client or the users may not fully grasp the entire situation. In the traffic example, for instance, the real problem may not be related to the design of the intersection but to the timing of the signal lights or to the tendency of students to jaywalk. *Implied solutions*, that is, a client's best guesses at solutions, frequently appear in problem statements. While implied solutions offer some useful insight into what the client is thinking, they may wind up restricting the design space in which the engineer searches for a solution. Also, the implied solution sometimes doesn't solve the problem at hand. For example, it is not obvious that reconfiguring the intersection will solve the student traffic problem. If students jaywalk, reconfiguring the intersection will do little or nothing to mitigate this. If the problem is that students are crossing a dangerous street, we may want to relocate the destination to which they are headed. The point is that we must carefully examine project statements in order to identify and deal with errors, biases, and implied solutions. Only then do we get to the real problem.

A better understanding of the problem that is shared by both the client and the designer results from clarification of the original problem statement.

We want to focus on developing a clearer understanding of what the client wants because it will help us see the lines along which a design might emerge. That is, we want to clarify what the client wants, account for what potential users need, and understand the technological, marketing, and other contexts within which our device will function. In so doing, we will be *defining* or *framing* the design problem clearly and realistically.

3.1.2   Asking questions and brainstorming about the client's problem

In Section 2.1 we talked about the role of questioning in the design process. In fact, there are two kinds of activities that design teams can initiate, sometimes in parallel, after being given the original design problem statement. The first is asking questions of the client(s) and of stakeholders who might have varying degrees of interest in the design, for example, potential users and experts in the field. The experts might be people versed in any relevant technology or other technical aspects, or marketing experts who are familiar with the market of users toward which the design is aimed.

It is useful to be well prepared when asking questions. If we know what we are looking for, we can guide the conversation and get more information. It also helps ensure that the people who are answering the questions feel their time is not being wasted. This is very important if a program of structured interviewing of experts and/or users is envisioned because such interviews, or similar detailed survey forms, will not produce useful or serious responses *unless* the respondents feel that it's worth their time to answer a lot of questions.

Brainstorming is the second activity that design teams can initiate in problem framing. Brainstorming is a group effort in which new ideas are elicited, retained, and perhaps organized into some problem-relevant structure. It is important that brainstorming sessions remain focused, so that when trying to identify objectives and constraints, the team's focus does not shift to something else, for example, functions. While there will inevitably be some "non-responsive" suggestions, the team should try to stick to the topic at hand, namely identifying goals and objectives, and perhaps constraints. The team can do this if its leader presents statements of ideas with phrases such as, "A desirable characteristic of the device would be . . . ," which reminds others that the team's focus is on the objectives. The best outcome of brainstorming is a list of characteristics that can be pruned and refined into an indented list of objectives for the design.

3.1.3 From lists of desired object attributes to lists of objectives

Now imagine that we are on a design team that is consulting for a company that makes both low- and high-quality tools (with a corresponding range of prices!). The company's management, seeking to penetrate a new market, has given the team a charter more specific than "Design a safe ladder," namely, "Design a new ladder for electricians or other maintenance and construction professionals working on conventional job sites." This is a "routine" design task, but to fully understand the goals of this design, we have to talk with management, some potential users, some of the company's marketing people, and some experts. We also need to conduct our own brainstorming sessions. We will get a better understanding of what our design project is really about by asking questions such as:

- What features or attributes would you like the ladder to have?
- What do you want this ladder to do?
- Are there already ladders on the market that have similar features?

And while asking these three questions, we might also ask:

- What does that mean?
- How are you going to do that?
- Why do you want that?

As a result of our discussions and brainstorming, we might generate the list of attributes and characteristics of a safe ladder design shown in List 3.1.

List 3.1. SAFE LADDER Attributes List

Ladder should be useful

Used to string conduit and wire in ceilings

Used to maintain and repair outlets in high places

Used to replace light bulbs and fixtures

Used outdoors on level ground

Used suspended from something in some cases

Used indoors on floors or other smooth surfaces

Could be a stepladder or short extension ladder

A folding ladder might work

A rope ladder would work, but not all the time

Should be reasonably stiff and comfortable for users

Step deflections should be less than 0.05 in

Should allow a person of medium height to reach and work at levels up to 11 ft

Must support weight of an average worker

Must be safe

Must meet OSHA requirements

Must not conduct electricity

Could be made of wood or fiberglass, but not aluminum

Should be relatively inexpensive

Must be portable between job sites

Should be light

Must be durable

Needn't be attractive or stylish

Note that the entries in List 3.1 reflect differences between fundamentally different concepts, in fact, many of the concepts we had defined in Chapter 1. Thus, let us review, illustrate and comment on the relevant concept definitions:

- **objective** *n*: something toward which effort is directed; an aim or end of action
  "Should be relatively inexpensive"
  "Should be light"

Objectives are the desired attributes and behaviors of a design.

Objectives or *goals* are expressions of the attributes and behavior that the client or potential users would like to see in a designed system or device. They are normally expressed as "being" statements that say what the design will *be*, as opposed to what the design must *do*. For example, saying that a ladder *should* be portable is a "being" term. "Being" terms identify attributes that make the object "look good" in the eyes of the client or user, expressed in the natural languages of the client and of potential users.

Objectives are also often written as statements that "more (or less) of [the objective]" is better than "less (or more) of [the objective]." For example, lighter is usually better than heavier if our goal is portability. In fact, as we explain in Section 3.4, we will introduce metrics for objectives that enable us to measure or quantify whether (or not) the objectives are met, which will then aide us in choosing among alternative designs. Another hint of how we measure the achievement of objectives is clear from the first objective above: Ladders costing $15 or $20, respectively, are both "relatively inexpensive," but which would be more desirable?

- **constraint** *n*: the state of being checked, restricted, or compelled to avoid or perform some action
  "Must not conduct electricity"
  "Must be durable"

Constraints are restrictions or limitations on a behavior or a value or some other aspect of a designed object's performance. Constraints are typically stated as clearly defined limits whose satisfaction can be framed into a binary choice; for example, the ladder material is a conductor or it is not, or the step deflection is less than 0.05 in or it is not. Constraints are important to the designer because they limit the size of a design space by forcing the exclusion of unacceptable alternatives. For example, a ladder design that fails to meet OSHA standards will be rejected.

Constraints are strict limits that a design must meet to be acceptable.

Objectives and constraints are closely related and sometimes seem to be interchangeable, but they are not. (As we will note on p. 3–27, there are circumstances where an objective may be converted into a constraint, but that does not make them interchangeable.) Constraints limit the size of the design space, while objectives permit the exploration of the remainder of the design space. That is, constraints are formulated to enable the rejection of alternatives that are unacceptable, while objectives enable a selection among design alternatives that are at least acceptable, or, in other words, that *satisfice*. Designs that satisfice may not be optimal or the best, but at least they satisfy all constraints. For example, we could minimally satisfy OSHA standards or we could significantly surpass those standards by making a "super-safe" ladder in order to obtain a marketing advantage. Or, on the price side, a goal that a ladder should be "relatively inexpensive" might also have a constraint that the ladder's cost cannot exceed $25. If we have *both* a low-cost objective and a $25 constraint, we may exclude some initial designs based on the constraint, while choosing among the remaining designs based on cost and other, non-economic objectives.

It is important to keep in mind that objectives and constraints *refer to the device or system being designed,* not to the design process. A "low-cost ladder," for example, has a low manufacturing or production cost. Costs incurred during the design process (engineering salaries, market surveys, prototype development, etc.) may be high, but that's a separate matter altogether.

Functions are actions that a successful design must perform.

- **function** *n*: the action for which a person or thing is specially fitted or used or for which a thing exists; one of a group of related actions contributing to a larger action
  "Must support weight of an average worker"
  "Must insulate user"

Simply put, *functions* are those things the designed device, system or process is supposed to *do*, the actions that it must perform. In an initial attributes list, functions are usually expressed as "doing" terms, such as the first function above. Often they refer to *engineering functions*, such as the second function highlighted above (which is also a constraint), which says that the flow of electrical charge must be prevented.

- **means** *n*: an agency, instrument, or method used to attain an end
  "Could be made of wood or fiberglass, but not aluminum"

Means or *implementations* are ways of executing those functions that the design must perform. On the attributes list these entries provide specific suggestions about what a final design will look like or be made of (e.g., the ladder will be made of wood or of fiberglass), so they often appear as "being" terms. However, it is generally obvious which "being" terms are objectives to be achieved and which "being" terms point to specific properties. Means and implementations are very much *solution-dependent* in that they are often selected to implement the functions that are to be performed by an already chosen design.

Implementations are specific choices of design options.

We can now pare the list of attributes (List 3.1) by removing or pruning the constraints, functions, and implementations, leaving only objectives on the list. Thus, our pruned list of objectives for the ladder is given in List 3.2.

While List 3.2 is useful as a list of objectives to be achieved, there is much more that we can do with it. In particular, if our list was much longer, we might find it difficult to use the list without organizing it in some way. Consider the several uses that we have identified for the ladder. While this is not an exhaustive list of ways to use a ladder, we may want to group or *cluster* these uses together in some coherent way. And one way to start grouping entries on the list is to ask ourselves why we care about them. For example, why do we want our ladder to be used outdoors? The answer is probably because that's part of what makes a ladder useful, which is another entry on our list. Similarly, we could ask why we care whether the ladder is useful. In this case, the answer is not on the list: We want it to be useful so that people will buy it. Put another way, usefulness makes a ladder marketable.

List 3.2. SAFE LADDER *Pruned Objectives List*

Ladder should be useful

Used to string conduit and wire in ceilings

Used to maintain and repair outlets in high places

Used to replace light bulbs and fixtures

Used outdoors on level ground

Used suspended from something in some cases

Used indoors on floors or other smooth surfaces

Should be reasonably stiff and comfortable for users

Should allow a person of medium height to reach and work at levels up to 11 ft.

Must be safe

Should be relatively inexpensive

Must be portable between job sites

Should be light

Must be durable

This suggests that we need an item on our list about marketing, for example, "The ladder should be marketable." This turns out to be a very helpful objective, since it tells us why we want the ladder to be cheap, portable, etc. (On the other hand, we should also be careful identifying "super-objectives" like marketability since almost any new or interesting product feature could fit under that rubric.) If we go through *thoughtful*, clustering questioning of this sort, we will find a new list that we can represent in an *indented outline*, with *hierarchies* of major headings and various levels of subheadings (e.g., List 3.3).

The revised, indented outline in List 3.3 allows us to explore each of the top-level objectives further, in terms of the sub-objectives that tell us how to realize it. At the highest level, our objectives turn us back to the original design statement we were given, namely to design a safe ladder that can be marketed to a particular group.

Now, we have certainly not exhausted all of the questions we could ask about the ladder, but we can identify in this outline some of the answers to the three questions mentioned just above. For example, "What do you mean by safe?" is answered by two sub-objectives in the cluster of safety issues, that is, that the designed ladder should be both stable and relatively stiff. We have answered "How are you going to do that?" by identifying several sub-objectives or ways in which the ladder could be useful within the "The ladder should be useful" cluster and by specifying two further "sub-sub-objectives" about how the ladder would be useful indoors. And we have answered the question "Why do you want that?" by indicating that the ladder ought to be cheap and portable in order to reach its intended market of electricians and construction and maintenance specialists.

List 3.3. SAFE LADDER Indented Objectives List

0. *A safe ladder for electricians*

1. The ladder should be safe

    1.1 The ladder should be stable

        1.1.1 Stable on floors and smooth surfaces

        1.1.2 Stable on relatively level ground

    1.2 The ladder should be reasonably stiff

2. The ladder should be marketable

    2.1 The ladder should be useful

        2.2.1 The ladder should be useful indoors

            2.2.1.1 Useful to do electrical work

            2.2.1.2 Useful to do maintenance work

        2.2.2 The ladder should be useful outdoors

        2.2.3 The ladder should be of the right height

    2.2 The ladder should be relatively inexpensive

    2.3 The ladder should be portable

        2.3.1 The ladder should be light in weight

        2.3.2 The ladder should be small when ready for transport

    2.4 The ladder should be durable

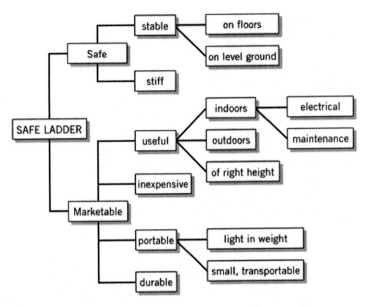

FIGURE 3.1 The objectives tree for the design of a safe ladder. It shows the first fruits of problem definition. Note the hierarchical structure and the clustering of similar ideas.

3.1.4 Building objectives trees

We now represent the indented outline of List 3.3 in graphical form by laying out a *hierarchy* of boxes, each of which contains an objective for the object being designed, as shown in Figure 3.1. Each layer or row of objective boxes corresponds to a level of indentation (indicated by the number of digits to the right of the first decimal point) in the outline. Thus, the indented outline becomes an *objectives tree*: A graphical depiction of the *objectives* or *goals for the device or system* (as opposed to goals for a design project or process). The top-level goal in an objectives tree — the node at the peak of the tree — is *decomposed* or broken down into subgoals at differing levels of importance or including progressively more detail, so the tree reflects an *hierarchical structure* as it expands downward. An objectives tree also *clusters* together related subgoals or similar ideas, which gives the tree some organizational strength and utility.

Objectives trees are ordered lists of the desired attributes of a design.

The graphical display of the tree is a very useful prop when the design team discusses its ideas with clients and other participants in the design process. It is also useful for determining what things we need to measure, since we will use these objectives to decide among alternatives. And the graphical tree format corresponds to the mechanics of the process that many designers follow. Often, the most useful way of "getting your mind around" a large list of objectives is to put them all on Post-It[TM] notes, and then move them around until the design team is satisfied with the tree. We will discuss some of the mechanics of tree building and problem definition in Section 3.1.5.

The process just outlined — from lists to refined lists to outlines to trees — has a lot in common with one of the fundamental skills of writing, being able to construct an outline. A topical outline provides an indented list of topics to be covered, together with the details of the subtopics corresponding to each topic. Since each topic represents a goal for the material to be covered, the identification of an objectives tree with a topical (or an indented) outline seems logical.

One final point about this simple example. Note that as we work *down* the tree, or move further in on the levels of indentation, we are doing more than just getting into more detail. We are also answering a generic *how* question for many aspects of the design, that is, the question of "*How* are you going to do that?"

We answer "how" by digging deeper into an objectives tree; we answer "why" by searching higher on an objectives tree.

Conversely, as we move *up* the tree, or further out toward fewer indentations, we are answering a generic *why* question about a specific (needed) function, that is, the question of "*Why* do you want that?" This enables us to track why we want some feature or other fine point in our design, which may be very important if we have to trade off features one against the other, because the values of these features may be directly attributable to the importance of the goals they are intended to serve. We will say more about this in Section 3.3.

3.1.5   How deep is an objectives tree? What about pruned entries?

Where do we end our list or tree of objectives? The simple answer is to stop when we run out of objectives or goals and implementations begin to appear. That is, within any given cluster, we could continue to parse or decompose our subgoals until we are unable to express succeeding levels as further subgoals. The argument for this approach is that it points the objectives tree toward a *solution-independent* statement of the design problem. That is, we know what characteristics the design has to exhibit, without having to make any judgment about how it might get to be that way. In other words, we determine the attributes of the designed object without specifying the way the objective is realized in concrete form.

Another way of limiting the depth of an objectives tree is to look out for verbs or "doing" words because they normally suggest functions. Functions do not generally appear on objectives trees or lists.

A second tree-building issue has to do with deciding what to do with the things that we have removed from the list. In the case of the functions and implementation, we simply put them aside (recording them in case they are good ideas), and pick them up again later in the process. In the case of constraints, however, it is often reasonable to reenter them into an appropriate place in the objectives tree, while being very careful to distinguish them from the objectives. For example, in an outline form of the objectives tree, we might use italics or a different font to denote constraints (see List 3.4 in Section 3.1.7). In a graphical form, we may wish to highlight constraints using differently shaped boxes. In either case, it is important to recognize that constraints are related to but are different from objectives, and they are used in different ways.

3.1.6   On the logistics of building objectives trees

When do we build an objectives tree? Right away? As soon as the client has offered us the design job? Or, should we do some homework first and perhaps try to learn more about the design task we're undertaking?

There's no hard and fast answer to these questions, in part because building an objectives list or tree is not a mathematical problem with an attendant set of initial conditions that must be met first. Also, building a tree is not a one-time, let's-get-it-done kind of activity. It's an iterative process, but one that surely should begin after the design team has at least some degree of understanding of the design domain. Thus, some of the questioning of clients, users, and experts should have begun, and some of the tree building can go on episodically while more information is being gathered.

Build an objective tree early, and modify it often while defining the problem.

Now, the various lists and the objectives tree are significant parts of the information gathered while we are defining or framing the design problem. How do we organize all of that information, particularly if we're sitting around a conference room table and doing some intensive brainstorming? Surely we'd use blackboards or whiteboards, but how do we do all the clustering and the hierarchical organization while team members are throwing out ideas in a rapid, stream-of-consciousness fashion? One way is to use Post-It notes, which come in various sizes nowadays, and write up individual notes for each entry on the list or in the tree. The notes can then be pasted on a board or display and later moved around as the team begins to organize the list of design attributes. Incidentally, this technique is also used in meeting environments to bring break-out panels to a shared understanding or set of outcomes by having a moderator sort each panel's Post-It notes into *affinity groups* that are identified by key ideas.

Two minor but important points. First, it is important that someone take notes during brainstorming sessions, in order to ensure that *all* suggestions and ideas are captured, even those that seem silly or irrelevant at the moment. It's always easier to prune out and throw away things than to recapture spontaneous ideas and inspirations. Second, after a rough outline of an objectives tree has emerged, it can be formalized and made to look pretty (and presentable) simply by using any standard, commercially available software package for constructing organization charts or similar graphical displays.

3.1.7   The objectives tree for the beverage container design

In the beverage container design problem, our design team is working for one of the two competing food product manufacturers, in this instance NBC. (We note parenthetically an interesting ethical problem that we will address in Chapter 12, that is, Could our design team, or our firm, take on the same or similar design tasks for both, or for two competing clients?) However, for now, let us suppose that we're dealing with a single client and that our client's project statement is as stated in Section 2.5.2: "Design a bottle for our new juice product."

In order to clarify what was wanted from this design, our design team questioned many people in NBC, including the marketing staff, and we talked to some of their potential customers or users. As a result, we found that there were several motivations driving the desire for a new "juice bottle," including: plastic bottles and containers all look alike; the client, as a national producer, has to deliver the product to diverse climates and environments; safety is a big issue for parents whose children might drink the juice; many customers, but especially parents, are concerned about environmental issues; the market is very competitive; parents (and teachers) want children to be able to get their own drinks; and, finally, children always spill drinks.

These motivations emerged during the questioning process, and their effects are displayed in the augmented attributes list for the container given as List 3.4. Some of the entries in this list are shown in italics because they are constraints. Thus, these constraint entries can be removed from a final list of the attributes that are objectives (to be re-inserted later, as discussed above).

List 3.4. *BEVERAGE CONTAINER Augmented Attributes List*

Safe ⟶	DIRECTLY IMPORTANT
Perceived as Safe ⟶	Appeals to Parents
Inexpensive to Produce ⟶	Permits Marketing Flexibility
Permits Marketing Flexibility ⟶	Promotes Sales
Chemically Inert ⟶	*Constraint* on Safe
Distinctive Appearance ⟶	Generates Brand Identity
Environmentally Benign ⟶	Safe
Environmentally Benign ⟶	Appeals to Parents
Preserves Taste ⟶	Promotes Sales
Easy for Kids to Use ⟶	Appeals to Parents
Resists Range of Temperatures ⟶	Durable for Shipment
Resists Forces and Shocks ⟶	Durable for Shipment
Easy to Distribute ⟶	Promotes Sales
Durable for Shipment ⟶	Easy to Distribute
Easy to Open ⟶	Easy for Kids to Use
Hard to Spill ⟶	Easy for Kids to Use
Appeals to Parents ⟶	Promotes Sales
Chemically Inert ⟶	*Constraint* on Preserves Taste
No Sharp Edges ⟶	*Constraint* on Safe
Generates Brand Identity ⟶	Promotes Sales
Promote Sales ⟶	DIRECTLY IMPORTANT

The augmented list (List 3.4) also shows how, after additional brainstorming and questioning, some of the listed goals are either expanded into sub-objectives (or subgoals) and others are connected to existing goals at higher levels. In one case a brand new top-level goal, Promote Sales, is identified. The objectives tree corresponding to (and expanded from) this augmented attribute list is shown in Figure 3.2, and a tree combining objectives and constraints is shown in Figure 3.3. The detailed subgoals that emerge in these trees clearly track well with the concerns and motivations identified in the clarification process.

As a result of the thought and effort that went into List 3.4 and the objectives trees of Figures 3.2 and 3.3, the design team rewrote and revised the problem statement for this design project to read: "Design a safe method of packaging and distributing our new children's juice product that preserves the taste and establishes brand identity to promote

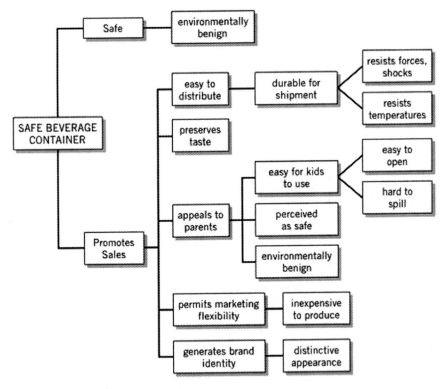

FIGURE 3.2 The objectives tree for the design of a new beverage container. Here the work on problem definition has lead to a hierarchical structuring of the needs identified by the beverage company and by the potential consumers — or at least the consumers' parents! — of the new children's juice drink.

sales to middle-income parents." Thus, as we noted in Chapter 2, one of the outputs of the design preprocessing (or problem definition) phase is a revised statement that reflects what has been learned about the goals for a design project. That is, the emergence of a clearer understanding of the client's design problem results in an objectives tree that points toward the expression of the features and behaviors wanted from the designed object, and it often results in a simultaneous revision or restatement of the client's original problem statement.

3.1.8 Revised project statements

We have assumed all along that design projects would be initiated with a relatively brief statement drafted by the client to indicate what he seems to want. All of the methods and outputs we have described in this chapter are aimed at understanding and elucidating these wants, as well as accounting for the wants of other potential stakeholders. As we gather information from clients, users, and other stakeholders, our views of the design problem will shift as we expose implicit assumptions and perhaps a bias toward an implied solution. Thus, it is important that we recognize the impact of the new

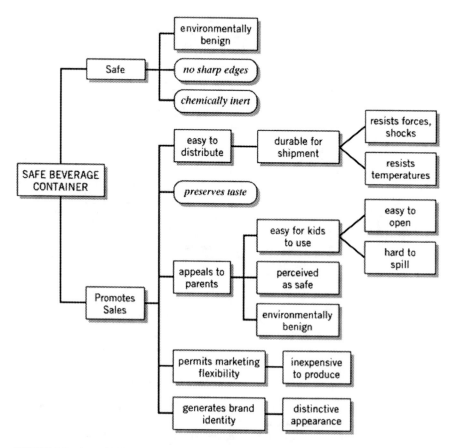

FIGURE 3.3 A combined tree (objectives in rectangles and constraints in ovals) for the design of a new beverage container. Here the goals for the new product are shown together with the constraints that apply to the object being designed.

Share revised problem statements with the client — they may just be right! information we've developed and that we formalize it by drafting a revised problem statement that clearly reflects our clarified understanding of the design problem at hand. We saw such a revised problem statement as one of the emergent products of the beverage container design (viz., Section 3.1.7), and a comparison of the initial and revised problem statements for this project speaks very clearly to the notion of exposing more precisely what the client wants. We will see a similar result in Section 3.6.

3.2 ON MEASURING THINGS

Having now identified the client's objectives for a design, we wonder: How will we know the objectives have been achieved? Also, does the client have any priorities, that is, are some objectives more important than others? These two questions imply another: How do we measure and compare design objectives? It is not clear that there is a way to "plot" design objectives along an axis, although it does make sense to plot evaluation points earned by objectives. Then evaluation points can be compared and some design decisions can be taken. But how do we award such evaluation points? And is there a scale on which we can lay out the evaluation points for each design?

Engineers are used to measuring all sorts of things: beam lengths, surface areas, hole diameters, speeds, temperatures, pressures, and so on. In each of these cases, there is a ruler or scale involved that shows a zero and has marks that show units, whether they be inches, microns, mm of Mercury, or degrees Fahrenheit or Centigrade. The ruler establishes a common basis for comparison. Without rulers, how would we meaningfully quantify the assertion that *"August* is taller than *Ulysses"*? Simply standing August and Ulysses against each other, back to back, doesn't work (especially if August and Ulysses are not easily moved or won't stand still!). However, by using a measuring stick that has a zero and is marked with fixed intervals of length that can be counted, we can establish real numbers to represent the heights of August and Ulysses.

The important concept here is that of having a *ruler* or *scale* with (1) a *defined zero*, and (2) a *unit* that is used to define the markings scribed onto the ruler. In mathematical terms, these properties enable *strong measurement*, as a consequence of which we can treat measured mathematical variables (say L for length, T for temperature, and so on) as we would any variable in calculus. Thus, strong measurements could be used as would any of our "normal" physical variables in a mathematical model.

Six different types of scales have been used to evaluate and test product designs (see Table 3.1). These different types of "scales" and their associated units of measure can each be used in different situations, but *there are limits to what can be done with these "measurements"* because some of them are not "real" measurements. *Nominal scales*, for example, are used to distinguish among categories. We can count the number of colors available, but there is no measure of color difference. The same can be said of *partially ordered scales*, such as hierarchies of families. Thus, these scales are of little use for examining most design choices, even if the distinctions drawn are of interest to the client, the users, or the designer!

Ordinal scales are used to place things in rank order, that is, in first, second, or *n*th place. This seems straightforward enough, but it is precisely here that measurement gets more complicated because it suggests assessing the *subjective preferences* of

TABLE 3.1 Measuring scales for testing and evaluating designs in the field of product design. Adapted from (Jones 1992)

Nominal scales, such as colors, smells, or even professions (e.g., teachers, lawyers, engineers).

Partially ordered scales, such as grandparent, parent, and child, which array themselves somewhat hierarchically.

Ordinal scales, such as first, second, third, etc.

Ratio scales, such as inches, seconds, or dollars. Ratio scales have natural reference points or base points.

Interval scales, such as degrees Centigrade, that have arbitrarily defined reference points or base points.

Multidimensional scales or *index numbers*, such as miles per gallon or kilometers per maintenance event, that are compounds of other scales of measurement.

individuals. That is, when we ask the client which design objectives are most important, we are typically asking for a subjective ranking of their perceived importance. To ask whether cost or portability is the more important objective in the design of a ladder is to ask a question to which the answer is different than the statement that "Joan Dym is 5 ft 9 in tall." We can indicate a preference for portability over price, but there is no sensible way to assess the degree or amount of that preference. For example, there is no meaningful way to say that "portability is five times more important than cost" because there is no scale or ruler that defines both a zero and a unit with which to make such measurements. Or for another example, by how much more do you prefer vanilla ice cream to chocolate? We shall address the important issue of assessing priorities for objectives in Section 3.3.

Ratio scales have naturally defined base points that have physical meaning (i.e., zero money, of zero height, etc.) and can be measured. In the case of objectives, ratio scales for design objectives would have specific values that can be understood as "zero." For example, the notion that a product will emit no pollution is straightforward.

Interval scales have defined reference points or base points, from which all others are referenced (or, to which all others are related). Interval scales are the closest of the traditional scales to strong measurement.

Assessing objectives often involves measurements for which there is no ruler. If "simplicity" is a design objective for a product, how would it be measured? The answer is that a *metric* would be introduced, for example, count the number of parts. A minimum number of parts would be identified as a base point, and other designs could be assessed by the number of parts they contain, with simpler designs having fewer or smaller numbers of parts. Developing and using metrics to assess the achievement of objectives will be discussed at length in Section 3.4.

Given the disparate nature of design objectives or a resulting set of designs for a product, it is far from clear that an identifiable scale or ruler can be meaningfully used to assess and evaluate either objectives or designs. It would be easier, for example, to evaluate designs by their estimated manufacturing costs, which are hard numbers that can be measured on a standard ratio scale, although doing that suggests that our only objective is minimizing manufacturing costs! But, for both design alternatives and design objectives, we are often trying to assess subjective preferences that are not easily rendered into quantitative terms.

3.3 SETTING PRIORITIES: RANK ORDERING THE CLIENT'S OBJECTIVES

We have been rather insistent in this chapter that we properly identify and list all of the client's objectives, while taking great care not to confuse constraints, functions or means with the goals set for the object being designed. But do we know that all of the identified objectives have the same import or value to the client or the users? Because we have made no effort to see whether there is any variation in the objectives' perceived value, it appears that we have implicitly assumed that each of the top-level objectives has the same value to all concerned. It is almost certain that some objectives are more important than others, so we ought to be able to recognize that and measure it. How are we going to do that?

3.3.1 Pairwise comparison charts: Individual rank orderings

Suppose we have a set of goals for a project whose relative values we want to *rank*, that is, we want to identify their value or importance relative to one another, and to order them accordingly. Sometimes we get really lucky and our client expresses strong and clear preferences, or perhaps the potential users do, so that the designer doesn't have to determine an explicit ranking. More often, however, we do have to do some ranking or we have to place some values ourselves. Thus, we propose here a fairly straightforward technique that can be used to rank goals that are at the same level in the hierarchy of objectives and are within the same grouping or cluster; that is, they have the same parent or antecedent goal within the objectives tree. It is very important that we make our comparisons of goals with these clustering and hierarchical restrictions firmly in mind in order to ensure that we're comparing apples with apples and oranges with oranges. For example, does it make sense to compare the subgoals of having a ladder be useful for electrical work and having it be durable? On the other hand, rank ordering the importance of the ladder's usefulness, cost, portability, and durability would be useful design information.

Suppose we are designing just such a ladder for which four high-level goals have been established: It should be inexpensive, useful, portable and durable. We further suppose that we can easily choose between any given pair of them. For example, we prefer cost over durability, portability over cost, portability over convenience, and so on. Where does leave us in terms of ranking all four goals? We can determine one answer to that question by constructing a simple chart or matrix that allows us to (1) compare every goal with each of the remaining goals individually and (2) add cumulative or total scores for each one of the goals.

We show in Table 3.2 a *pairwise comparison chart* (PCC) for our four-objective ladder design. The entries in each box of the chart are determined as binary choices, that is, every entry is either a 1 or a 0. Along the row of any given goal, say Cost, we enter a zero in the columns for the goals Portability and Convenience that are preferred over Cost, and we enter 1 in the Durability column because Cost is preferred over Durability. We also enter zeros in the diagonal boxes corresponding to weighting any goal against itself, and we enter ratings of 1/2 for goals that are equally valued. The scores for each goal are determined simply by adding across each row. We see that in this case, the four goals can be ranked (with their scores) in order of decreasing value or importance: Portability (3), Convenience (2), Cost (1), Durability (0).

Note, too, that the score of 0 earned by Durability does *not* mean that we can or should drop it as an objective! Durability earned the 0 because it was ranked as *least*

TABLE 3.2   A pairwise comparison chart (PCC) for a ladder design

Goals	Cost	Portability	Convenience	Durability	Score
Cost	••••	0	0	1	1
Portability	1	••••	1	1	3
Convenience	1	0	••••	1	2
Durability	0	0	0	••••	0

important, that is, it placed last in the line of the four objectives ranked. If it were of *no importance*, it would not have been listed as an objective to begin with. Thus, we *cannot* drop objectives that score zeros.

It should also be kept in mind that the pairwise comparison (also known as the Borda count), if done correctly, preserves the property of *transitivity*: Thus, in the ladder design we preferred Portability to Convenience, and Convenience to Cost, then the PCC produced a consistent result when it said that we preferred Portability to Cost. To ensure such consistency, however, we must be sure to distribute entries that are only 1s or 0s, or integral multiples of 1 (and 0!). Other than in the case of exactly equal preference, we cannot award fractions to different objectives without sacrificing transitivity and correctness.

Now, the simple PCC process just described, is a valid way of ordering things, but its results should be taken as *no more than a straightforward rank ordering*, or an ordering of place in line. The scores assembled in Table 3.2 do not constitute what we had defined as (mathematically) strong measurement because there is no rational scale on which we can measure the four objectives, and the zero is only implied, not defined. Thus, these ranking scores should not be used in subsequent calculations (and so the zero won't get us into trouble)! They are a useful guide for further thought and discussion, but they are *not* a basis for further calculations. In particular, PCC rankings *cannot* be used to *weight* or *scale* objectives.

3.3.2   Pairwise comparison charts: Aggregate rank orderings

Life is still more complicated when assessing the preferences of groups. We have been working in the framework of the single designer or decision maker who is making a subjective assessment, determined to obtain a meaningful and useful ranking. The group situation — in which members of a design team vote on their preferences so that their individual votes can be gathered into an aggregated set of preferences for the entire team — is still more complicated and a subject of both research and discussion. (See the section notes for further reading.) The well-known sticking point derives from the well-known *Arrow Impossibility Theorem* of decision theory, for which Kenneth J. Arrow won the Nobel Prize in Economics in 1972. It states, in essence, that it is impossible to run a "fair" election — or to select a "fair" objective or attribute — and preserve transitivity if there are more than two candidates from which to choose! There is corresponding discussion in the design community as to the role that decision theory plays in the design process, but we believe that the PCC (or Borda count) can be used to indicate the collective ordering of the preferences of a design team.

Suppose a team of twelve designers is asked to rank order three designs: *A*, *B*, and *C*. In doing so, the twelve designers have, as individuals, produced the following twelve sets of orderings:

$$
\begin{array}{ll}
\text{1 preferred } A \succ B \succ C & \text{4 preferred } B \succ C \succ A \\
\text{4 preferred } A \succ C \succ B & \text{3 preferred } C \succ B \succ A
\end{array} \tag{3.1}
$$

where $\succ$ is the ranking symbol used to write "*A* is preferred to *B*" as $A \succ B$.

The collective will of the design team is worked out through the aggregated PCC shown in Table 3.3. A point is awarded to the winner in each pairwise comparison, and

TABLE 3.3 An aggregated pairwise comparison chart (PCC) for 12 designers

Win / Lose	A	B	C	Sum/Win
A	••••	$1 + 4 + 0 + 0$	$1 + 4 + 0 + 0$	10
B	$0 + 0 + 4 + 3$	••••	$1 + 0 + 4 + 0$	12
C	$0 + 0 + 4 + 3$	$0 + 4 + 0 + 3$	••••	14
Sum / Lose	14	12	10	••••

then the total points earned by each alternative from all of the designers is summed. The aggregate ranking of preferred designs is

$$C \succ B \succ A \tag{3.2}$$

That is, the group consensus was that C was ranked first, B second and A last. Thus, the 12 designers choose design C as their collective first choice, though it was not the unanimous first choice. In fact, only 3 of 12 designers ranked it first! However, as pointed out by Arrow, there is no such thing as a "fair" election, no matter how many voters (assuming at least two!), if there are more than two names on the ballot. However, the PCC as applied here provides as good a tool as there is for these purposes, so long as its results are used with the same caution noted for individual PCCs.

Pairwise comparison voting by members of a design team must be used carefully because it can be problematic.

How would we use this in a design setting, if not as a direct decision driver? One approach would be to recognize that each design must have had elements or features that were attractive, else the points awarded each would have been far different. For example, had the points awarded in a PCC vote been $C = 24$, $B = A = 6$, we could conclude that the voters didn't see much overall merit in designs A and B. On the other hand, had the votes turned out $C = 18$, $B = 16$, and $A = 2$, then it would be reasonable to assume that there were two designs that were perceived to be nearly equal, in which case combining their best features would be a good design strategy.

3.3.3 Using pairwise comparisons properly

The pairwise comparison method should be applied in a *constrained*, *top-down* fashion, so that (1) objectives are compared only when emanating from a common node at the same level of abstraction or level on the objective tree, and (2) the higher-level objectives are compared and ranked before those at lower, more detailed levels. The second point seems only a matter of common sense to ensure that more "global" objectives (i.e., those more abstract objectives that are higher up on the objectives tree) are properly understood and ranked before we fine tune the details. For example, when we look at the objectives for the safe ladder (see Figure 3.1), it is more important to decide how we rank safety against marketability than its use for electrical work or for maintenance. Similarly, for the beverage container (Figure 3.2), it is again more meaningful to rank safety against sales promotion before worrying about whether the container is easier to open than it is harder to spill. Also, depending on the nature of the design task, it is quite possible that only top-level objectives need be so ranked. Only when complex subsystems (within large and complex systems) are being designed would it make much sense to rank objectives below the top level.

Goals	Environ. Benign	Easy to Distribute	Preserve Taste	Appeals to Parents	Market Flexibility	Brand ID	Score
Environ. Benign	••••	0	0	0	0	0	0
Easy to Distribute	1	••••	1	1	1	0	4
Preserve Taste	1	0	••••	0	0	0	1
Appeals to Parents	1	0	1	••••	0	0	2
Market Flexibility	1	0	1	1	••••	0	3
Brand ID	1	1	1	1	1	••••	5

(a) ABC´s weighted objectives

Goals	Environ. Benign	Easy to Distribute	Preserve Taste	Appeals to Parents	Market Flexibility	Brand ID	Score
Environ. Benign	••••	1	1	1	1	1	5
Easy to Distribute	0	••••	0	0	1	0	1
Preserve Taste	0	1	••••	1	1	1	4
Appeals to Parents	0	1	0	••••	1	1	3
Market Flexibility	0	0	0	0	••••	0	0
Brand ID	0	1	0	0	1	••••	2

(b) NBC´s weighted objectives

FIGURE 3.4 Pairwise comparison charts for the design of the new beverage container. Here, goals for the product are ranked one against another by designers working for (a) ABC and (b) NBC. The relative rankings of these objectives varies considerably in each chart, thus reflecting the different values held by each company.

In addition, given the subjective nature of these rankings, we should ask whose values are being assessed when we use such a ranking tool. Marketing values can easily be included in different rankings. In the ladder design, for example, the design team might need to know whether it's "better" for a ladder to be cheaper or heavier. On the other hand, there could be deeper issues involved that, in some cases, may touch upon the fundamental values of both clients and designers. For example, consider how the objectives for the beverage container design might be ranked at the two competing companies, ABC and NBC. We show PCCs for the ABC- and NBC-based design teams in Figures 3.4(a) and (b), respectively. These two charts and the scores in their right-hand columns show that the folks at ABC were far more interested in a container that would generate a strong brand identity and be easy to distribute than in one that was environmentally benign or would appeal to parents. At NBC, on the other hand, the environment and the taste preservation ranked more highly. Thus, subjective values show up in PCCs and, consequently, in the marketplace!

It is also tempting to place our *ranked* or ordered objectives on a *scale* so that we can manipulate the rankings in order to attach relative weights to objectives or to do some other calculation. It would be nice to be able to answer questions such as,

How much more important is portability than cost in our ladder? Or, in the case of the beverage container, *How much more* important is environmental friendliness than durability? A little more? A lot more? Ten times more? We can easily think of cases where one of the objectives is substantially more important than any of the others, such as safety compared to attractiveness or even cost in an air traffic control system, and other cases where the objectives are essentially very close to one another. Sadly, however, there is no mathematical foundation for scaling or normalizing the rankings obtained with tools such as the PCC. The numbers obtained with a PCC are *subjective* preferences about relative value or importance. They do not represent strong measurement. Therefore, we should not try to make these numbers seem more important by doing further calculations with them or by giving them unwarranted precision.

Lastly, and continuing the spirit of the foregoing, it is also tempting to want to construct *weighted* objectives trees that explicitly show relative scores for every goal and subgoal by integrating scores from the PCCs into the objectives trees. But we cannot weight objectives without repeating the error of building an appealing numerical edifice on a mathematically unsound foundation.

3.4 DEMONSTRATING SUCCESS: MEASURING THE ACHIEVEMENT OF OBJECTIVES

Having determined what our client wants in a design in terms of rank-ordered objectives, we now take up the question of assessing how well a particular design *actually does* all these things. As we noted in Section 3.2, such assessment requires *metrics*, that is, standards that measure the extent to which a design's objectives are realized. In principle it is easy to devise metrics, as all we need are units and a scale of something that can be *measured* about an objective, and a way to *assign* a value to the design in terms of those units. In practice it is often hard to devise (and then apply) an appropriate metric. So how will we know that the metrics we are developing are good and appropriate metrics?

Metrics are used to measure how well objectives are met.

3.4.1 Establishing good metrics for objectives

First and foremost, a *metric* should *actually measure the objective* that the design is supposed to meet. Often designers try to measure some phenomenon that, while interesting, is not really on point for the desired objective. If the objective is to appeal to consumers, for example, measuring the number of colors on the package may be a poor metric. On the other hand, sometimes we need to invoke *surrogate metrics* because there are no obvious measures appropriate to the objective of interest. For example, to assess the durability of a cell phone, we might subject it to a drop test wherein we evaluate its survival in drops from different heights. Similarly, the simplicity (or, conversely, the complexity) of a product might be assessed in terms of the number of parts that are needed to make the product, or perhaps in terms of the product's estimated assembly time. Thus, surrogate metrics are quite useful when they are measurable properties that strongly relate to the objective of interest.

Having decided on what to measure, the next step is to *determine the appropriate units* with which to make the measurement. For an objective of low weight for a ladder, for example, we could use units of weight or mass, that is, kg, lb, or oz. For an objective of low cost, our metric would be measured in currency, that is, $US in the United States. Having determined the appropriate units, we must also be sure that the metric enables the *correct scale* or *level of precision*. For a low-weight ladder, weight should not be measured in tons or milligrams.

The next step in the process of developing metrics is to *assign points* within the context of a scale or a range expressed in the right *units of interest* or *figures of merit*. For example, if we want a fast car, we might use speed in km/hr as the figure of merit and assume that the range of speed of interest is 50 km/hr $\leq$ speed $\leq$ 200 km/hr. Then we could assign points linearly distributed over the range, that is, from 0 points at the low end (50 km/hr) up to 10 points at the high end (200 km/hr). Thus, a design alternative that has a projected speed of 170 km/hr would earn or be awarded 8 points. With reference to assessing the durability of a cell phone, we might drop the phone over a range of heights of 1 m $\leq$ height $\leq$ 10 m and then assign points from 0 points at the low end (1 m) up to 10 points at the high end (10 m).

Note that in awarding points (just above) for speeds or drop test heights, we are implicitly assuming that we have a plan for measuring performance that is compatible with the type of scale and units selected. Such a "measurement" plan could include laboratory tests, field trials, consumer responses to surveys, focus groups, etc. However, while some things are relatively easy to measure directly (e.g., weight on a balance scale) or indirectly (e.g., weight by computing volume), others must be *estimated* (e.g., the top speed of a planned airplane using a back-of-the-envelope model), and other things are neither easily measured nor easily estimated (e.g., cost can be hard to estimate without knowing the manufacturing techniques to be employed, the number of units to be made, the components to be included in the design, etc.).

Often the appropriate "units" are general categories (e.g., "high," "medium," or "low") or subjective or qualitative rankings (e.g., "great," okay," or "lousy"). In Table 3.4

TABLE 3.4   Scales or rulers for awarding points depending on perceived value of a solution (*Use-Value Analysis*) or perceived value of the idea or concept (*VDI 2225 Guidelines*)

Use-Value Analysis		VDI 2225 Guidelines	
Solution Value	Points Awarded	Perceived Value	Points Awarded
absolutely useless	0	unsatisfactory	0
very inadequate	1		
weak	2	just tolerable	1
tolerable	3		
adequate	4	adequate	2
satisfactory	5		
good, w/drawbacks	6	good	3
good	7		
very good	8	very good (ideal)	4
exceeds requirements	9		
excellent	10		

we display two means of quantifying qualitative rankings or assigning "measurement" points for such categories or rankings. Eleven rankings of the value of a solution are offered in *Use-Value Analysis*, with points then being awarded on a scale ranging from 0 (absolutely useless) to 10 (ideal). There are five rankings in the German *VDI 2225* standard, with points awarded on a scale ranging from 0 (unsatisfactory) to 4 (very good/ideal) depending on the degree to which an idea or a concept or something else is considered valuable.

In the context of qualitative rankings, consider once again the objective of a low-cost ladder. The information needed to accurately assess the manufacturing costs for ladders may not be available without a significant and expensive study. One alternative might be to estimate the manufacturing cost by summing the costs of the ladder's components when purchased in given lot sizes. This does disregard some relevant costs (e.g., assembling components, company overheads), but it allows the design team to distinguish between designs with expensive elements and designs with cheaper elements. Alternatively, the designers might seek expert input from the client and then rank the designs into ordinal categories such as "very expensive," "expensive," "moderately expensive," "inexpensive," and "very cheap."

It is important that the measurements of the achievements of all of the objectives of a design alternative be made consistently, on the same ruler or scale, so that some objectives do not dominate the overall assessments by virtue of being measured on scales that awarded more points than earned by other objectives. In fact, the use-value analysis and VDI 2225 guidelines of Table 3.4 can also be used to ensure that we are assessing quantitative performance ratings on similar, consistent scales. In Table 3.5 we show how two different sets of quantitative performance ratings in their figures of merit, for mass per unit power (measured in kg/kW) and service life (measured in km), both arrayed against the *Use-Value Analysis* and German *VDI 2225* scales.

TABLE 3.5 Measuring quantitative performance levels for figures of merit of mass per unit power (kg/kW) and for service life (km) measured on the *Use-Value Analysis* and *VDI 2225* scales or rulers

Measured/Estimated Values		Value Scales	
Mass/Power (kg/kW)	Service Life (km)	Use-Value Points	VDI 2225 Points
3.5	20×10^3	0	0
3.3	30×10^3	1	
3.1	40×10^3	2	1
2.9	60×10^3	3	
2.7	80×10^3	4	2
2.5	100×10^3	5	
2.3	120×10^3	6	3
2.1	140×10^3	7	
1.9	200×10^3	8	
1.7	300×10^3	9	4
1.5	500×10^3	10	

It is also important to determine whether or not the information derived from using a metric is worth the cost of actually performing a measurement. The value of the metric may be small in comparison with the resources needed to obtain the measurement. In such cases we can either develop a new metric, find another means for measuring the expensive metric, or look for an alternative way of assessing our design. There may be other metrics that provide equivalent information, in which case we may be able to choose a less expensive measurement. In other cases, we may decide to use a less accurate method to assess our designs. As a last resort, we may decide to convert the hard-to-measure objective into a constraint, which allows us to consider some designs and reject others. (Recall that on p. 3–6 we noted the distinction between *converting* objectives to constraints and *confusing* objectives with constraints.) In the case of designing a low-cost ladder without adequate cost information, perhaps that objective could be converted into a constraint, such as, "Contains no parts costing more than $20." This constraint indirectly works toward the original objective, while enabling the dismissal of designs that seem certain not to be low in cost.

Good metrics measure the right thing, have clear units, and are cost effective. If a good metric is not affordable, consider making the corresponding objective into a constraint.

A few closing comments about metrics:

- A metric should be *repeatable*. That is, others conducting the same test or measurement would obtain the same results, subject to some degree of experimental error. This characteristic can be met either by using standard methods and instruments, or, if no such methods are available, by carefully documenting the protocols being followed. It also makes it incumbent upon the design team to use sufficiently large statistical samples where possible.

- The outcomes of metrics assessment should be expressed in *understandable units of measure*.

- A metrics assessment should elicit only *unambiguous interpretation*. That is, the results of a metrics assessment should lead all members of a design team (as well as all other stakeholders) to the same conclusion about the measurement. We certainly don't want a post-assessment debate about the meaning of the assessment or measurement of a given metric.

It is clear that judgment is called for in selecting and applying a good metric. The scale and units should be appropriate to the design objectives, and means of measuring them must be available and affordable. In general, good metrics result from careful thought, extensive research, and ample experience — which suggests that the selection of metrics can certainly be enhanced by the synergy derived from a cooperative, well functioning team.

3.4.2 Establishing metrics for the beverage container

Let us now establish metrics for the six objectives of the beverage container problem that are identified in Figure 3.3. As we will see immediately, the six metrics will be, to different degrees, *qualitative* metrics simply because there is no direct measurement that we can make for any of the objectives. Thus, we will establish metrics that are analogous to

Use-Value Analysis and the VDI 2225 Guidelines. We will also keep in mind what we've said before: Metrics should be solution-independent, that is, metrics should be established without any reference to the kinds of design solutions or alternatives that may emerge from the design process.

Consider that we want the beverage container to be *environmentally benign*. Products that are environmentally benign must at worst do no harm to the environment, that is, they should produce no hazardous waste or residue. At best, containers should be easily reused, or — and almost as good — their materials should be recyclable. Thus, we might propose the following qualitative metric:

Objective: *Beverage container should be environmentally benign.*

Units: Rating assessment of most environmentally desirable alternative from 0 (worst) to 100 (best).

Metric: Assign points according to the following scale:

Completely reusable:	100 points
Material is recyclable:	90 points
Material is easily disposable:	50 points
Material is disposable with difficulty:	25 points
Material is a hazardous waste:	0 points

It is also quite likely that a surrogate metric of *environmental costs* could be established for this objective. This means ascertaining the cost of washing bottles so that they can be relabeled and reused, as they once were for sodas and other drinks for many years. Similarly, we might estimate the cost of recycling the materials, since glass bottles and aluminum cans can be deconstructed (i.e., broken down, melted, etc.) into their constituent materials. Finally, the social and opportunity costs of disposal can probably be estimated. This would be apply to disposing of relatively benign materials (such as cardboard) or more hazardous materials and products (such as the plastic bags that confound the lives of those who dwell under the sea, or the small seed-like detritus that is eaten by unsuspecting birds). Then a quantitative surrogate metric can be established using known or estimated environmental costs. (See Exercise 5.9.) Finally, it is also worth noting that environmental, life-cycle and sustainability issues are increasingly central in product design, as we will discuss in Chapter 11.

We want the beverage container to be *easy to distribute*. There are a variety of issues that could enter here, including: whether the container can be easily *packed*, in terms of both shape and size; whether it is *breakable*; and whether the juice product or its container are sensitive to *temperature*. It is also likely that standard container shapes make it easier for store owners to provide shelf space for the new juice. These are clearly multidimensional objectives, so in fact it might be worth building a mathematical model that encapsulated the above three dimensions (and perhaps more), or perhaps the constraint list might be modified. For the present exercise, we propose the following use-value analog:

Objective: *Beverage container should be easy to distribute.*

Units: Rating of design team's assessment of the ease of packing and stacking the container, from 0 (worst) to 100 (best).

Metric: Assign points according to the following scale:

Very easy to pack and stack:	100 points
Easy to pack and stack:	75 points
Can be packed and stacked:	50 points
Hard to pack and stack:	25 points
Very hard to pack and stack:	0 points

This is also a metric for which a beverage company likely has a lot of experience and a lot of data about what works and what doesn't. Clearly, a data-driven metric is much more meaningful — and persuasive — than the qualitative "approximation" we are adopting here.

We also want the beverage container to *preserve taste*. One of the constraints given (in Figure 3.3) for this beverage container is that it should be *chemically inert*. This suggests that the objective of preserving taste depends on perception, that is, on whether people believe that drinks taste differently if they are in different containers. Some people do not like drinking coffee from Styrofoam or paper cups (which means they won't get take-out coffee very often!), and beer drinkers have long been known to prefer bottles to cans. Thus, for this objective the design team will once again go with a value-use kind of metric:

Objective: *Beverage container should preserve taste.*

Units: Rating of design team's assessment of how good the new juice will taste as a function of the container, from 0 (worst) to 100 (best).

Metric: Assign points according to the following scale:

Won't change the taste at all:	100 points
Will change the taste a little bit:	75 points
Will change the taste noticeably:	50 points
Will change the taste a lot:	25 points
Will render the juice undrinkable:	0 points

This is another instance in which the beverage company is almost certain to have both experience and data. In fact, for this objective and the remaining three objectives (*appeal to parents*, *permit marketing flexibility*, and *generate brand identity*), the design team is almost certainly going to turn to the beverage company's marketing teams and other in-house resources for information that would be relevant in assessing the attainment of these objectives. As we noted in Section 2.3.3.1, this is an occasion to use established techniques for determining market demands, including focus groups and structured questionnaires and surveys. Likely, too, the senior management of ABC and NBC will have preferences of their own that will enter into the picture.

3.5 CONSTRAINTS: SETTING LIMITS ON WHAT THE CLIENT CAN HAVE

There are limits to everything. That is why constraints are extremely important in engineering design, as we noted in Section 3.1.2 when we articulated some differences between constraints and objectives.

As a practical matter, many designers use constraints as a sort of "checklist" for designs in order to prune the set of designs to a more manageable size. Such constraints, which can be included in properly marked trees that include both objectives and constraints, are usually expressed in terms of specific numbers. By way of contrast, objectives are normally expressed as verbal statements that sometimes can be formulated in terms of continuous variables or numbers that may allow for a range of values of interest to the designer. To reiterate our earlier illustration of this point, a goal that a ladder should be cheap could be stated in terms of having a materials or manufacturing cost that does not exceed a fixed limit or constraint, say $25. On the other hand, we could have *both* an objective that the ladder be cheap *and* a constraint that puts a limit on the cost. In that case we are able to choose among a set of designs whose costs to build are different, as long as they are all below the limit set by the constraint alone. This, again, is the strategy of "satisficing" wherein we select among design alternatives that are acceptable.

Constraints enable us to identify and exclude unacceptable designs.

There is still another approach to dealing with objectives that can be cast in terms of "continuous variables." There are many design domains in which we can formulate mathematical relationships between many of the design variables. For example, we might know how the cost of the ladder depends on its weight, its height, the size of its projected market, and other variables. In such cases, we may try to *optimize* or get the best design, say the minimum-cost ladder, using procedures much the same as we use to find the maxima or minima of multi-variable calculus problems. Similarly, *operations research* techniques allow calculations to be performed when the design variables are discrete in nature, for example, when the ladder cost depends on the number of steps or connections, or if a ladder is restricted to be made in fixed-interval lengths, say that the ladder must be 5, 6, or 7 feet long. Optimization techniques are clearly beyond the scope of our discussions, but the underlying idea that design variables, and design objectives, interact and vary with one another is also a theme that we will further elaborate in the following section when we discuss ways of assessing the comparative values of design goals.

3.6 DESIGNING AN ARM SUPPORT FOR A CP-AFFLICTED STUDENT

In their first engineering course at Harvey Mudd College, *E4: Introduction to Engineering Design*, first-year engineering students are assigned the task of developing a conceptual design for a device or system. The projects are typically done for the benefit of a nonprofit or educational institution, and they provide the students with the insight that good engineering design may be (and is) done in nontraditional, noncorporate settings. The course also stresses the formal design methods we are presenting in this book. In order to illustrate student design within the E4 environment, we now begin to describe the design of a device to support and stabilize the arm of a young student afflicted with cerebral palsy (CP) as she writes or draws. The sponsor of this project, the Danbury School, is a special education school within the Claremont (California) Unified School District that serves children with severe orthopedic and medical problems. Students may be as young as three years old, and Danbury School offers classes through the sixth grade. Danbury School has a long history of working with the students in Harvey Mudd's E4 course, dating back to E4's first offering in the spring 1992 semester. Among the E4 projects that have been done for Danbury School are designs for a robotic arm to feed handicapped children, a computer input device for handicapped children, and school restrooms for students who have been diagnosed with orthopedic disabilities.

In the design problem at hand, E4 teams were asked to design a device for Jessica, a third-grade student who had been diagnosed with CP. While a particular user for the design had been identified, it was the hope of the Danbury School's principal — and, as it would turn out, of some of the student designers — that a design might be worth refining and developing into a product that could be offered to similarly disabled students elsewhere. The full problem statement is (an abbreviated version was given in Section 2.5):

> *The Danbury Elementary School of the Claremont Unified School District has a number of students with the diagnosis of cerebral palsy (CP), a neuro-developmental impairment that causes disturbances of voluntary motor function. For these students, activities that require fine muscle movements (e.g. writing) are particularly difficult because of impaired motor control and coordination as a result of CP. There is ample evidence indicating that these students write more effectively when an instructor physically stabilizes either the hand or the elbow to reduce extraneous movement. A device that can achieve the same physical effect by counteracting the involuntary movement would be desirable since this would increase the students' functional independence.*

A reading of the above initial problem statement makes it clear that the design teams had many questions to answer before they could begin to specify the ultimate form of an arm support. Among the most pressing of these questions is, What exactly do the client (Danbury School) and the user (Jessica) want (and need)? To answer this question, the students had to undertake research into cerebral palsy, the personal and classroom environments in which Jessica would be working, and into existing designs for arm supports and/or restraints. In addition, teams had to determine what terms such as "more effectively" and "increase . . . independence" meant to the client and to Jessica. This was accomplished by a combination of library research, web searches, interviews with Jessica, and repeated interviews with Danbury School staff. The end result of this was that objectives lists and refined client statements were developed.

3.6.1　Objectives and constraints for the Danbury arm support

The objectives elicited by two different teams are shown as an objectives tree (Figure 3.5) and as a list of objectives (Figure 3.6) *as they presented them in their final reports*. Also, in both instances, lists of constraints were developed along with the objectives. Both lists may have some errors or problems that are worth your attention. However, even with their shortcomings, there are a number of interesting points to be made about these two sets of objectives and constraints. First, neither of the sets of objectives or constraints are identical. While this is not surprising, given that the trees reflect the work of two different teams, it highlights the fact that many of the objectives and constraints that occur to designers are subject to analysis, interpretation, and revision. Thus, it is very important that designers carefully review their findings with the client before proceeding too far in the design process.

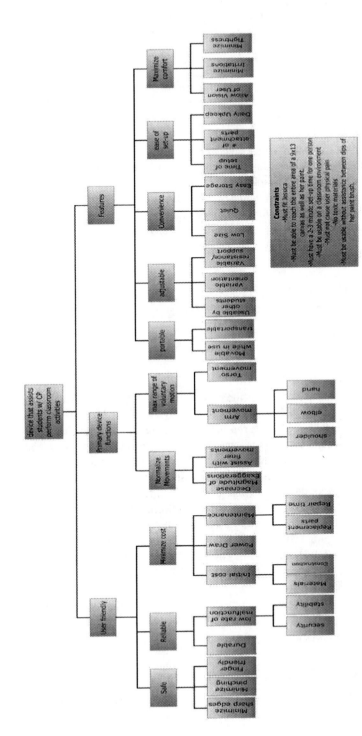

FIGURE 3.5 An objectives tree done by a first-year design team for the Danbury arm support project. Are there any entries in this tree that don't belong?

The following lists outline primary and secondary design objectives as well as constraints identified by the design team through the analysis of the revised problem statement. These form the basis for the functions and means analysis and design selection process.

Design Objectives

- Design should minimize involuntary movement of the upper arm
  ○ Should be safe
  ○ Should be comfortable
  ○ Should be durable
  ○ Should not impair/restrict voluntary motion
- Design should be applicable to multiple individuals and wheelchairs
  ○ Size should be adjustable
  ○ Mounting mechanism should be adaptable
- Design should minimize the cost of production
- Restraint mechanism should be easy to install and maintain

Design Constraints

- Design must reduce and counteract involuntary movement of the upper arm
- Design must not require more than two to three minutes for setup by an adult

FIGURE 3.6 Another student team's lists of objectives and constraints for the Danbury arm support project. How does this set of objectives and constraints compare with those shown in Figure 3.5? Are all of its entries appropriate?

It is also worth noting that the objectives tree in Figure 3.5 has been given in great detail, so that the "How?" and "Why?" questions that can be answered by traversing an objectives tree are easily answered here with some specificity. Such a dense objectives tree, however, raises some other interesting questions, including: Do we need to develop metrics for each and every sub- and sub-sub-objective on an objectives tree? How many of these subsidiary objectives (and metrics) should we consider when we select a design from among a set of design alternatives? We shall respond to the first question in Section 3.6.2 immediately below, and to the second in Section 5.4.

A second point to note is that one of the teams has chosen to incorporate much more detail (Figure 3.5), perhaps already reflecting some additional research into the details of potential designs, while the other team's objectives are much more general (Figure 3.6), likely reflecting primarily what the client had indicated in personal interviews. The designer often informs and educates the client, providing the client with a better understanding of the problem as the process of clarifying objectives unfolds. This takes on a particular importance when we are considering functions and requirements (cf. Chapter 4).

3.6.2 Metrics for the objectives for the Danbury arm support

The teams developed and applied metrics for their own sets of objectives. Figure 3.7 shows some of the metrics results presented by one design team, and this table

Objectives	Metrics	Conclusion	Result
1. Minimize number of sharp edges	number of sharp edges	inherent sharp metal edges	Fail
2. Minimize pinching	number of pinching possibilities	user quite comfortable	Pass
3. Finger friendly	number of places on device to get finger caught	not safe to handle	Fail
4. Durable	disconfiguration, misalignment of device after regular use	insecure mount, misaligns	Fail
5. Remain secure on user	Conditions under which device remains securely attached to user	arm remains attached to device	Pass
6. Maintain stable position	Conditions where position, orientation of device maintain mounting setting	insecure mount	Fail
7. Minimize cost	estimate dollar amount	less than other products	Pass
8. Normalize arm movement	User ability to draw straight line compared to ability to do so without the device	failure in extensibility hurts use	Fail
9. Maximize range of voluntary motion	Degree of freedom in motion of wrist, elbow, arm, and torso	comfortable range of motion, except torso forward bend	Pass (except torso)
10. Movable while in use	Required assembly condition to move device	does not require disassembly	Pass
11. Transportable	Necessary level of disassembly for movement	does not require disassembly	Pass
12. Usable by multiple students	Range of permissible arm sizes	adjustable size permits various arm sizes	Pass

FIGURE 3.7   This table shows metrics for 12 of the 23 objectives in the fourth level of the objectives tree of Figure 3.5. This set of results was accompanied by the statement that "The specific units and scales for each metric are not presented due to size constraints." So, within their 62-page final report the team couldn't find room for the details that give meaning to their metrics!

presents some interesting issues. First, only *results* were given; there were no scales or units. One might wonder about the impression conveyed to a client (and other readers) by the apparent unwillingness to make room to credibly document the basis for design selection. In fact, the impact of some of the neat results obtained by this team (to be presented later) could be lessened because of such inattention to detail. Second, results were shown for each of the 23 sub-sub-sub-objectives displayed in the fourth level of the tree in Figure 3.5. Thus, the three objectives at the second level of the tree and the 10 at the next level were not directly assessed, likely because they were so abstract that meaningful measurements could not be made. Third and last, some of the formal metrics appear to be very qualitative. It may often be the case that only qualitative assessments are possible, but a client might find it easier to accept such

Objective	Metric
Safety	Measured by number of possible ways in which device can cause bodily harm. Scale: Total Points = 10–# ways to cause harm
Stabilization	Ability to resist sudden accelerations. Scale: 1 to 10 by Subjective Evaluation
Comfortable	Perceived comfort of device. Scale: Total Points = 10–# sources of discomfort
Non-Restrictive	Measured by the area of allowed motion. Scale: Total Points = 10 (Area/2 sq. ft.)
Ease of Installation	Measured by the number of minutes required for installation. Scale: Total Points = 10 – 2 (Minutes Required)
Durable	Measured by flimsiness, points of failure, ability to resist torques. Scale: Total Points = 10–# of points of failure
Adjustability	Measured by the device's ability to fit a range of wheelchairs and individuals. Scale: 1 to 10 by Subjective Evaluation
Low Cost	Determined by the production cost of one unit. Scale: Total Points: 10–(Cost/$200)

FIGURE 3.8 This table presents an alternative approach to Figure 3.7. This team has developed metrics and corresponding scales for each of the major objectives and sub-objectives given in Figure 3.6. This team left no doubt about what was to be measured and how. It is a minor point, but it is not helpful that the order in which the metrics are listed does not correspond to the order in which the objectives are listed in Figure 3.6.

judgments when complete details are given for those objectives whose metrics can be measured.

Figure 3.8 shows the metrics and their corresponding scales and units developed by the other design team. These metrics are basically solid, testable measures of the achievement of the objectives given. This approach also leads to the rapid adoption of a very small set of design alternatives, within which the selection of components remained somewhat larger. It can be argued that in both cases, the teams might have been more successful had they gone through several thoughtful iterations.

3.6.3 Revised project statements for the Danbury arm support

After conducting their research and extensive interviews with their client, including several reviews of their objectives lists and trees, the design teams revised the original problem statements. One of the teams produced the following revised problem statement:

> The problem presented to the team involves Jessica, a third-grader at Danbury Elementary School. Jessica has recently begun painting, but because she suffers from cerebral palsy, she has difficulty pursuing her new interest. Jessica paints with her left hand, with her elbow held above the rest position, using a combination of arm and torso movement. While painting, Jessica exhibits exaggerated movements, and lack of control of finer movements, in all directions. These problems are amplified when her arm is fully extended. Currently, when Jessica wants to paint, she requires

a teacher or staff member to hold her left elbow stable. The staff at Danbury school has asked the team to try to design a device that would decrease the magnitude of the exaggerations and assist Jessica in controlling her finer movements. The device must permit the same range of voluntary motion currently employed while painting. Thus, the device would take the place of the teacher or staff member and increase Jessica's functional independence while painting in a classroom environment. The Danbury staff must be able to set-up the device in a classroom environment in eight minutes or less. Optimally, the device could be used by other students with cerebral palsy or other functionally similar conditions at Danbury Elementary school.

A second design team produced a revised problem statement with less detail:

The Danbury Elementary School of the CUSD has a student diagnosed with Cerebral Palsy (CP), a neuro-developmental impairment which causes disturbances of voluntary motor function. For this student, activities that require fine muscle movements, such as painting, writing, and eating, are particularly difficult because of impaired motor control and coordination. There is ample evidence indicating that this student paints more effectively when an instructor holds onto the lower portion of the upper arm (right above the elbow) and thus minimizes extraneous movements of the shoulder. The school desires a device that can minimize the student's involuntary shoulder movements and thus allow her to paint semi-independently. Such a device would ideally be applicable in other CP cases and must be easily implemented by an adult.

3.7 NOTES

Section 3.1: Insights into the need to find and remove bias and implied solutions in assessing problem statements have been provided by Collier (1997). More examples of objectives trees can be found in (Cross 1994), (Dieter 1991, and (Suh 1990). (Cross 1994) and (Dieter 1991) also show weighted objectives trees. The very important notion of *satisficing* is due to Simon (1981).

Section 3.2: Measurements and scales are very important in all aspects of engineering, and not just design. Our discussion takes on a positivist approach (Jones 1992, Otto 1995).

Section 3.3: Some aspects of measurements have recently become controversial in the design community, to a degree beyond our current scope. Some of the critiques derive from an attempt to make design choices and methods emulate long-established approaches of economics and social choice theory (Arrow 1951, Hazelrigg 1996, Hazelrigg 2001, Saari 1995, Saari 2001a, Saari 2001b). The PCCs outlined in the text are exactly the same as the best tool offered by the social choice theorists, the Borda count (Dym, Scott and Wood 2002).

Section 3.4: Our discussion of metrics is strongly influenced by the German approach to design (Pahl and Beitz 1997).

Section 3.5: Constraints are discussed in (Pahl and Beitz 1997).

Section 3.6: The results for the Danbury arm support design project are taken from final reports ((Attarian et al. 2007) and (Best et al. 2007)) submitted during the Spring 2007 offering of Harvey Mudd College's first-year design course, E4: Introduction to Engineering Design. The course is described in greater detail in (Dym 1994b).

3.8 EXERCISES

3.1 Explain the differences between biases, implied solutions, constraints, and objectives.

3.2 The HMCI design team established in Exercise 2.5 has been given the problem statement shown below. Identify any biases and implied solutions that appear in this statement.

Design a portable electric guitar, convenient for air travelers, that sounds, looks, and feels as much as possible like a conventional electric guitar.

Revise the problem statement so as to eliminate these biases and implied solutions.

3.3 Develop an objectives tree for the portable electric guitar. (Some team members will have to play the roles of client and users for this design project.)

3.4 Design a strategy for obtaining the weights for the objectives tree of Exercise 3.3.

3.5 The HMCI design team established in Exercise 2.5 has been given the problem statement shown below. Identify any biases and implied solutions that appear in this statement.

Design a greenhouse for a women's cooperative in a village located in a Guatemalan rain forest. It would enable cultivation of medicinal preventive herbs and aid the villagers' diets. It would also be used to grow flowers that can be sold to supplement villagers' income. The greenhouse must withstand very heavy daily rains and protect the plants inside. The greenhouse must be made of indigenous materials because the villagers are poor.

Revise the problem statement so as to eliminate these biases and implied solutions.

3.6 Develop an objectives tree for the rainforest project. (Someone will have to play the roles of client and users for this design project.)

3.7 Correct and revise the objectives tree developed by one of the arm support teams and shown in Figure 3.5.

3.8 Correct and revise the lists of objectives and constraints developed by the other arm support team and shown in Figure 3.6.

3.9 Develop a set of metrics for the portable electric guitar of Exercise 3.2. If a metric is likely to be difficult to measure, indicate how it can be reframed as a constraint.

3.10 Develop a set of metrics for the rain forest project of Exercise 3.5. If a metric is likely to be difficult to measure, indicate how it can be reframed as a constraint.

CHAPTER 4

FUNCTIONS AND REQUIREMENTS

How do I articulate the design in engineering terms?

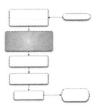

SO FAR we have focused on defining the client's design problem. Now we move from the client's perspective into engineering practice as we translate the client's needs into quantitative, engineering terms that enable us to ensure that those needs are met. In engineering terminology, then, we first identify *functions* that the design must perform and then formulate *requirements* that specify how the performance of those functions can be assessed. We will also note that the label requirements also incorporates specifications for other behaviors or attributes (in addition to functions) required of the design. Thus, we focus here on the fifth and sixth of the design tasks identified in Figure 2.3.

4.1 IDENTIFYING FUNCTIONS

Asked what a bookcase does, a child might answer that, "It doesn't *do* anything, it just sits there." An engineer, however, would say that the bookcase does a number of different things (and does them well if it's a good design!). In this view the bookcase: resists the force of gravity exactly, so that books neither fall to the floor nor are forced into the air; and separates the books into categories chosen by the owner, either with dividers or some limits on individual shelf length. Thus, there are two ways in which this bookcase performs functions or *does things*, even as it appears to "just sit there." Understanding what a designed device must do is essential to creating a successful design. In this section we will explore what we mean when we talk about a design doing something, and we will describe techniques for identifying and listing functions.

It is well worth noting that an engineer must be able to specify functions properly because there are consequences for failing to understand and design for *all* of the functions in a design. The forensic engineering literature is rich with cases in which engineers failed to realize some additional, unmet function(s), often with tragic results.

4.1.1 Functions: Energy, materials, and information flow and are transformed

We can think of functions in a number of different ways, beginning with our dictionary definition from Chapter 1:

- **function** *n*: the action for which a person or thing is specially fitted or used or for which a thing exists; one of a group of related actions contributing to a larger action

Thus, in the simplest terms, *functions* are those actions the designed device or system is supposed to take, the things that that system or device is meant to do.

But what does it mean to do something? For our work as designers, we can relate *doing* something to *transforming* an *input* into an *output*. Recall that in elementary calculus we write $y = f(x)$ to denote how the *input* of an independent variable x is *transformed* into the *output* of the dependent variable y by the function $f(x)$. In multivariate calculus this notion is extended to include multiple inputs and multiple outputs. Similarly, management studies use *transformation functions* to transform a vector of inputs (labor, materials, technology, etc.) into a set of outputs (products, services, etc.). In all of these cases we are highlighting the existence of a relationship between some independent variables (i.e., *inputs*) and some dependent or response variables (i.e., *outputs*), and characterizing that relationship in a formal way.

And what is being transformed? For most of our purposes, *engineering functions* involve the flow or transformation or transfer of *energy*, *materials*, and *information*. The kinds of energy we see include mechanical, thermal, fluid and electrical, and these forms of energy are transformed as they are transmitted, converted or dissipated. Energy can also be stored and supplied. We also view energy transfer quite broadly to include forces used to support, forces that are transmitted, current flows, charge flows, and so on. And as we shall repeat later, *energy must be conserved*, that is, all of the energy going into a device or a system must come out. This does not mean that the device or system is ideal and that no energy is lost. Rather, it means that we must account for all of the energy — it can't simply disappear — including that which is dissipated.

Similarly, materials flow occurs in a variety of ways, including: moving or flowing through some conveyance, being transferred or located in a container, being separated into constituents, or added to, mixed in with, or located within one or more materials. Thus, cement, aggregate and water are mixed to create concrete, which is then typically moved (while being mixed), poured, finished, and allowed to set and harden.

Finally, information flow can include the transfer of data in any of several forms, including tables and charts on paper through data transmitted over the Internet or by wireless, as well as electrical or mechanical signals transmitted to sense or measure behavior and to control response. The transformation of information occurs when, for example, a room temperature measured by a thermometer is transmitted electronically to a wall thermostat, as well as when a person in the room uses that thermostat to instruct the heater or air conditioner to change what it has been doing. We might even think of the energy that is transformed when data is accumulated and processed into information, and a similar transformation of energy when information is processed to become knowledge.

4.1.2 Expressing functions

Given that functions are the things that a designed device must do in order to be successful, the statement of a function usually consists of an "action" verb and an object or noun. For examples, lift a book, support a shelf, mix two fluids, measure the temperature, or switch on the light are action verb–object couples.

The object or noun in the statement of function may start off with a very specific reference to a particular design project, but experienced designers look for more general cases. For example, we might characterize one of a bookcase's functions as "support books," but this implies that the bookcase would only hold books. Clearly, shelves in bookcases often support trophies, art, or even piles of homework. Thus, a more basic and more useful statement of the function to be served here is that shelves should "resist forces due to gravity," which could then be associated with any objects weighing less than some predetermined weight. That is, our statement of function is that shelves should support some number of kg (or lbs). When describing functions, then, we should use a verb-noun combination that best describes the most general case.

Functions are often expressed as verb-object pairs.

We also want to avoid tying a function to a particular solution. If we were designing a cigarette lighter, for example, we might be tempted to consider "applying flame to tobacco" as a function. This might imply that the only way to light the tobacco is by using a flame (and that tobacco is the only material to be lit). Car lighters, however, use electrical resistance in a wire to achieve this function. Thus, a better statement of this function might be "ignite leafy matter," or even "ignite flammable materials." (Somewhat parenthetically, we could consider the following questions. In light of the well-documented health hazards associated with smoking, is there an ethical issue for an engineer who is asked to design a better cigarette lighter? Is it an appropriate design task? We discuss ethics in engineering and design in Chapter 12, but we note here that that particular issue might be missed if the lighter is viewed as a camping tool.)

We can also categorize functions as being either *basic* or *secondary* functions. A *basic function* is defined as "the specific work that a project, process, or procedure is designed to accomplish." *Secondary functions* would be (1) any other functions needed to do the basic function or (2) those that result from doing the basic function. Secondary functions can themselves be either required or unwanted functions. *Required secondary functions* are clearly those needed for the basic function. Consider, for example, an overhead projector. Its basic function is to project images, and it has required secondary functions that include converting energy, generating light, and focusing images. *Unwanted secondary functions* are undesirable byproducts of other (basic or secondary) functions. For the overhead projector, generating heat and generating noise are such unwanted secondary functions. Such undesirable byproducts often generate new required functions, such as quieting noise or dissipating generated heat. This last point also suggests the importance of ensuring that all secondary functions are anticipated, lest they turn into undesirable *unanticipated side effects* that may significantly affect how a new design is perceived and accepted.

4.1.3 Functional analysis: Identifying functions

We now turn to *functional analysis* to identify the functions that are to be performed by the device or system we're designing. One starting point for analyzing the functionality of a proposed device is a "black box" that clearly delineates the boundary between the device and its surroundings. Inputs to and outputs from the device occur across that boundary, so we can assess the inputs and outputs by (1) tracking the flow of energy, materials, and information through the device's boundary, and (2) detailing how energy is used or converted and how materials and/or information are processed to produce the desired functions. We now describe the analysis of both "black" and "transparent" boxes, as well as three other methods used to determine functions: enumeration, dissection or reverse engineering, and the construction of function-means trees.

4.1.3.1 *Black boxes and transparent boxes*   Recall that our earlier discussions of mathematical and management functions featured inputs and outputs, both singly or in groups. This input-output model is also useful in modeling system designs and their associated functions. One tool that helps relate inputs and outputs, and the transformations between them, is the *black box*. A black box is a graphic representation of the system or object being designed, with inputs shown entering the box on its left-hand side and outputs leaving on the right. *All* of the known inputs and outputs should be specified, even undesirable byproducts that result from unwanted secondary functions. In many cases functional analysis helps us identify inputs or outputs that have been overlooked. Once a black box has been drawn, a designer can ask questions such as "What happens to this input?" or "Where does this output come from?" We can answer such questions by removing the cover of the black box, thus making it into a *transparent box*, to see what is going inside. That is, we expose the transformations of inputs to outputs by making a box transparent. We can also link more detailed "sub-inputs" to (smaller) internal boxes that produce related "sub-outputs" within any given box.

Black box analyses connect outputs to inputs.   As a brief illustration, let's look at the basic function of a power drill. In fact, a power drill is a mildly complex system (see Section 4.1.3.2), but at the top level it has three inputs: a source of electrical power (energy), a supporting force (energy) that holds or grasps the drill, and the control of speed and direction (information) of the drill chuck's rotation, which is also the drill's output. In Figure 4.1 we show this power drill (system) as a simple black box that transforms the controlled power input into a rotating chuck, in which we can insert a drill bit to drill a hole or a screwdriver blade to drive a screw. How does this actually happen? What functions are performed in a power drill? Can we identify all of the (many) sub-functions performed inside of the power drill's black box? We can answer these questions, and we will in Section 4.1.3.2 where we *dissect* or take apart such a drill.

Now consider another familiar system, a radio. Experience suggests that a radio has three inputs: an airborne signal within that part of the frequency spectrum containing the

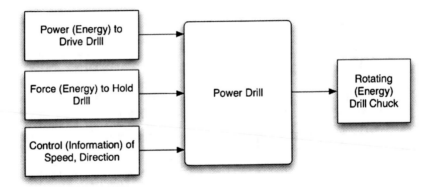

FIGURE 4.1 This is a black box for a power drill. Notice that both of the inputs and the single output are encapsulated in a single, top-level function, the power drill's *basic function*: provide power to a screw. To know how these two inputs are actually transformed into the single output, we will remove the cover on the black box when we dissect or reverse engineer the power drill (see Section 4.1.3.2 and Figure 4.3).

radio frequencies (RF), a controllable source of electrical power, and a vector of desired outputs (such as particular stations and volume levels). And a radio has three obvious outputs: sound, heat, and a display that indicates whether the user's desired frequency and volume level were realized. We show in Figure 4.2(a) a black box for a radio that transforms an incoming RF signal into audio signal that could include music, talk, and even some noise! We also see there that the basic, top-level function of the radio is multi-dimensional. In addition to an input RF signal (information), we need the energy supplied through the radio's power cord to make the radio run, and we need the information contained in the vector of user choices to choose a particular sound. The radio transforms the information and energy input into several outputs. Two of the outputs are desired: the audio signal (energy and information) from the (electromechanical) speaker and a vector of status indicators (information and energy) that confirm the station, volume, bass-treble balance, and so on. One output is unwanted: heat (energy) is generated by the radio circuitry as the power input is converted and used to identify and amplify the selected signal (information) and to drive the speaker (energy and information).

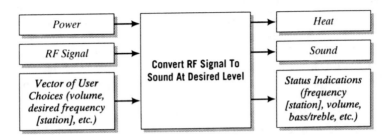

FIGURE 4.2(a) This is a black box for the radio. Again, we see that all of the inputs and outputs are somehow related to the radio's basic or top-level function. When we take the cover of this black box we will see how the energy and information inputs are actually transformed into corresponding outputs (see Figure 4.2(b)).

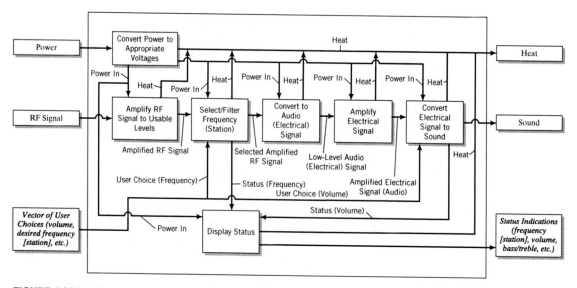

FIGURE 4.2(b)   The cover on the black box of Figure 4.2(a) has been removed or made transparent. Notice that in order to transform the inputs into outputs, a large number of secondary functions are required. If our design responsibilities (or our curiosity) demanded it, we could also remove the covers on some or all of these functions. We also see that this design calls for the heat to be allowed out of the box on its own. In many designs, we would add a specific function, "dissipate heat," and then decide on a strategy (i.e., a set of sub-functions) for doing so.

If we take the cover off this box (Figure 4.2(b)), we see several new black boxes within. These boxes include transforming the power from that of a wall outlet, 110V, to a level appropriate for the radio's internal circuitry (likely 12V). Other internal functions include filtering out unwanted frequencies, amplifying the signal, and converting the RF to an electrical signal that drives speakers. Thus, making our black box cover transparent revealed a number of additional functions. If we had to design the radio, we would probably remove the covers of even more of the boxes than we now see. If, on the other hand, we were assembling a radio from known parts, we might stop at this level. This method of making internal boxes transparent and analyzing their internal functions is also called the *glass box method*. No matter which term we use, the effect is the same: We keep opening internal boxes until we fully understand how all inputs are transformed into corresponding outputs and identify any side effects that are produced by these transformations.

The black box method can be a very effective way to determine functions, even for systems or devices that do not have a *physical* box or housing. The only requirement for using a black or transparent box is that *all* of the inputs and outputs be identified. For example, to design a playground for a rainy climate, our inputs would include children, their parents or caregivers, and the rain. Our outputs would include entertained children, satisfied parents, and water. If we forget the water, our playground design may suffer for lack of proper drainage. (As an aside, it is generally not sufficient to include general terms such as "weather" unless we are willing to consider how the weather is translated into water, ice, wind, and heat *within* our box.)

Black boxes become transparent when we ask how inputs are transformed into outputs.

A final point on the black or glass box method is that we must be very careful when we define the *boundaries* of a system or subsystem whose functions we are identifying, as there is a tradeoff. If we set boundaries too broadly, we may incorporate functions that are beyond our control (or design), for example, generating the household electric current for the radio. If we draw boundaries too narrowly, we may limit the scope of the design. For example, the radio output could be an electrical signal that is fed to speakers, or it could be the acoustic signal coming from speakers. Thus, the boundary drawn here decides whether or not speakers are included in the radio. Such decisions are really about the scope of the design problem and should be resolved as the design problem is framed.

4.1.3.2 *Dissection, or reverse engineering* Most engineers, and indeed, most curious people, ask the question, "What does this do?" when confronted with a button, knob or dial. Natural follow-ups may be "How does it do that?" or "Why would you want to do that?" When we follow up these questions with remarks on how we might do it better or differently, we are engaging in the art of *dissection* or *reverse engineering*. Reverse engineering means taking a device or system that does some or all of what we want our design to do and dissecting — or deconstructing or disassembling — it to find out, in great detail, just how it functions or works. We may not be able to use that design for any number of reasons: it may not do all the things we want, or do them very well; it may be too expensive; it may be protected by a patent; or it may be our competitor's design. But even if all of these reasons apply, we often can gain insight into our own design problem by looking at how other people have thought about the same or similar problems. (Remember that design problems are (see Section 1.3) "*open-ended because they typically have several acceptable solutions.*")

The process is actually quite simple. We begin with a means that has been used by a previous designer, then determine what functions are realized by that given means. We then explore alternative ways of doing the same thing. For example, to understand the functioning of an overhead transparency projector, we might find a button that, when pushed, turns on the projector. A projector button controls the function of turning the projector on or off. This can be done in other ways, including toggle switches and bars along the front of the projector. It is an interesting exercise to consider just how many functions can be thought of for this commonplace device.

A more complicated example is the power drill, for which we showed a top-level black box analysis in Figure 4.1. Let us take the cover off that black box while, simultaneously, dissecting such a power drill. In Figure 4.3 we show an exploded view of a DeWalt™ corded power drill (Model D21008K), and in Figure 4.4 we show a glass box that results from "drilling down one level" to expose the major subsystems. For example, the power cord (8) transmits electrical power into the drill, where a switch (6) both directs that power and transmits information about its level to a *Universal Motor*. The Universal Motor converts the electrical power into mechanical power and then transmits that power and information about its level to the *Transmission*. The Transmission increases the torque output by reducing the speed of power transmission from the 30,000 rpm of the rotor (1) in the Universal Motor to a (peak) output of 2,500 rpm for the rotating chuck (13). Note that even in this brief description, it is often difficult to describe all that happens at the same level of detail.

Several cautions should be noted about using dissection or reverse engineering to find functions. First, the devices being dissected were developed to meet the goals of a particular client and a target set of users. This audience may have had very different

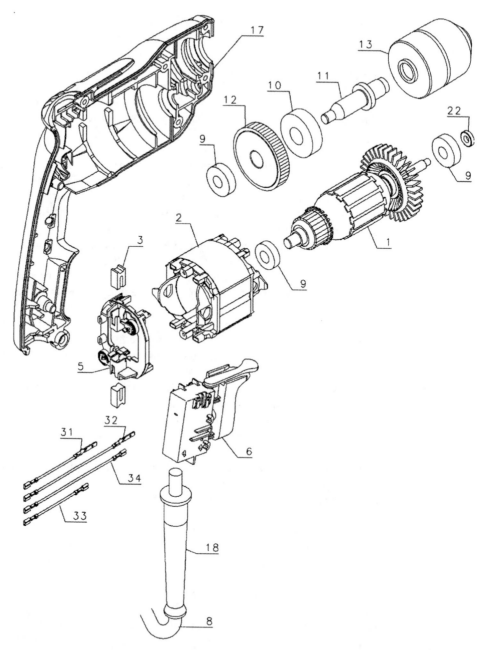

FIGURE 4.3 This is an exploded view of the major subsystems in a DeWalt™ corded power drill, Model D21008K. We can identify the following major subsystems: *Switch* (6); *Universal Motor* consisting of brushes (3, 5), stator (2), rotor and armature (1), helical pinion; *Transmission* consisting of helical pinion, bearings (9, 10), spur gear (12), spindle (11) and chuck (13); and the *Clam Shell* cover in the upper left. (Courtesy of Black & Decker Corporation.)

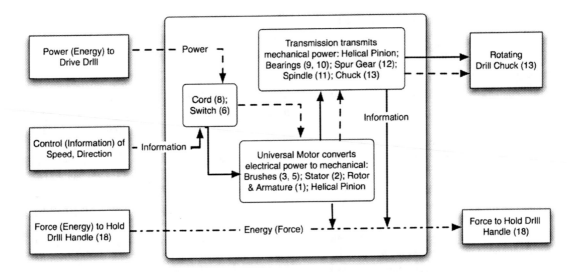

FIGURE 4.4 This is a transparent box for the DeWalt™ corded power drill (Model D21008K), for which we showed a black box in Figure 4.1 and an exploded view in Figure 4.3. Here we have removed the cover of that black box and "drilled down one level" by reverse engineering the corded drill in order to identify the major sub-functions performed by the major sub-systems (and their sub-inputs and sub-outputs) needed to realize the basic drill function identified in Figure 4.1.

concerns than are called for in the current project. Thus, a designer should be sure to stay focused on the current client's needs. Second, there is often a temptation to limit the new means to those that work in the context of the object that is being dissected. For example, all of the means for turning on and off the power to the classroom projector are more or less compatible with a stand-alone device. In some settings, however, it might be more appropriate to remove these controls from the device itself and make them part of some more general room controls. In theaters, for example, lights and other controls are often located in the projection room, rather than in wall switches. It is important that we do not become captive to the design being used to assist our thinking.

A third caution is that while we treat the terms dissection and reverse engineering as equals, they may not always refer to exactly the same process. This is because dissection is sometimes viewed just as it is in a high school biology laboratory, wherein a frog is dissected to reveal its anatomical structure. Here, dissection is more descriptive than analytical. In reverse engineering we go a step further as we try to determine means for making functions happen, which means that we are trying to analyze both the functional behavior of a device and how that functional behavior is implemented.

There is a fourth consideration, one that we stated earlier but reiterate here. We need to define functions in the broadest possible terms and only focus down when it is necessary. Restricting functions to the most immediate terms found on the object being reverse engineered may lead us to mimic someone else's design, rather than fully appreciating opportunities for new ideas. Further, there are serious intellectual property and ethical issues tied to reverse engineering. It is never appropriate to claim as our own the ideas of others. In some cases this can be a violation of law. We will discuss intellectual property in Chapter 9 and ethics in Chapter 11, but it is always important to respect the ideas of others at least as stringently as any other (tangible) property they might hold. After all, wouldn't we want the same protections for our own ideas?

Dissection should enhance our appreciation of the ideas of others.

4.1.3.3   Enumeration

Another basic method of determining functions for a designed object is to simply *enumerate* or list all of the functions that we can readily identify. This is an excellent way to begin functional analysis for many objects. It leads us to consider what the basic function of the object is and it may prove useful for determining secondary functions. However, we might get "stumped" very early in this process. Consider, for example, a bridge. If the bridge is used for highway traffic, we might note that its basic function is to act as a conduit for cars and trucks, and then we might scratch our heads before being able to add much to this initial, single-entry list. However, there some useful "tricks" we can use to extend an enumerated list.

One trick is to imagine that an object exists and ask what would happen if it suddenly vanished. (Philosophers call this *St. Anselm's Riddle.*) If a bridge disappeared entirely, for example, any cars on the bridge would fall into the river or ravine over which the bridge crosses. This suggests that one function of a bridge is to support loads placed on the bridge. If the abutments ceased to exist, the deck and superstructure of the bridge would also fall, which suggests that another function of the bridge is to support its own weight. (This may seem silly until we recall that there have been more than a few disasters in which bridges collapsed because they failed to support even their own weight during their construction. Among the most famous of such infelicitous bridges is the Quebec Bridge over the St. Lawrence River, which collapsed once in 1907 with the loss of 75 lives and again in 1916 when its closing span fell down.) If the ends of a bridge that connect to various roads disappeared, traffic would not be able to get on the bridge, and any vehicles on the bridge would be unable to get off. This suggests that another function of a bridge is to connect a crossing to the road network. If road dividers on our bridge were removed, vehicles headed in one direction could collide with vehicles headed in the other. Thus, separating traffic by direction is a function that many bridges serve, and it is a function that can be accomplished in several ways. For example, New York's George Washington Bridge assigns different directions of traffic to each its two levels. Other bridges use median strips.

Effective function enumeration goes beyond making a list.

Another way to determine functions is to consider how an object might be used and maintained over its lifetime. In the case of our bridge, for example, we might note that it is likely to be painted, so that one function is to provide the bridge's maintenance workers with access to all parts of the structure. This function might be served with ladders, catwalks, elevators, and so on.

Consider once again our beverage container design problem. Here, because we have ample experience with such containers, we can readily name or list the functions served by a beverage container, including at least the following:

- contain liquid
- get liquid into the container (fill the container)
- get liquid out of the container (empty the container)
- close the container after opening (if it is to be used more than once)

- resist forces induced by temperature extremes
- resist forces induced by handling in transit
- identify the product

Note that the functions of getting liquid into and out of the container are distinct. This is evident after a brief reflection on canned beverages: Liquid is sealed in by a permanent top, while access is obtained through a pull tab. We might have noticed this distinction between the filling and emptying functions had we considered the "life cycle" of a beverage container.

At the heart of our approaches to function enumeration lies the need for the designer to list the verb-noun pair that corresponds to each and every function of the designed object. However, since enumeration is often difficult, we must turn to other methods.

4.1.3.4 *Function-means trees*

We often have ideas about how a designed device or system might work early in the design process. While we warn against "marrying your first design" and caution against trying to solve design problems until they are fully understood, it is often true that early design ideas suggest different functional aspects. Consider the hand-held (cigarette) lighter. Clearly, if we use a flame to ignite leafy materials we will encounter different secondary functions than if we use hot wires or lasers. One such difference could be the shielding of the igniting element if the device is to be pocket-sized. A function-means tree can help us sort out secondary functions in cases where means or implementations can lead to different functions.

A *function-means tree* is a graphical representation of a design's basic and secondary functions. The tree's top level shows the basic function(s) to be met. Each succeeding level alternates between showing means by which the primary function(s) might be implemented and displaying the secondary functions made necessary by those means. Some graphical notation is employed to distinguish functions from means. For example, functions and means can be shown in boxes with different shapes or written in different fonts. Figure 4.5 shows part of a function-means tree for the hand-held cigarette lighter. Note that the top level function has been specified in the most general terms possible. At the second level, a flame and a hot wire are given as two different means. These two means imply different sets of secondary functions, as well as some common ones. Some of these secondary functions and their possible means are given in lower levels.

Once a function-means tree has been developed, we can list all of the functions that have been identified, noting which are common to all (or many) of the alternatives and which are peculiar to a specific means. Functions that are common to all of the means are likely to be inherent to the problem. Others are addressed only if the associated design concept is adopted after evaluation.

A function-means tree has another useful property because it begins the process of associating what we must do with how we might do it. We will return to this issue in Chapter 5 when we present a tool to help us generate and analyze alternatives. That tool, the morphological chart, lists the functions of the designed device and the possible means for realizing each function in a matrix format. The effort we put into the function-means tree really pays off then.

Two cautions should be noted regarding function-means trees. The first, and perhaps most obvious, is that a function-means tree *is not* a substitute for either formulating

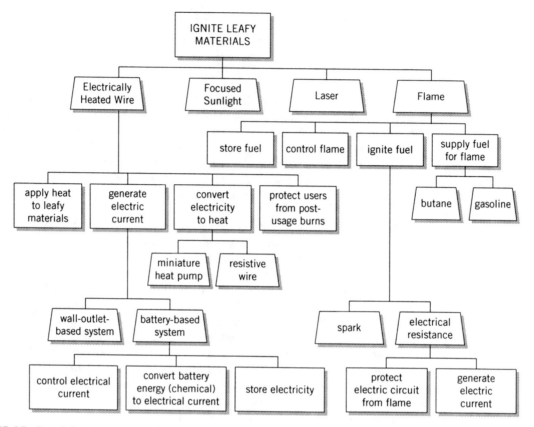

FIGURE 4.5   Part of a function-means tree for a cigarette lighter. (Functions are shown in rectangles, while means are shown in trapezoids.) Notice that there are different sub-functions that result from different means. It is often the case that conceptual design choices will result in very different functions at the preliminary and detailed design stages.

the problem or for generating alternatives. It may be tempting to use the outcome of the function-means tree as a complete description of the available alternatives, but doing so will likely restrict the design space much more than need be. The second caution is that function-means trees should not be used without using some of the other tools described above. One mistake commonly made by novices (or students) is that they adopt a tool because it somehow "fits" with their preconceived ideas of a solution. This transforms the design process from a creative, goal-oriented activity into just a mechanism for making choices that a designer wanted. That is, because the function-means tree allows us to work with appealing means or implementations, we may overlook functions that might have turned up with a less "solution-oriented" technique.

Beware of using function-means trees to reinforce preconceived ideas.

4.1.4   A caution about functions and objectives

Novice designers often make lists of objectives when functions are appropriate, and vice versa. This happens because objectives sometimes express a functional need. For example, a bookshelf design might have as an objective that it hold the complete *Harry Potter* (by J. K. Rowling) and *Lord of the Ring* (by J. R. Tolkein) series of fables, whereas the requirements include both a functional requirement that it supports the weights of those collected sets and a requirement that the shelf has the attribute of being long enough to accommodate all ten volumes in hardback editions (seven of *Harry Potter*, three of *Lord of the Ring*).

It is also the case, as noted in Section 3.1.4, that one of the signs that we are nearing the end of the objectives tree is that the "why" flavor turns toward "how," meaning that functions may be emerging as ways in which objectives can be achieved. Confusion between objectives and functions can also be eased by keeping in mind whether the focus is on "being" participles or on "doing" verbs. As we noted in Section 3.1.3:

Objectives are adjectives; functions are verbs!

- *Objectives* describe what the designed device will be like, that is, what the final object will *be* and what qualities it will have. As such, *objectives detail attributes* and are usually characterized by *linking verbs* such as "are" and "be."
- *Functions* describe what the object will *do*, with a particular focus on the input-output transformations that the device or system will accomplish. As such, *functions transform inputs to outputs*, and are usually characterized by active verbs.

The distinction between objectives and functions is terribly important, but its centrality is often only grasped fully after a great deal of serious practice.

4.2 DESIGN REQUIREMENTS: SPECIFYING FUNCTIONS, BEHAVIOR, AND ATTRIBUTES

In Chapter 2 we noted that design requirements specify in engineering terms the functions, behaviors and attributes of a design. Such requirements, also called "specs" because they used to be called *specifications*, provide a basis for determining a design because such specs become the targets of the design process against which we ensure our success in performing them. Design requirements are presented in three forms that represents different ways of formalizing a design's functional performance and behavior for engineering analysis and design:

Prescriptive requirements specify **values** *for attributes of the designed object.* For example, "A step on a (safe) ladder shall be made from Grade A fir, have a thickness of at least 0.75 in, have a length that does not exceed 80 in, and be attached to the side rails through a full-width groove at each end."

Procedural requirements specify **procedures** *for calculating attributes or behavior.* For example, "The maximum bending stress σ_{max} in a step in a (safe) ladder shall be calculated from $\sigma_{max}=Mc/I$ and shall not exceed the allowable stress σ_{allow}."

Performance requirements specify **performance levels** *that must demonstrate successful functional behavior.* For example, "A step on a (safe) ladder shall support an 800 lb gorilla."

Thus, *prescriptive* requirements specify values of attributes that a successful design must meet (e.g., a beverage container must be made of recyclable plastic). *Procedural* requirements mandate specific procedures or methods to be used to calculate attributes or behavior (e.g., a beverage container must be disposable as stipulated by EPA standards.) *Performance* requirements characterize the desired functional behavior of the designed object or system (e.g., a beverage container must contain 75 ml).

As noted in Section 4.1, determining what a designed object or system must do is essential to the design process. Functional requirements don't mean much if we don't consider *how well* the design must perform its functions. For example, if we want a device that produces musical sounds, we should specify how loudly, how clearly, and at what frequencies the sounds are produced. Thus performance or functional requirements must be specified or defined.

In addition, if a system or device has to work with other systems or devices, then we must specify how those systems interact. We call these particular requirements *interface performance requirements.*

4.2.1 Attaching numbers to design requirements

It is normally up to the designer to cast functions in terms that facilitate the application of engineering principles to the design problem at hand. Thus, the designer has to translate functions into measurable terms in order to be able to develop and assess a design. We must find a way to measure the performance of a design in realizing a specific function or objective and then establish the range over which that measure is relevant to the design. And designers must determine the extent to which ranges of improvements in perform-ance really matter.

Determining the range over which a measure is relevant to a design and deciding how much improvement is worthwhile are interesting problems. Our conceptual starting point for assessing the value of a gain in performance of a design (at some unspecified cost) is the curve shown in Figure 4.6. It is similar to what economists call a *utility plot* with which the benefit of an *incremental* or *marginal* gain in performance can be found. The utility or value of that design gain is plotted on the ordinate on a normalized range

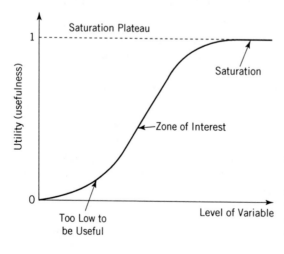

FIGURE 4.6 A hypothetical performance specification curve. Notice that until some minimal level is realized, no meaningful benefit is achieved. Similarly, above some saturation plateau, there is no meaningful benefit in gaining still more. The actual shape of the curve is likely to be uncertain in most cases.

from 0 to 1. The level of the attribute being assessed is shown on the abscissa. For example, consider using processor speed as a measure of a laptop computer performance. At processor speeds below 100 MHz, the computer is so slow that a marginal gain from, say, 50 MHz to 75 MHz provides no real gain. Thus, for processor speeds below 100 MHz, the utility is 0. At the other end of the utility curve, say, above 5 GHz, the tasks for which this computer is designed cannot exploit additional gains in processor speed. For example, browsing the World Wide Web may be more constrained by typing speed or communication line speeds, so that an incremental gain from 5 GHz to 5.1 GHz still leaves us with a normalized utility of 1. Thus, the utility plot is *saturated* at high speeds.

What happens at performance levels between those that have no value and those on the saturation plateau, say between 100 MHz and 3 GHz for the computer design? In this range we expect that changes do matter, and that increases in processor speed do improve the incremental or marginal gain. In Figure 4.6 we show an *S-* or saturation curve that displays *qualitatively* what is happening. There are clearly gains to be made as we move toward faster speeds, and the value of those gains can be determined from the curve. Thus, the utility of the entire S-curve is initially flat (or 0) at low processor speeds, increases measurably over a range of interest, and then plateaus at 1 because added gains are not valued.

This sort of behavior seen in a utility plot is rather common. Economists refer to the *law of diminishing returns*; however, it isn't a "law." We don't usually know the actual shape or precise details of the S-curve (it may not look nearly as smooth as what we have sketched in Figure 4.6), so we choose to approximate it by a collection of straight lines, such as that shown in Figure 4.7. Here there are still regions where gains no longer interest us, as indicated by the horizontal lines at levels 0 and 1. In the middle range, however, we suppose that we are just as happy to increase our levels of the design variable (e.g., processor speed) to obtain a corresponding linear gain in the utility, anywhere within this range of interest. Perhaps most important here is the point that, qualitatively, we are simply saying that the straight line defines a range within which we expect to achieve design gains by tuning the design variable in question.

Consider another example. Suppose we are asked to design a Braille printer that is quiet enough to be used in office settings. None of the competing designs are quiet enough to be so used. How quiet does this design really have to be? To answer this question, we

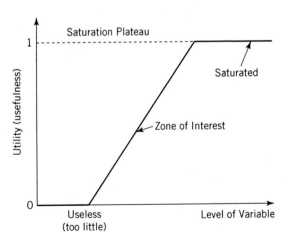

FIGURE 4.7 A linear approximation of the hypothetical performance specification curve shown in Figure 4.3. In this case, the design team has agreed on the lower (minimum) and upper (saturation) levels of interest, and is assuming that an equal increase anywhere along the sloping line brings the user an equal gain.

must determine the relevant units of noise measurement and the range of values of these units that are of interest. We would also find out how much noise is generated by current printer designs and whether or not listeners can distinguish among different designs. If one printer produces the same noise level as made by a pin dropping on a carpet, while another generates the noise level of a ticking watch, we would likely view both as quiet enough to be fully acceptable. Similarly, if one printer is as loud as a gas lawn mower and another is as loud as an un-muffled truck, there is no utility gained by distinguishing between these two designs as neither design will be used in an office setting. (Note that this example shows a *reverse* S-curve in which we start at saturation because there is no gain to be made at such low levels of quietness, and then we degrade to a level of no utility for printers that are uniformly too loud.)

Since sound intensity levels are usually measured in decibels (dB), we might conclude that some range of dB is likely to be of interest. Carrying this further, we might look for some indication of how much noise is produced by other devices and within different environments. We show sound intensities for various devices and environments in Table 4.1. For reference, we show the noise exposure levels to which workers may be exposed in Table 4.2. These levels, expressed in hours of exposure, are defined by OSHA, the federal agency concerned with the safety of work environments. With such environmental and exposure information in hand, the designer can identify a range of interest for a performance specification for the Braille printer. New printer designs must generate less than 60 dB noise in an office environment. Further, smaller values of generated noise

TABLE 4.1 Sound intensity levels that are produced by various devices and are measured in various environments

Level (dB)	Qualitative Description	Source/Environment
10	Very Faint	Hearing Threshold; Anechoic Chamber
20	Very Faint	Whisper; Empty Theater
30	Faint	Quiet Conversation
40	Faint	Normal Private Office
50	Moderate	Normal Office Background Noise
60	Moderate	Normal Private Conversation
70	Loud	Radio; Normal Street Noise
80	Loud	Electric Razor; Noisy Office
90	Very Loud	Band; Unmuffled Truck
100	Very Loud	Lawn Mower (Gas); Boiler Factory

Sound intensity levels are measured in decibels (dB) and are a logarithmic expression of the square of acoustic power. Thus, a 3 dB shift corresponds to a doubling of the energy produced by the source, while the human ear cannot distinguish between levels that differ by only 1 dB (or less). After (Glover 1993)

TABLE 4.2 Permissible noise exposures in American work environments expressed in intensity levels (dB) permitted during various daily durations (hours)

Daily Duration (Hours)	Sound Level (dB)
0.5	110
1	105
2	100
3	97
4	95
8	90

These levels and durations are defined by the Occupational Safety and Health Administration (OSHA), If workers are exposed to levels above these or for times longer than indicated, they must be given hearing-protection devices. After (Glover 1993).

levels are considered gains, down to a level of 20 dB. All designs that generate less than 20 dB are equally good. All design that produce more than 60 dB are unacceptable. Note that any realistic designs will generate noise levels that are so far below the OSHA exposure values that occupational safety is not an issue here.

4.2.2 Setting performance levels

We now extend the above discussion to setting performance levels. First, we determine design parameters that reflect the functions or attributes that must be measured and the units in which those parameters are to be measured. Then we establish the range of interest for each design parameter. For desirable design variables (i.e., qualities or attributes), utility values *below a threshold* are treated as equals because no meaningful gains can be made. Utility values above a *saturation plateau* are also indistinguishable because no useful gains can be achieved. (We are assuming a standard S-curve in which the threshold comes first and the plateau last.) The range of interest lies between the threshold and the plateau. It is within this zone that the design gains should be matched to and measured with respect to the design parameters that are the subject of a given performance specification. This process works well when we exercise judgment in setting performance requirements based on: sound engineering principles, an understanding of what can and cannot be reasonably measured, and an accurate reflection of both client's and users' interests.

Performance requirements require sound engineering, reasonable measurements, and clarified client interests.

Consider once again the beverage container. Each of the functions that were specified in Section 4.1.2.4 has a range of values that must be specified. Some of those functions and some relevant questions associated with each function are:

- *contain liquid*: How much liquid must the container contain, at what temperatures? Is there a range of fluid amounts that we can put into a container and still meet our objectives?

- *resist forces induced by temperature extremes*: What temperature ranges are relevant? How might we measure the forces created by thermal stresses on the container designs?
- *resist forces induced by handling in transit*: What are the range of forces that a container might be subject to during routine handling? To what degree should these forces be resisted in order for the container to be acceptable?

Note that similar but distinct problems arise for the second and third functions on this list, as they both relate to forces.

We can now specify a set of performance requirements that the container designs should meet by addressing these and similar questions. For example, we might indicate that each container must hold 12 ± 0.01 oz. In this case the requirement has become a constraint because the corresponding utility plot is a simple binary switch: We either meet this design specification or we don't. (Of course, it is possible to study the container design problem as one in which a variable single-size serving is possible, in which case there may exist a linearized S-curve for the container size where smaller is better.) Still another performance requirement could emerge from a production concern, namely, that the containers can be filled by machines at a rate of 60–120 containers per minute. Thus, any container that cannot be filled at least this fast creates a production problem, while a faster rate might exceed current demand projections.

We might also specify that the designs should allow the filled containers to remain undamaged over temperatures from –20 to +140°F. Temperatures lower than a threshold of –20°F are unlikely to be encountered in normal shipping, while temperatures higher than a plateau of +140°F indicate a storage problem. It may be that some designs that appeal in other ways are limited here by either temperature extreme. A judgment will have to be made about the importance of this function and its associated performance requirement.

It is also worth noting that the specification of the performance of a device is often published *after* it has been designed and manufactured because users and consumers want to know whether the product is appropriate for *their* intended use. End users, however, are usually not parties to the design process, and so they depend on published performance requirements that set out the performance levels that can be expected from a device or system. In fact, in many instances designers examine the performance requirements of similar or competing designs to gain insight into issues that may affect end users.

4.2.3 Interface performance requirements

As previously noted, performance requirements also specify how devices or systems must work together with other systems. Such requirements, called *interface performance requirements*, are particularly important in cases where several teams of designers are working on different parts of a final product, and all of the parts are required to work together smoothly. For example, a designer must ensure that the final design of a car radio is compatible with the space, available power, and wiring harness of the car. Thus, a design team that has divided a project into several parts must ensure that the final parts will work together. In such cases, the boundaries between the subsystems must be clearly defined, and anything that crosses the boundaries must be specified in sufficient detail to allow all teams to proceed.

Interface performance requirements are increasingly important for large firms that, in a highly competitive international arena, are trying to minimize the total time needed to design, test, build, and bring to market new products. Most of the world's major automobile companies, for example, have reduced their design and development times for new cars to one-half or less of what they were a decade ago by having design teams work *concurrently*, or at the same time, on many systems or products, all of which must work together and be suitable for manufacture. This puts a premium on the ability to understand and work with interface performance requirements.

Developing interface performance requirements is easy theoretically, but is extremely hard in practice. During conceptual design, the boundaries or interfaces between the systems that must work together must be specified, and then requirements for each item that crosses a boundary must be specified. These requirements might be a range of values (e.g., 5V, ±2V), or logical or physical devices that enable the boundary crossing (e.g., pinouts, physical connectors), or simply an agreement that a boundary cannot be breached (e.g., between heating systems and fuel systems). In every case, the designers of systems on both sides of a boundary must have reached a clear agreement about where the boundary is and how it is to be crossed, if at all. This part of the process can be difficult and demanding in practice since teams on all sides are, in effect, placing constraints on all of the others. A black box functional analysis could be helpful in developing interface requirements because it allows all of the parties to identify the inputs and outputs that must be matched and to deal with any side effects or undesired outputs.

4.2.4 A caution on metrics and requirements

A *metric*, in its broadest meaning, is a rule by which meaningful measurements can be made. Thus, at this level of abstraction, requirements — including both functional requirements and behavioral requirements — are also metrics. In the present context, however, we use metrics to scale the achievement of objectives, and we use requirements to scale the performance of functions and behaviors (or attributes):

- *Metrics apply to objectives (only).* They allow both designers and clients to assess the extent to which an objective has been achieved by a particular design.
- *Requirements are applied to functions and behaviors.* They specify required performance levels for functions and behaviors of a design.

Metrics are measures of achievement of objectives; requirements are measures of functional and behavioral performance.

Metrics are needed for *all* of the objectives that are being considered in the design selection process, although that doesn't mean that each objective has to earn high marks on its metric. Requirements are required, and therefore a design must meet *each and every one* of them.

It is also worth noting that metrics and requirements are used in different fashions. Metrics are applied *past* tense, to assess whether objectives *have been* achieved. Requirements, like constraints, are developed for *future* application, to specify the functional or behavioral performance that *must be* achieved in order for a design to be considered to be successful.

4.2.5   A note on customers' requirements

A more advanced tool, *quality function deployment* (QFD), builds on the performance requirements method with the goal of achieving higher-quality products. Used widely in product manufacturing, QFD calls for charting *customer requirements* and engineering attributes in a *matrix* format that makes it possible to relate and weigh them, one against another. The intent is to erect a *house of quality* that exposes both positive and negative interactions of the engineering requirements, thus enabling a designer to anticipate and weed out performance conflicts. QFD is briefly described in Section 11.5.

4.3   FUNCTIONS FOR THE DANBURY ARM SUPPORT

In Section 3.6 the objectives for the design of an arm support/restraint for a Danbury School student diagnosed with cerebral palsy (CP) were clarified by two student design teams that, in turn, developed lists of objectives and constraints and wrote revised problem statements. Now we present some of their results on functions and on functional requirements or specifications.

A number of functions must be performed by the designed arm in order to support Jessica's writing and drawing. Both design teams used the enumeration method to develop the list of functions shown in Tables 4.3 and 4.4. Notice that many of the functions in Table 4.3 are of the form "enable...". While this is not an unreasonable way to begin functional analysis, it is better to cast functions in more active forms. For example, "Enable size adjustability" might be stated as "Adjust size" or as "Adjust to a (specified) range of sizes," depending on which of these two is the correct interpretation of what the design team intended. Similarly, "Enable adjustable orientation" can be more succinctly stated as "Adjust orientation" or as "Adjust to a (specified) range of orientations."

Some functions listed in Table 4.3 are so vague that it is hard to know what means are available for performing the function, or how to specify or write a requirement for the performance of that function. For example, the function "Resist environmental damage," which could also be an objective, does not identify particular environmental threats.

TABLE 4.3   One list of functions for the Danbury arm support, as developed by one of the first-year design teams

Attach to something secure

Stay secure on arm

Support Jessica's arm

Decrease the magnitude of exaggeration

Enable size adjustability

Enable adjustable resistance/support

Enable adjustable orientation

Resist damage due to mishandling

Resist environmentally induced damage

Prevent physical pain

Provide comfort

TABLE 4.4   One list of functions for the Danbury arm support, as developed by one of the first-year design teams

Attach to arm

Attach to stabilizing point

Dampen motion

Allow for range of motion

Provide comfort

Provide adjustability

(Similar concerns could be raised about "Prevent physical pain" and "Provide comfort.") To be fair, the team did provide further commentary on the meaning of their functions. With regard to possible environmental damage, the team stated that "The device will sustain minimal damage if exposed to water, air moisture, dirt, dust, or other environmental factors in order to maximize the lifetime of the device." This does provide more detail, but still leaves open "other environmental factors," and it also introduces another objective, namely, "to maximize the lifetime of the device." Further, it also does not quantify how much exposure water, air, etc., the device is expected to survive. We do not mean to nitpick or be unduly critical, but just as it is important that designers specify functions in general terms that do not imply solutions, it is also important that they be sufficiently specific that their meaning is clear and can be translated into meaningful statements of requirements.

The second team enumerated the list of functions are displayed in Figure 4.4. This is a sharper, more concise list of functions. It is interesting to note that most of the functions listed in Table 4.4 appear in Table 4.3. One that does not appear to be so repeated, "Dampen motion," is likely intended as the function "Decrease the magnitude of exaggeration," although it could be included in the function "Enable adjustable resistance/ support" in Table 4.4. Similarly, the three adjustment functions of Table 4.3 (on size, resistance and orientation) might be viewed as three articulated sub-functions of a single top-level function, "Enable adjustability" — which is more properly cast in Table 4.4 as the more active "Provide adjustability." Likewise, the two functions "Prevent physical pain" and "Provide comfort" of Table 4.3 can also be viewed as more detailed articulations of the function "Provide comfort" of Table 4.4.

While neither team was specifically asked to produce a formal list of design requirements or specifications, the first design team did produce such a set of requirements for their eleven (11) functions; they are displayed in Table 4.5. While this set of requirements served the team adequately for this project, they would clearly not be accepted as a formal set of requirements because they do not provide the "hard" engineering statements associated with the need to specify (and then measure or test) exactly what the performance of a function means. It is also clear from Table 4.5 (as well as the student team's actual final report) that the team viewed the requirements more as a test of acceptability of its chosen, final design. This is an understandable line of thought, but it is important to remember that the requirements are used prospectively to figure out precisely what has to be done to achieve a function, e.g., just how much structure do we need to support a given weight, and not just retrospectively to assess whether the design works.

TABLE 4.5 **The requirements (or specifications) that correspond to the 11 functions listed in Table 4.3**

Functions	Performance Specifications (Requirements)	Specification Met?
Attach to something secure		No
Attach securely to arm	Must not permit user to remove arm without assistance or use of free arm	Yes
Support Jessica's arm	Once user raises arm to a specific height above arm rest, must maintain this height without dropping or requiring user to apply muscle tension unless user chooses to change	Yes
Decrease the magnitude of the exaggeration of Jessica's arm movements	As the user attempts to move her arm from point A to point B, the distance she ends at from point B should be smaller than this distance if she were not using the device	No
Enable adjustability of size	Component containing/supporting user arm should have two setting: (1) fit to arm (to not allow play), (2) allow a little play	Yes
Enable adjustability of resistance/support	3 resistance settings: (1) none, (2) resist sudden jerks, (3) completely resist movement	No
Enable adjustability of orientation	2 stationary positions: (1) arm rest (just above arm rest of wheelchair), (2) work position (above table, next to canvas)	Yes
Resist damage due to mishandling	Maintain structure and form during mounting, dismount, and transport	Yes
Resist environmentally induced damage	Performance is not sensitive to dust, water, or paint	Yes
Prevent physical pain	User should not receive bruises, cuts, or experience strain	Yes
Provide comfort	Components containing/supporting arm contain some form of cushion or soft padding	Yes

Note that the requirements are qualitative, rather than 'hard,' quantitative statements. The requirements were used to assess (in a pass/fail manner) whether the selected design did perform its intended functions.

4.4 MANAGING THE REQUIREMENTS STAGE

For many inexperienced designers, the requirements (or requirements) stage is the most difficult. It requires practice and effort to learn to think in terms of functions and to regard them as intellectual objects distinct from objectives and constraints. To this end, team-based approaches can be quite helpful.

There are some engineering tasks that are better done by individuals (e.g., calculating a stress in a bridge cable), while there are others that are best done by groups (e.g., designing a major suspension bridge). Setting requirements is a task that has both individual and group components. Many teams find it useful to have its members initially use the methods as individuals, and subsequently have the entire team review, discuss, and revise these individual results. This sort of divide-and-conquer approach ensures that team members are prepared for meetings and that there is an experienced practitioner for each tool or method. Group review and revision allows team members to build on each other's ideas in developing and setting functional requirements, so the team benefits from fresh viewpoints and further critical thinking.

4.5 NOTES

Section 4.1: Further details about engineering functions and requirements can be found in (Ullman 1997). The black box for the radio used in Section 4.1.1 was developed by our colleague, Carl Baumgaertner, for use in HMC's first-year design course (E4). The term glass box method was coined in (Jones 1992). The function-means tree used was developed by our former HMC colleague, James Rosenberg, to illustrate an example originally proposed in (Akiyama 1991).

Section 4.3: The results from the Danbury arm support design project are from the final reports (Attarian et al. 2007) and (Best et al. 2007).

4.6 EXERCISES

4.1 Explain the differences between functions and objectives.

4.2 Explain the differences between metrics and requirements.

4.3 Using each of the methods for developing functions described in Section 4.1, develop a list of the functions of the portable electric guitar of Exercise 3.2. How effective was each of these methods in developing the specific functions?

4.4 Using each of the methods for developing functions described in Section 4.1, develop a list of the functions of the rain forest project of Exercise 3.5. How effective was each of these methods in developing the specific functions?

4.5 Based on the results of either Exercise 4.3 or 4.4, discuss the relationship between methods of determining functions, the nature of the functions being determined, and the nature of the device or system being designed. Is the designer's level of experience also likely to affect the outcome of functional analysis?

4.6 Do the research necessary to determine whether there are any applicable standards (e.g., safety standards, performance standards, interface standards) for the design of the portable electric guitar of Exercise 3.2.

4.7 Describe the interfaces between the portable electrical guitar of Exercise 3.2 and, respectively, the user and the environment. How do these interfaces constrain the design?

4.8 Developing countries often have different safety (and other) standards than are typically found in countries such as Canada and the United States. How could this affect the design of both the portable electric guitar of Exercise 3.2 and the rain forest project of Exercise 3.5?

4.9 What are the interface design boundaries and issues for the design and installation of a new toilet for a building?

CHAPTER 5

GENERATING AND EVALUATING DESIGN ALTERNATIVES

How do I create feasible designs? Which one is preferred or "best"?

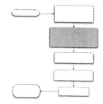

WE DESCRIBED the process of problem framing in Chapter 3 and the establishment of functions and requirements in Chapter 4. Now we will finish our work on conceptual design by generating design concepts or schemes to meet our objectives and evaluating alternative designs to see which are worth pursuing. Thus, we focus here on design tasks 7–10 as identified in Figure 2.3.

5.1 USING A MORPHOLOGICAL CHART TO GENERATE A DESIGN SPACE

We've so far done the "easy work," that is, we have identified the objectives we want the design to achieve, the constraints within it must work, and the functions it must perform. But how do we *generate* or where do we find the actual designs? For now we will say that the *design alternatives* or candidates can be found in a *design space* — an imaginary "intellectual space" that contains or envelops all of the potential solutions to our design problem. A design space is a useful context to the extent that it conveys a *feel* for the design problem at hand: A *large design space* invokes the image of a design problem that has (1) a very large, perhaps even infinitely large, number of designs, or (2) a large number of design variables that may each take on a wide range of values. We will say more about design spaces in Section 5.2. For now we will focus on one technique — the *morphological chart* — for creating and visualizing a design space and identifying the design alternatives within that space. In fact, *morph charts*, as they are affectionately called, show the design space in a way that enables us to identify potential designs while also giving us a sense of the design space's size.

5.1.1   Creating a morph chart

How do we actually create or build a morph chart? We start by constructing a list of the functions that our design must perform and the attributes or features it must have. The list should be of a reasonable and manageable size, with all functions or features being at the same level of detail. This helps ensure consistency. We then list all of the different *means* of implementing each function or attribute identified. Thus, for example, a list of functions and features for the beverage container problem consistent with the objective Promote Sales (see Figure 3.2) might look like:

> Contain Beverage
>
> Material for Beverage Container
>
> Provide Access to Juice
>
> Display Product Information
>
> Sequence Manufacture of Juice and Container

We then list all of the means by which each of these functions and attributes can be achieved and attach them to their corresponding entries, as in this table:

Contain Beverage:	Can, Bottle, Bag, Box
Material for Beverage Container:	Aluminum, Plastic, Glass, Waxed Cardboard, Lined Cardboard, Mylar Films
Provide Access to Juice:	Pull-Tab, Inserted Straw, Twist-Top, Tear Corner, Unfold Container, Zipper
Display Product Information:	Shape of Container, Labels, Color of Material
Manufacturing Sequence of Juice and Container:	Concurrent, Serial

This tabulated information provides what we need to construct the morph chart shown in Figure 5.1, where the same information is displayed as a matrix that is both usefully organized and visually appealing. The required functions and features are listed in the matrix's leftmost column, while for each function and attribute the means are identified and listed in cells in the corresponding row. A conceptual design can be constructed by linking one means, any means, for each identified function, subject only to interface constraints that may prevent a particular combination. For example, one design could consist of a Mylar bag with a zipper that is made (and stored) in advance of the beverage's expected delivery and has a color chosen to correspond to a particular beverage (see Figure 5.2(a)). We thus assemble designs in the classic "Chinese menu" style, choosing one means from each of rows *A, B, C* ... to combine into a design scheme. This is also analogous to a spreadsheet that enables certain "calculations."

MEANS FEATURE/ FUNCTION	1	2	3	4	5	6
Contain Beverage	Can	Bottle	Bag	Box	••••	••••
Material for Drink Container	Aluminum	Plastic	Glass	Waxed Cardboard	Lined Cardboard	Mylar Films
Mechanism to Provide Access to Juice	Pull Tab	Inserted Straw	Twist Top	Tear Corner	Unfold Container	Zipper
Display of Product Information	Shape of Container	Labels	Color of Material	••••	••••	••••
Sequence Manufacture of Juice, Container	Concurrent	Serial	••••	••••	••••	••••

FIGURE 5.1   A morphological (morph) chart for the beverage container design problem. The *functions* that the device must serve are listed in the left most column. For each function, the *means* by which it can be implemented are arrayed along a row to the function's right. A conceptual design or scheme can be constructed by linking one means for each of the five identified functions, thus assembling a design in the classic "Chinese menu" style. (See Figure 5.2.)

How many potential solutions are identified in a morph chart? Or, in other words, just how big is our design space? The correct answer would account for the *combinatorics* that result from combining any single means in a given row with each of the remaining means in all of the other rows. Thus, for the beverage container morph chart of Figure 5.1, the number of design alternatives could be as large as $4 \times 6 \times 6 \times 3 \times 2 = 864$.

While it seems that the design space for this simple example has suddenly become very large, it is important to recognize that not all of these 864 connections are, in fact, feasible or candidate designs. For example, we are unlikely to design a glass bag with a zipper! (See Figure 5.2(b).) Thus, while building a morphological chart provides a way to create a design space and identify alternatives, it also provides the opportunity to *prune* that design space by identifying and excluding incompatible alternatives. We cannot assume that all of the connections made by following out the combinatorial arithmetic are valid connections: There are clearly other combinations of container designs that cannot work (e.g., glass cans with pull-tabs or inserted straws) and therefore must be excluded from the design space. To exclude such alternatives we can apply design constraints, physical principles, and plain common sense. We should also remember that technologies and, consequently, available means do change over time. For example, the waxed paper container has evolved from being one that only supported the containment function to a modern one that incorporates a twist-cap at its top, so that this container supports both containment and intermittent mixing (after it has been opened) of the contents.

MEANS FEATURE/ FUNCTION	1	2	3	4	5	6
Contain Beverage	Can	Bottle	Bag		Box	••••
Material for Drink Container	Aluminum	Plastic	Glass	Waxed Cardboard	Lined Cardboard	Mylar Films
Mechanism to Provide Access to Juice	Pull Tab	Inserted Straw	Twist Top	Tear Corner	Unfold Container	Zipper
Display of Product Information	Shape of Container	Labels	Color of Material	••••	••••	••••
Sequence Manufacture of Juice, Container	Concurrent	Serial	••••	••••	••••	••••

(a)

MEANS FEATURE/ FUNCTION	1	2	3	4	5	6
Contain Beverage	Can	Bottle	Bag	Box	••••	••••
Material for Drink Container	Aluminum	Plastic	Glass	Waxed Cardboard	Lined Cardboard	Mylar Films
Mechanism to Provide Access to Juice	Pull Tab	Inserted Straw	Twist Top	Tear Corner	Unfold Container	Zipper
Display of Product Information	Shape of Container	Labels	Color of Material	••••	••••	••••
Sequence Manufacture of Juice, Container	Concurrent	Serial	••••	••••	••••	••••

(b)

FIGURE 5.2 The morphological (morph) chart for the beverage container design problem (Figure 5.1) is used to show (a) one feasible design alternative whose means are shaded in blue and (b) one infeasible combination whose means are shaded in red.

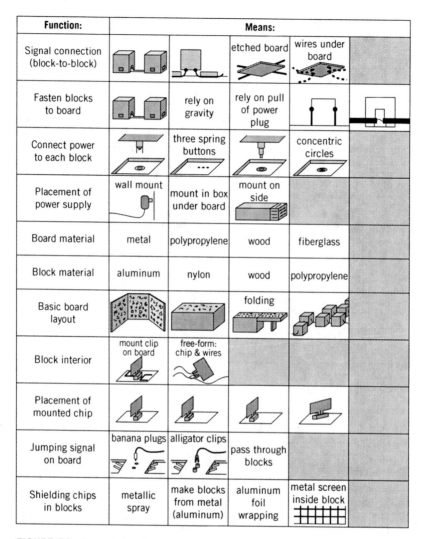

Function:	Means:			
Signal connection (block-to-block)			etched board	wires under board
Fasten blocks to board		rely on gravity	rely on pull of power plug	
Connect power to each block		three spring buttons		concentric circles
Placement of power supply	wall mount	mount in box under board	mount on side	
Board material	metal	polypropylene	wood	fiberglass
Block material	aluminum	nylon	wood	polypropylene
Basic board layout			folding	
Block interior	mount clip on board	free-form: chip & wires		
Placement of mounted chip				
Jumping signal on board	banana plugs	alligator clips	pass through blocks	
Shielding chips in blocks	metallic spray	make blocks from metal (aluminum)	aluminum foil wrapping	metal screen inside block

FIGURE 5.3 A morph chart for a 'building block' analog computer that was done in Harvey Mudd's E4 design course (Hartmann, Hulse et al. 1993).

Morph charts include all potential alternatives; infeasible ones must be pruned.

It is important to list features and functions at the same level of detail when building a morph chart because we don't want to compare apples to oranges. Thus, for the beverage container, we would not include means of **Resisting Temperature** and means for **Resisting Forces** and Shocks within the morph chart of Figure 5.1 because they are more detailed functions that derive from sub-goals that are much further down in the objectives tree of Figure 3.2. Similarly, when doing a complex design task (e.g., designing a building), we don't want to worry about means for identifying exits or for opening doors while developing different concepts for moving between floors, which might include elevators, escalators, and stairways.

As a further illustration, we show in Figure 5.3 a morph chart constructed in a first-year project done at Harvey Mudd that was aimed at designing a "building block" analog computer. This morph chart is appealing in its use of graphics and icons to illustrate many of the means that could be applied to achieve its functions. (This style of morph chart may also be useful in the kind of C-sketch discussions described in Section 5.2.3.2.)

Similarly, morph charts can also be used to expand the design space for large, complex systems by listing principal subsystems in a starting column and then identifying various means of implementing each of the subsystems. For example, to design a vehicle, one subsystem would be its Source of Power, for which the corresponding means could be Gasoline, Diesel, Battery, Steam, and LNG. Each of these power sources is itself a subsystem that needs further detailed design, but the array of power subsystems expands the range of our design choices.

5.1.2 Decomposing complex design spaces

Large design spaces are complex because of the combinatorial possibilities that emerge when hundreds or thousands of design variables must be assigned. Design spaces are also complex because of interactions between subsystems and components, even when the number of choices is not overwhelming. In fact, one aspect of design complexity is that collaboration with many specialists is often critical because it is rare that a single engineer knows enough to make all of the design choices and analyses.

Two designed objects that have large design spaces are passenger aircraft (e.g., the Boeing 747), and major office buildings (e.g., Chicago's Sears Tower). A 747 has six million different parts, and we can only imagine how many parts there are in a 100-story building, from window frames and structural rivets to water faucets and elevator buttons. With so many parts, there are still more design variables and design choices. Yet, while both the 747 and the Sears Tower have very large design spaces, these devices differ from one another because their performance presents different challenges and airplanes are constrained in ways that buildings are not. Architects and structural designers of a skyscraper have far more choices for the shape, footprint, and structural configuration of a high-rise than do aeronautical engineers who are designing fuselages and wings. While weight is important as the number of floors and occupants of a building mount, and while the shapes of high-rise buildings are analyzed and tested for their response to wind, they are subject to fewer constraints than are the payload and aerodynamic shape of aircraft. As we will discuss further in Section 5.2.3, constraints play an important role in limiting the size of a design space.

A *small* or *bounded design space*, on the other hand, conveys the image of a design problem in which (1) the number of potential designs is limited or small, or (2) the number of design variables is small and they, in turn, can take on values only within limited ranges. The design of individual components or subsystems of large systems often occurs within small design spaces. For example, the design of windows in both aircraft and buildings is so constrained by opening sizes and materials that their design spaces are relatively small. Similarly, the range of framing patterns for low-rise industrial warehouse buildings is limited, as are the kinds of structural members and connections used to make up those structural frames.

The "size" of a design is space reflects the number of possible design solutions and the number of design variables.

The complexity of large design spaces emerges because values of many design variables are highly dependent either on choices already made or on those yet to be made. How do we attack such complex design spaces? Our approach is to apply the idea of *decomposition*, or *divide and conquer*, the process of dividing or breaking down a complex problem into sub-problems that are more readily solved. Designs of airplanes, for example, can be decomposed into sub-problems: the wings; the fuselage; the avionics; the tail; the galley; the passenger compartment; and so on. In other words, the design problem and the design space are broken into manageable pieces that are taken on one at a time!

Divide and conquer, or decompose the problem.

The tools presented in this chapter are designed to assist in decomposing a design problem into solvable sub-problems and reassembling their solutions into coherent, feasible designs. The morphological chart described in Section 5.1.1 is particularly suited to (1) decomposing the overall functionality of a design into its constituent sub-functions; (2) identifying the means for achieving each of those functions; and (3) enabling the composition, or *recomposition*, of possible design solutions. The recomposition or *synthesis* of feasible or workable solutions is particularly important. Imagine how it would feel to take apart a very elaborate clock mechanism in order to fix something, and then find that it couldn't be put back together because a new part was too large or a fitting had a slightly different configuration! Similarly, when we are recomposing candidate designs, we have to be sure to exclude incompatible alternatives.

5.1.3   Limiting the design space to a useful size

On the more general subject of guiding the search for design solutions, there are some pragmatic guideposts for narrowing a search space. For example, the design of vehicles for a campus transportation system will be influenced by the *user needs* of potential users. Candidate vehicles might include low-end, simple bikes; nifty, high-tech bikes; recumbent bikes; tricycles; or even rickshaws. Among user needs that might affect the consideration and design of these vehicles would be the availability of parking facilities at lecture and residence halls, the need to carry packages, the need for access for the handicapped, and so on. User needs could have a major impact on the kinds of features called for in a campus vehicle, and thus on the very vehicle type.

Just as clearly, the *availability of different technologies* might stimulate the process of generating alternatives. Imagine the different materials of which a bike or bike-like device could be made: The choice of materials affects a bike's appearance, its manufacture, and its price.

Finally, *external constraints* may affect the design. Examples of such constraints include a team's competence in certain design fields (e.g., it may be more comfortable designing tricycles than high-tech performance bikes) or the availability of manufacturing capabilities (e.g., it should avoid designing a bike made of composite materials if the only available manufacturing facility forms and connects metals).

Similarly, there are practical considerations to keep in mind to ensure that the size and scope of a design space are manageable. Such considerations are often issues of *common sense*. In particular, lest the lists of issues and of candidate designs become too large or too silly, we can:

- *invoke and apply constraints*, in much the same way we did just above while assessing the influence and importance of user needs

- *freeze the number of attributes* being considered to avoid details that are unlikely seriously to affect the design at this point (e.g., the color of a bike or car is not worth noting in a design's early stages)

- *impose some order on the list,* perhaps by harking back to data gathered during problem definition that suggests that particular functions or features are more important
- *"get real!"* or, in other words, watch out when silly, infeasible ideas are repeated often, although, again, such common sense must be applied in ways consistent with our earlier admonitions about maintaining supportive environments for design brainstorming

5.2 EXPANDING AND PRUNING THE DESIGN SPACE

It is worth remembering that design generation is an exciting, creative activity, but it is a goal-directed creative activity: It is designed to serve a known purpose, not to search for one. The goal may be imposed externally, as is often the case in engineering design firms, or internally, as in the development of a new product in a garage. But *there is a goal toward which that creative activity is aimed.*

It is also worth remembering that *creative activity requires work.* As Thomas Edison famously said, "Invention is 99 percent perspiration and one percent inspiration." In other words, we have to be willing to do some serious work if we expect to be successful at generating design alternatives. Therefore, in order to do good, goal-directed design generation, we ask: Beyond the morph chart, what else can we do to generate design ideas? Or, how can we usefully expand the size and range of our design space?

Two central themes will emerge in our answers to these two questions. One is that we rarely gain any advantage by reinventing the wheel. In other words, and particularly when we identify functions and the means to execute them, we should be aware that other people may have already tried to implement some of the functions we want to make happen. It is only common sense to suggest that we should identify, study, and then use the formidable amount of existing information that is already available. Thus, it should not be a surprise that much of what we will now discuss has parallels with our prior discussion of means for acquiring information (cf. Section 2.3.3).

The second principal theme is that as we describe activities that a design team, sometimes including experts or other stakeholders, might undertake to generate design alternatives, we will also hark back to some of the ideas advanced in Section 2.3.3.1 about brainstorming. We will suggest both particular activities a group can undertake in Section 5.2.3 and ideas about how team members might think about things in Section 5.2.4.

5.2.1 Taking advantage of design information that is already available

In Section 2.3.3.1 we detailed the importance of conducting literature reviews to identify prior work in the field and determine the state of the art. This includes locating and studying previous solutions, product advertising, vendor literature, as well as handbooks, compendia of material properties, design and legal codes, etc. *The Thomas Register* is one valuable digest of product vendors. It lists more than one million manufacturers of the kinds of systems and components used in mechanical design. Annual updates of the 23-volume *Register* are found in most technical libraries. Further, while much more material is becoming available on the World Wide Web, it is risky to assume that *all* information is web-available, so web-surfing should not be seen as the only way to identify and retrieve design-related knowledge.

Two information gathering tasks that we discussed in Section 2.3.3.1 are central to product design. Competitive products are:

- *benchmarked* to evaluate how *well* such existing products perform certain functions, and

- *reverse engineered* to see *how* functions are performed and so identify other ways of performing similar functions (see also Section 4.1.3.2).

Another source of already available information is the notes taken during the problem framing phase because it is likely that some ideas and solutions emerged then. Thus, looking back at old notes allows us to recapture old or premature ideas that can now be used to help generate design concepts and alternatives.

5.2.2 Expanding a design space without reinventing the wheel: Patents

One related activity that we can undertake while generating alternatives is to search for relevant patents that have been awarded. We do this to avoid reinventing the wheel and to leverage our thinking by building on what we already know about a still emerging design. We might also do a patent search to identify available technology that we can use in our design, assuming we can negotiate appropriate licensing agreements with the patent holder(s).

Patents are a kind of *intellectual property*, by which we mean that holders of patents are identified as those who are given the credit for having discovered or invented a device or a new way of doing things. A designer's list of patent awards carries a lot of weight in engineering practice. This intellectual credit is awarded by the U.S. Patent Office (USPTO) to individuals and/or to corporations after they file an application that details what they believe to be the *new art* or originality of their invention or discovery. The USPTO awards two basic patents after patent applications are thoroughly evaluated by USPTO patent examiners:

- *Design patents* are granted on the *form* or appearance or "look and feel" of an idea. They clearly relate to an object's visual appearance, as a result of which they are relatively weak patents because only minor alterations in the appearance of a device are enough to create a new product.

- *Utility patents* are granted for *functions*, that is, on how to do something or make something happen. They are stronger and often hard to "work around" because they focus on function rather than form.

In either case, it should be kept in mind that a patent reflects the fact that the patent holder has been certified as the *owner* of this intellectual property.

A computer-based version of an index to patents, *Classification and Search Support Information System* (CASSIS), can be found at most libraries. Patents are listed by individual class and subclass numbers that are detailed in a rather complex classification index. The USPTO maintains its own web site (and search engine) that presents data about granted patents, and it publishes a weekly edition of the *Official Gazette* that lists, in numerical order by class number, all patents granted in the previous week.

Since patents result from (examined) *claims*, they can be — and often are — challenged because others feel that they have developed the relevant *prior art* that forms the foundation of that challenged patent. Filing patent applications is a mixed blessing for the designer. The award of a patent provides protection for new and innovative ideas. At the same time, however, patents may inhibit the development of ideas for second- and third-generation improvements of existing devices or processes.

5.2.3 The design team can do things to expand the design space

Brainstorming was earlier identified as a small-group activity done to clarify a client's goals. We pointed out that new ideas, some related, some not, might be generated, and we suggested that these new ideas be recorded but not evaluated. We also suggested that design team members show respect for the ideas and offerings of their teammates. We now extend those behavioral themes in the context of design generation, emphasizing some "rules of the game" for three different "games" that teams can "play" to generate design alternatives.

Successful design calls for two different kinds of thought, divergent thinking and convergent thinking:

- *Divergent thinking* is done when we try to remove limits or barriers, hoping instead to be expansive while trying to increase our store of design ideas and choices. Thus, we want to "think outside of the box" or "stretch the envelope" when we are trying to expand the design space and generate design alternatives.

- *Convergent thinking* describes what we do to narrow the design space to focus on the best alternative(s) after we have opened up the design space sufficiently. Our problem solving takes on a narrow focus so as to converge to a solution within known boundaries or limits.

As a goal-oriented activity, design combines divergent and convergent thinking. The design process seeks to converge toward the goal of a "best" solution, however "best" is defined. At the same time, the process's activities that lead us to converge on a solution require divergent thinking. For example, we want to open up the problem space to better understand the problem and identify other stakeholders. Similarly, we engage in divergent thinking when we expand our functions to cover the most general case. After completing such "divergent" activities we try to converge to an adequate representation of our understanding. Thus, setting and documenting priorities among objectives is a convergent process, as is selecting appropriate metrics for our objectives. This interplay of modes of thinking (and acting) is part of what gives design its intellectual richness.

Think outside the box, but within the physics!

We now describe three intuitive activities that work to encourage divergent "free thinking" to enhance our collective creativity, although we do *not* mean to encourage thinking that violates the axioms of logic or physical principles. People working in groups may interact more spontaneously, in a free-spirited way that calls forth associations from group members in ways that we cannot anticipate or logically force. These activities are also *progressive* in nature because there are iterative cycles within in each that result in the progressive emergence and refinement of new design ideas.

5.2.3.1 The 6–3–5 method

The first group design activity is the *6–3–5 method*. The name derives from having *six* team members seat themselves around a table to participate in this idea generation game, each of whom writes an initial list of *three* design ideas briefly expressed in key words and phrases. The six individual lists are then circulated past each of the remaining group members in a sequence of *five* rotations of *written* comment and annotation. Verbal communication or cross-talk is not allowed. Thus, each list makes a complete circuit around the table, and each member of the group is stimulated in turn by the increasingly annotated lists of the other team members. When all of the participants have commented on each of the lists, the team would use a common visualization medium (e.g., a blackboard) to list, discuss, evaluate, and record all of the design ideas that have resulted from a group enhancement of all of the team members' individual ideas.

We can generalize this method to the "*m–3–(m-1)*" method by starting with *m* team members and using *m–1* rotations to complete a cycle. However, the logistics of ever-lengthening lists written on increasingly crowded sheets of paper, and of providing tables that seat more than six, suggest that six may be a "natural" upper limit for this activity. And especially in an academic setting, we would prefer fewer than six — ideally no more than four — on a project team.

5.2.3.2 The C-sketch method

The *C-sketch method* starts with an initial sketch of a single design concept by each of the team members, and then proceeds as does the 6–3–5 method. Each sketch is circulated through the team in the same fashion as the lists of ideas in the 6–3–5 method, with all of the annotations or proposed design modifications being written or sketched on the initial concept sketches. Again, the only permissible form of communication is by pencil on paper, and the discussions that follow the end of a complete cycle of sketching and modifying follow those described in the 6–3–5 method. Research suggests that the C-sketch method can becomes unwieldy with even five team members due to the crowding of annotations and modifications on a given sketch. However, the C-sketch method is very appealing in an area such as mechanical design because there is strongly suggestive evidence that sketching is a natural form of thinking in mechanical device design. Research has also shown that drawings and diagrams facilitate the grouping of relevant information added in marginal notes, and they help people to better visualize the objects being discussed.

It is worthwhile at this point to comment on sketching as a technique. Sketching is a valuable tool for designers and is an integral part, for example, of the design generation approach described above. Being able to quickly and clearly convey ideas in the form of a sketch is a powerful skill in design. There are a few simple guidelines for making quick sketches that are easy to learn and put into practice:

- On *lettering*: Clear, easily read notes on sketches can be tremendously useful in conveying the meaning behind the sketched ideas. For example, it is generally useful to follow the famous architectural style of using evenly spaced block letters; they are generally much clearer than cursive scribbles. The sketches in Figure 5.4 are from one of the Danbury arm support teams. Note that the lettering is uneven and a little hard to read in places. Block lettering takes no more time to write, and it produces a document that is much easier to read. For example, look ahead to the sketch in Figure 8.1.

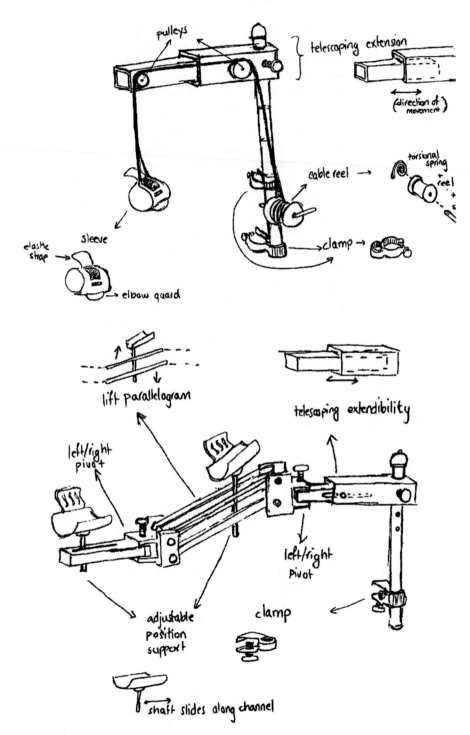

FIGURE 5.4   Sketches of two of the design alternatives produced by the first design team for the Danbury arm support.

- On *proportion control*: It is generally a good idea to sketch designs on graph paper. Graph paper makes it easier control the relative sizes of parts without taking the time to make measurements with a ruler. Further, it is a good idea to think ahead of the components to be sketched before drawing actually begins. Still further, sketching should begin with "blocking in" the overall length and width of a part, laying out the largest component first. Details should be added last.

There are several types of sketches that designers routinely use to convey design information, including *orthographic, axonometric, oblique* and *perspective* sketches (see Figure 5.5). Orthographic sketches lay out the front, right side and top views of a part, and we will discuss such sketches in some detail in Chapter 8. We will briefly describe the other types of sketches here:

- *Axonometric* sketches (Figure 5.5(b)) are done starting with an axis: typically a vertical line with two lines 30° from the horizontal. This axis forms the corner of your part. The object is then blocked in using light lines, with the overall size first. Then vertical lines are darkened, followed by other lines. All lines in these sketches are either vertical, or parallel to one of the two 30° lines. Details of the part are added last.

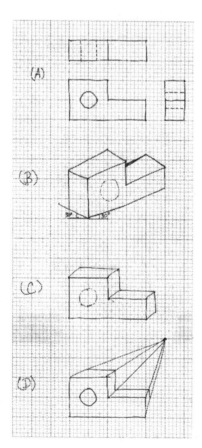

(A)

(B)

(C)

(D)

FIGURE 5.5 Four types of sketches of the same object: (a) orthographic, (b) axonometric, (c) oblique and (d) perspective.

- *Oblique* sketches (Figure 5.5(c)) are probably the most common type of quick sketch. In an oblique sketch, the front view is first blocked in roughly, then depth lines are added and finally details such as rounded edges are added.

- *Perspective* sketches (Figure 5.5(d)) are similar to oblique sketches in that the front view is blocked in first. Then a vanishing point is chosen and projection lines drawn from the points on the object to the vanishing point. The depth of the part is then blocked in using the projection lines. Finally, as in the other sketches, details are added to the part.

Note that each of these sketching techniques provides a structural framework for putting design thoughts on paper. Thus, mastering a few simple techniques means mastering some important design tools! The design sketches in Figure 5.4, done by one of the student design teams, show aspects of some of these techniques, most notably the oblique sketch. Chances are this carefully detailed design sketch took a much more time to produce than any sketches done early in the design process where the focus is on getting out concepts clearly and quickly.

A final note on design sketches. Design teams prepare clearer drawings, often using drawing software, for technical reports and presentations. Figure 5.6 shows examples of such drawings by one of the Danbury arm support teams. These drawings can provide a great deal of information and are nice for presenting the work to clients. Note that this team has used top and side views (think orthographic drawings!) to convey information about its designs. It should be noted, however, that these more formal sketches cannot substitute for the detailed design drawings that must be produced in order to manufacture the design. We will discuss these more formal types of drawings in Section 7.3.

You don't have to be an artist to be a visual thinker.

5.2.3.3 The gallery method

The *gallery method* is a third approach to getting team reactions to drawings and sketches, although the sketching and communication cycles are handled differently. In the gallery method, group members first develop their individual, initial ideas within some allotted time, after which all of the resulting sketches are posted, say on a corkboard or a conference room whiteboard. This set of sketches forms the backdrop for an open, group discussion of *all* of the posted ideas. Questions are asked, critiques offered, and suggestions made. Then each participant returns to her or his drawing and suitably modifies or revises it, again within a specified period of time, with the goal of producing a second generation idea. The gallery method is thus both iterative and progressive, and there is no way to predict just how many cycles of individual idea generation and group discussion should be held. Our only recourse would be to apply the common-sense dictum of the *law of diminishing returns*: We proceed until a consensus emerges within the group that one more cycle will not produce much (or any) new information, at which point we quit because it doesn't make sense to expend any further effort on this particular activity.

5.2.4 Thinking about thinking divergently

A *metaphor* is a figure of speech. It is a style in which attributes of one object or process are used to give depth or color to the description of a second object or process. For example, to describe engineering education as drinking from a fire hose is to suggest that

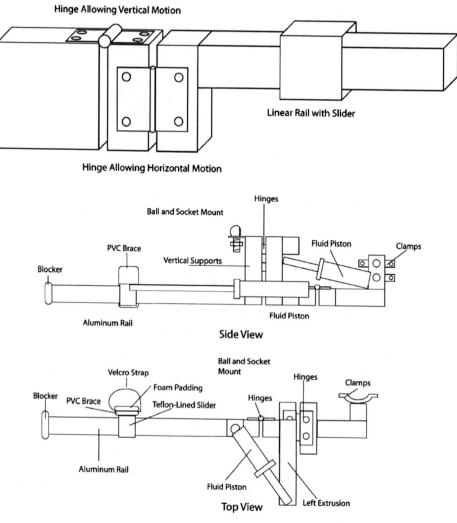

FIGURE 5.6 Design drawings for the design alternatives produced by the second design team for the Danbury arm support.

engineering students are expected to absorb a great deal of knowledge rapidly and under great pressure. We use metaphors to point out *analogies* between two different situations, that is, to suggest that there are parallels or similarities in the two sets of circumstances. Analogies can be very powerful tools in engineering design. One of the most often cited is the Velcro fastener, for which the analogy drawn was to those pesky little plant burrs that seem to stick to everything on which they're blown.

The Velcro fastener resulted from a *direct analogy* in which its inventor made a direct connection between the individual elements of the plant burr and the connecting fibres of the fastener. We also can use *symbolic analogies*, as when we plant ideas or talk about objectives trees. In these cases we are clearly drawing connections through some underlying symbolism. We might also apply *personal analogies* by imagining what it would feel like to be the object (or part of it) that we're trying to design. For example, how might it feel to be a tin beverage container with a pull-tab? We could also stray into the realm of *fantasy analogies* by imagining something that is literally fantastic or beyond belief. Of course, looking at our world at the beginning of the third millennium, just how much more fantastic can it get than space travel, instantaneous and reliable personal communication across the globe, and seeing clearly into the human body with CT scans and magnetic resonance imaging (MRIs)?

Fantasy analogies suggest another approach to "thinking out of the box." We are not very far past the time when many of the technologies we take for granted were thought to be outrageous ideas that were beyond belief. When Jules Verne published his classic *20,000 Leagues Under the Sea* in 1871, the idea of ships that could "sail" deep in the ocean was viewed as preposterous. Now, of course, submarines and seeing unfamiliar yet exciting life forms under water are part of everyday experience. We cannot escape the idea that design teams might imagine the most outrageous solutions to a design problem and then seek ways to make such solutions useful. For example, airplanes that are invisible to radar were once considered far-fetched. The arterial stents used in angioplastic surgery (Figure 5.7) are also devices once thought to be impossible. Who would have believed that an engineering structure could be erected within the narrow (3–5 mm in diameter) confines of a human artery?

The impossible is often the starting point for the excellent!

The stent suggests still another aspect of analogical thinking, namely, looking for *similar solutions*. The stent clearly is similar in both intent and function to the scaffolding

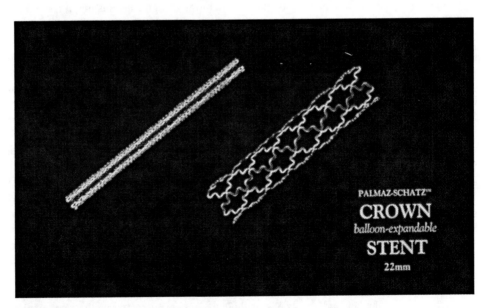

FIGURE 5.7 This is a PALMAZ-SCHATZ™ balloon expandable coronary *stent*, that is, a device used to maintain arterial shape and size so as to allow uninhibited and natural blood flow. Note how this structure resembles the kind of scaffolding often seen around building renovation and construction projects. (Photo courtesy of Cordis, a Johnson & Johnson Company.)

erected to support walls in mines and tunnels as they are being built. Thus, the stent and the scaffold are *like ideas*.

We could invert this idea by looking for *contrasting solutions* in which the conditions are so different, so contrasting, that a transfer of solutions would seem totally implausible. Here we would be looking for *opposite ideas*. Fairly obvious contrasts would be between strong and weak, light and dark, hot and cold, high and low, and so on. One example of using an opposite idea occurs in guitar design. Most guitars have their tuning pegs arrayed at the end of the neck. In order to make a portable guitar, one clever designer chose to put the tuning pegs at the other end of the strings, at the bottom of the body, in order to save space and thus accentuate the guitar's portability.

Finally, in addition to finding similar and contrasting solutions, we recognize a third category. *Contiguous solutions* are developed by thinking of *adjoining* (or *adjacent*) *ideas* in which we take advantage of natural connections between ideas, concepts and artifacts. For example, chairs prompt us to think of tables, tires prompt us to think of cars, and so on. Contiguous solutions are distinguished from similar solutions by their adjacency; that is, bolts are adjacent to nuts and are contiguous solutions, while bolts and rivets serve identical fastening functions and are thus similar solutions.

5.3   APPLYING METRICS TO OBJECTIVES: SELECTING THE PREFERRED DESIGN

If we've done a good job of generating design concepts, we'll likely have several feasible designs from which to choose — and choose we must because time, money, and personnel resources are always limited, so we can't fully develop all of our alternatives. No matter how our schemes have been developed, whether with a morph chart or a less structured approach, we have to "pick a winner" among the identified options and select one or (perhaps) two concepts for further elaboration, testing, and evaluation.

A variety of approaches are used to assess and select among design alternatives, some formal and seemingly rigorous, some as simple as picking the one preferred the most or "liked" the best. The client or a customer may make a choice for reasons unspecified, or an executive or a champion may make a decision based on personal criteria.

We will discuss three methods for choosing from among a set of alternative designs or concepts, each a variant of the *Pugh selection chart*: the numerical evaluation matrix, the weighted benchmark chart, and the best-of-class chart. We will also discuss the idea of concept screening. The three selection methods explicitly link design alternatives to *unweighted design objectives* (recall the discussion of Section 3.3.1), but they do not have the mathematical rigor that goes along with finding maxima and minima in calculus. Rather, we are striving only to bring some order to judgments and assessments that are subjective at their root. Just as professors give grades to encapsulate judgments about how well students have mastered concepts, ideas, and methods, designers try to integrate the best judgments of stakeholders and design team members and exploit these judgments in a sensible and orderly manner. We must use common sense when we look at the results of any such method.

No matter what kind of chart or other decision support technique we apply, our first step should always be to evaluate each alternative in terms of all of the constraints that apply, since alternative designs must be rejected if the constraints are not met. As we outline our three selection methods, we will show how the relevant constraints are being applied and how the design space has been narrowed accordingly.

5.3.1 Numerical evaluation matrices

We show a *numerical evaluation matrix* for the beverage container problem in Figure 5.8. This chart show sboth constraints (upper rows) and objectives (lower rows) in the left-hand columns, while the scores assigned to each objective are shown in design-specific columns on the right. By applying the constraints we can rule out glass bottles and aluminum containers because of potential sharp edges. This reduces the number of designs to two, the Mylar bag and the polyethylene bottle. These two alternatives are evaluated against the metrics for the objectives, as detailed in Section 3.4.2.

There are two questions of interest for us. First, how do we actually do the assessment of attainment for each objective? Simply put, we apply the solution-independent metrics developed in Section 3.4.2. In the present case, since the metrics for the last three objectives (*appeal to parents, permit marketing flexibility*, and *generate brand identity*) require marketing information that is not available to us (and may well be applied independently by management as the final arbiter), we would suggest that the design teams choose a design based only on the first three objectives. This suggestion is made in part because we have established metrics for the goals that the new beverage container should *be environmentally benign, be easy to distribute*, and *preserve taste*. But we also recommend limiting the number of decisive objectives to the top two or three because anyone

DESIGN CONSTRAINTS (C) AND OBJECTIVES (O)	Glass bottle, with twist-off cap	Aluminum can, with pull-tab	Polyethylene bottle, with twist-off cap	Mylar bag, with straw
C: No sharp edges	✖	✖		
C: No toxin release				
C: Preserves quality				
O: Environmentally benign			80	40
O: Easy to distribute			40	60
O: Preserves taste			90	100
TOTALS			210	200

FIGURE 5.8 A *numerical evaluation matrix* for the beverage container design problem. Note that only three of the six objectives originally identified for this design are utilized here, in part because we think these three objectives are more important than the other three, and in part because we have metrics (and presumably data) for these three objectives.

faced with a list of unweighted objectives would find it hard to balance or mediate among more than two or three at a time. Thus, on the presumption that these three objectives are the most important ones, we should do our evaluation by applying the appropriate corresponding metrics.

In the present instance, and taking at face value the data presumably taken from applying the metrics, it would appear that the choice is a plastic or polyethylene bottle with a twist-off cap (210) points over the Mylar bag with a straw (200 points). But we should recognize immediately that the difference between these two sums is but a small fraction of their individual values. Thus, as a practical matter, these totals can be taken as essentially equal.

This last point raises the second question of interest: How do the client's or the designer's values enter into the design selection process? We have assumed so far that these three objectives are seen as essentially equally important by both ABC and NBC. However, it is not hard to imagine that NBC might value an environmentally friendly container above all else, while ABC might value ease of distribution. With these values in mind, NBC could choose the glass bottle with the same degree of rationality as ABC choosing the Mylar bag.

Beyond these "calculated" results about these two hypothetical candidate designs, the most important feature in Figure 5.8 is that each chart shows (or uses) the same results as the metrics are applied to the two design alternatives. Recall from our discussion in Section 3.4 that metrics are measurable indicators of how well specific objectives are met. Thus, if our metrics had different values for different design alternatives, we would have to wonder whether there was a defect in the testing process. Ideally, the metrics and the testing procedures should not change simply because the design team has a different client.

It is worth noting that this might not be the case if the companies were independently doing their designs and, consequently, rating each product on its different dimensions. That is, it is not at all hard to imagine that some companies might find a Mylar bag significantly more expensive to produce or distribute than they would a polyethylene bottle. In such a case, the metric's result for ease of distribution of the bottle might jump from 40 to 80, for example, in which case ABC might make the same choice as did NBC.

5.3.2 The priority checkmark method

The *priority checkmark method* is a simpler, qualitative version of a numerical evaluation matrix. We simply rank the objectives as high, medium, or low in priority. Objectives with high priority are given three checks, those with medium priority are given two checks, while objectives with low priority are given only one check, as shown in Figure 5.9. Similarly, metric results are assigned as 1 if they are awarded more than 50 points (on a scale of 0–100), and as 0 if their award is less than 50. Thus, if a design alternative meets an objective in a "satisfactory" way, it is then marked with one or more checks, as shown in Figure 5.9. Finally, the number of checks are summed over all of the valid alternatives (the constraints having already been applied). This method is easy to use, makes the setting of priorities rather simple, and is readily understood by clients and by other parties. On the other hand, the priority checkmark approach lacks detailed definition and it sets up all of the metrics as binary variables that are either checked (satisfactory) or not. This makes it easy to succumb to temptation and "cook the results" to

DESIGN CONSTRAINTS AND OBJECTIVES	Priority (✔)	Glass bottle, with twist-off cap	Aluminum can, with pull-tab	Polyethylene bottle, with twist-off cap	Mylar bag, with straw
C: No sharp edges		✘	✘		
C: No toxin release					
C: Preserves quality					
O: Environmentally benign	✔ ✔ ✔			1 x ✔ ✔ ✔ ✔ ✔ ✔	0 x ✔ ✔ ✔ • • • •
O: Easy to distribute	✔			0 x ✔ • • • •	1 x ✔ ✔
O: Preserves taste	✔ ✔			1 x ✔ ✔ ✔ ✔	1 x ✔ ✔ ✔ ✔
TOTAL CHECKS				5 ✔	3 ✔

FIGURE 5.9 A *priority benchmark chart* for the beverage container design problem. This chart qualitatively reflects NBC's values in terms of the priority assigned to each objective, so it is a qualitative version of the ordering in the PCC of Figure 3.4.

achieve a desired outcome. Note, in fact, how much more convincing it is for the glass bottle to beat the Mylar bag by a vote of 5 ✔ to 3 ✔, rather than being first with "only" 210 points to 200 points.

5.3.3 The best-of-class chart

Our last method for ranking alternatives is the *best-of-class chart*. For each objective, we assign increasing scores to each design alternative that range from 1 for the alternative that meets that objective best, 2 to second-best, and so on, until the alternative that met the objective worst is given a score equal to the number of alternatives being considered. If, for example, there are five alternatives, then the best at meeting a particular objective would receive a 1, and the worst a 5. Ties are allowed (e.g., two alternatives are considered "best" and so are tied for first) and are handled by splitting the available rankings (e.g., the two "firsts" would each get a score of $(1 + 2)/2 = 1.5$). The remaining calculation would proceed as was done for the numerical evaluation matrix, and the *lowest* summed score would be considered to be the best alternative design under this scheme.

The best-of-class approach also has its advantages and its disadvantages. One advantage is that it allows us to rank alternatives with respect to a metric, rather than simply treat as a binary, "yes or no" decision, as we did with priority benchmarks. It, too, is relatively easy to implement and explain, and it can be done by individual team members or by a design team as a whole to make explicit any differences in rankings or approaches. The disadvantages of this approach are that it encourages evaluation based on opinion rather than testing or actual metrics, and it may lead to a moral hazard akin to that attached to priority checkmarks, that is, the temptation to fudge the results or cook the books.

5.3.4   An important reminder about design evaluation

No matter which of the three selection methods is used, design evaluation and selection demand careful, thoughtful judgment. First and foremost, as we have cautioned earlier, the ordinal rankings of the objectives obtained using PCCs *cannot* be meaningfully scaled or weighted. To draw a crude analogy, think of being at the finishing line of a race without a clock: we can observe the order in which racers finish, but we cannot measure how fast (i.e., how well) they finish the course. Similarly, while we can measure *ranking* with a PCC, we cannot measure or scale objectives' weights from their PCC order of finish.

A PCC's ordinal rankings cannot be weighted or scaled.

Further, *common sense must always be exercised* when we are evaluating results. If the metrics results for two alternatives are relatively close, they should be treated as effectively tied, unless there are other unevaluated strengths or weaknesses. Second, if we are surprised by our evaluations, we should ask whether our expectations were simply wrong, whether our measurements were consistently applied, or whether our rankings and our metrics are appropriate to the problem. Third, if the results meet our expectations, we should ask whether they represent a fair application of the evaluation process, or whether we have just reinforced preconceived ideas or biases. Finally, if some alternatives have been rejected because they violated constraints, it might be wise to ask whether those constraints are truly binding.

There is no excuse for accepting results blindly and uncritically.

5.3.5   Concept screening

It is also worth mentioning that the relative ease of using the foregoing methods suggests that they might also be used to do informal *concept screening*, perhaps earlier in the design process, as a way to easily narrow the field of candidate designs. We could further facilitate a rapid screening with the priority benchmark method, and the other selection tools, by taking the weights out of the matrices or calculations. Such screening could also be made a group process by clustering the checkmarks or some other symbols (e.g., dots) according to the number of people who vote for a concept. In this case, the number of votes would be easily evident from a visual display of the group members' votes.

5.4   GENERATING AND EVALUATING DESIGNS FOR THE DANBURY ARM SUPPORT

We now return to following the two design teams working on the arm support for the CP-afflicted student at the Danbury School. The lists of functions that the teams developed are shown in Tables 4.3 and 4.4, and Table 4.5 shows a set of requirements corresponding to the functions in Table 4.3. Now we follow the teams as they develop design alternatives.

Both teams used morph charts based on the functions they'd identified to build a design space of meaningful alternatives. In addition, although to different degrees, the teams did research on the availability of devices that were intended to serve the same functions for the same kinds of users. (It is worth noting, too, that teams in HMC's E4 environment are routinely told that if they identify an existing product that meets the

Functions	Means			
Attach to something secure	Clamp to chair armrest	Strap to chair backrest	Clamp to table	Strap to user's arm using Velcro
Attach securely to arm	Velcro straps	Buckles	Clamps	Sleeves
Support Jessica's arm	Stationary support cup attached to an arm under student's elbow	Mobile support cup attached to an arm under student's elbow	Support frame with sliding bar	Free-swinging sling
Decrease the magnitude of the exaggeration of Jessica's arm movements	Dashpots	Torsion springs	Elastic wires/cables	Brake pads on hinges
Enable adjustability of size	Adjustable straps	"Baseball cap" snaps	Elastic material	Telescoping extensions
Enable adjustability of resistance/support	Tightening screw	Adjustable brake pads	Pads to compress torsion springs	Reel to shorten elastic wires/cables
Enable adjustability of orientation	Telescoping rods	Lockable hinged arm	Sliding rail	Lockable pivot disks
Resist damage due to mishandling	Slip covers	Rubber padding	Foam padding	Non-deformable materials
Resist environmentally induced damage	Covers over parts with small crevices	Waterproof material	Rustproof material	
Prevent physical pain	Emergency release	Cover moving parts	Cover sharp edges	
Provide comfort	Soft coverings	Soft padding	Air holes	Breathable material

FIGURE 5.10 A significant fraction of the morph chart developed by one of the student teams designing an arm support for the Danbury School.

Function				Means		
Attach to arm	Air pressure cuff	Elbow Pad	Velcro straps	Rope, string	Arm rest	Sleeve, rings
Attach to stabilizing point	Clamps (one or two)	Nut and Bolt	Harness	Magnets	Chemistry-style clamps	
Dampen motion	Pneumatics	Flywheel	Seatbelt	Elastics	Viscous fluid	Friction
Allow range of motion	Telescopic rods	Rails				
Provide comfort	Pillows	Air cushion	Foam padding	Gel cushioning		
Provide adjustability	Jack	Nuts and bolts	Pressure	Slide rails	Piston	

FIGURE 5.11   The morph chart from the second design team working on the Danbury School arm support.

client's objectives and satisfies the client's constraints, then an acceptable design report is to recommend that the client buy the existing product. The teams are told this in part because to let them know that this is a legitimate outcome, and in part to encourage the teams to do their research so as to avoid reinventing the wheel!) Figure 5.10 shows most of the morph chart developed by one team, and Figure 5.11 shows the other. A comparison of the two morph charts reinforces the earlier observations about focused thinking in that Figure 5.11 reflects a more bounded morph chart (and design space) because that team had a sharper, more concise list of functions. Similarly, the three "Enable adjustability of . . ." functions and their associated means of Figure 5.10 might again be viewed only as three articulations of sub-functions and means of the top-level function, "Enable adjustability." In particular, if the overall objective is to provide adjustability, then is there a need now to consider details of the different kinds of adjustments at this time?

A further problem to consider is that very large morph charts suggest a very large number of possible combinations. Some 13,310 (i.e., $11 \times 11 \times 11 \times 10$) alternatives are conceivably available in the partial chart of Figure 5.10. The smaller morph chart of Figure 5.11 has only 7,200 conceivable combinations! Thus, the total number of outcomes is overwhelming for this design problem. Clearly, some strategy is needed for grouping and organizing the functions and the resultant design alternatives.

In fact, both teams followed similar approaches of blending possibilities derived from their morph charts with information gained from their research and with their experience-based judgments and gut-level feelings. Some of the design sketches and drawings for the design concepts produced by the teams are shown in Figures 5.4 and 5.6, and photos of the prototypes produced by both teams are shown in Figure 5.12. We will show more sketches, drawing and photos in Chapter 8 when we discuss drawing and prototyping in significant detail, but is interesting to show some of the results here as they are the most immediately tangible fruits (and joys) of successful conceptual design. The particular designs recognized in the lower sketch of Figure 5.4 and in the two photos of Figure were those chosen, respectively, by the two teams after applying

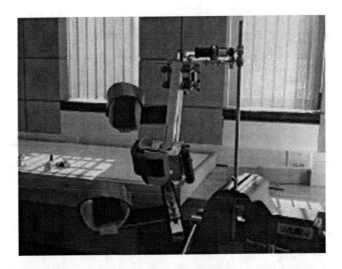

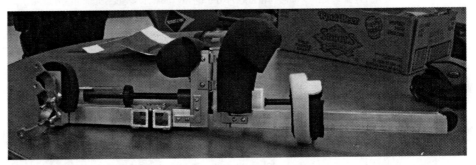

FIGURE 5.12 Photos of two prototypes of the Danbury arm support, taken from the final reports of two student design teams.

the metrics they had developed to their own sets of objectives (see Section 3.6.2 and Figures 3.7 and 3.8).

5.5 MANAGING THE GENERATION AND SELECTION OF DESIGN ALTERNATIVES

Generating designs is an exciting activity for both experienced and inexperienced designers, although selecting the "best" design may not be so easy. But as with requirements development, design generation is an activity best done by groups (e.g., choosing among suspension, arch or cable-stayed bridges). In fact, we have identified several group activities that were designed to foster divergent and creative thinking. In these activities, team members can build on each other's ideas and continue to benefit from diverse views and enhanced critical thinking.

5.6 NOTES

Section 5.1: Zwicki (1948) originated the idea of a morphological chart. Further discussion and examples of morph charts can be found in (Cross 1994), (Jones 1992), and (Hubke 1988).

Section 5.2: The address of the USPO's Web site is www.uspto.gov. Another often-used Web site is www.ibm. com/patents. Group methods of idea generation are explored and described in (Shah 1998) and synectics are defined in *Webster's Ninth New Collegiate Dictionary* (Mish 1983) and described in (Cross 1994). Approaches to creativity and analogical thinking in a group setting are described in (Hays 1992).

Section 5.3: Pugh's concept selection method is discussed in (Pugh 1990), (Ullman 1992, 1997), and (Ulrich and Eppinger 1995, 2000).

Section 5.4: The results for the Danbury arm support design project are taken from final reports ((Attarian et al. 2007) and (Best et al. 2007)) submitted during the spring 2007 offering of Harvey Mudd College's first-year design course, E4.

5.7 EXERCISES

5.1 Explain what is meant by the term "design space" and discuss how the size of the design space might affect a designer's approach to an engineering design problem.

5.2 Using the functions developed in Exercise 5.4, develop a morphological chart for the portable electric guitar.

5.3 Organize and apply a process for selecting means for realizing the design of the portable electric guitar.

5.4 Using Web-based patent lists (identified in Section 5.2.2), develop a list of patents that are applicable to the portable electric guitar.

5.5 Using the functions developed in Exercise 4.6, develop a morphological chart for the rain forest project.

5.6 Organize and apply a process for selecting means for realizing the design of the rain forest project.

5.7 Using Web-based patent lists (identified in Section 5.2.2), develop a list of patents that are applicable to the rain forest project.

5.8 Describe an acceptable proof of concept for the rain forest project. Would a prototype be appropriate for this project? If so, what would be the nature of such a prototype?

5.9 Look up or estimate the costs of disposing of the beverage containers made by the ABC and NBC, as detailed in Figures 5.8 and 5.9.

CHAPTER 6

DESIGN MODELING, ANALYSIS, AND OPTIMIZATION

Where and how do math and physics come into the design process?

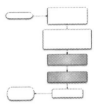

WE HAVE now completed our conceptual design in that we have one or two design alternatives that we've selected as our "final" design. Now we have to do preliminary design (design tasks 11 and 12 in Figure 2.3) and detailed design (tasks 13 and 14).

Unlike much of what we have done so far, the preliminary and detailed design phases require that we exercise mathematical models and techniques that describe and enable us to analyze physical behavior. We can't possibly introduce here all of the models and techniques needed to model all the kinds of designs that engineers do. However, we can illustrate the major points and "habits of thought" by analyzing the possible design of part of a ladder. Thus, we will model a ladder so that we can do preliminary and detailed design of a step on a ladder. After discussing some fundamental mathematical modeling ideas, we will model, analyze, refine, and optimize an archetypal ladder rung.

6.1 SOME MATHEMATICAL HABITS OF THOUGHT FOR DESIGN MODELING

What is a model? Our trusty dictionary states the following:

- **model** *n*: a miniature representation of something; a pattern of something to be made; an example for imitation or emulation; a description or analogy used to help visualize something (e.g., an atom) that cannot be directly observed; a system of postulates, data and inferences presented as a mathematical description of an entity or state of affairs

This definition suggests that *modeling* is an activity in which we think about and make models or representations of how devices or objects of interest behave. Since there are many means by which devices and behaviors can be described — words, drawings or sketches, physical models (to be discussed in Chapter 7), computer programs, or mathematical formulas — for our present purposes it is worth focusing the dictionary definition by thinking of a mathematical model as a *representation in mathematical terms* of the behavior of real devices and objects.

As engineers and designers, we describe and analyze objects and devices with mathematical models so that we can *predict* their behavior. Every new airplane or building, for example, represents a model-based prediction that the plane will fly and the building will stand without producing dire, unintended consequences. Thus, it is important for designers to ask: How do we create mathematical models or representations? How do we validate such models? How do we use them? And, are there any limits on their use?

6.1.1   Basic principles of mathematical modeling

Mathematical modeling is an activity that has underlying principles and a host of methods and tools. The overarching principles are almost philosophical in nature. We describe our intentions and purposes for building a mathematical model in terms of a list of questions (and their answers) as follows:

- *Why* do we need a model?
- What do we want to *find* with this model?
- What data are we *given*?
- What can we *assume*?
- *How* should we develop this model, that is, what are the appropriate physical principles we need to apply?
- What will our model *predict*?
- Are the predictions *valid*?
- Can we *verify* the model's predictions?
- Can we *improve* the model?
- How will we *use* this model?

Good habits of thought are essential for good design modeling.

We should note that we often invoke the last principle, *use*, early in the modeling process, along with *why* and *find*, because how we use a model is often intimately connected with the reason we created it. More importantly, we point out that this list of questions is *not* an algorithm for building a good mathematical model. The underlying ideas are key to mathematical modeling, and to problem formulation generally. Thus, the individual questions will recur often during the modeling process, and the list should be regarded as a general approach to *habits of thought* for mathematical modeling.

6.1.2 Abstractions, scaling, and lumped elements

An important decision in modeling is choosing the right level of detail for the problem at hand, which thus dictates the level of detail for the attendant model. We call this part of the modeling process *abstraction*. It requires a thoughtful approach to identifying the phenomena to be emphasized, that is, to answering the fundamental question about why a model is being developed. Thinking about finding the right level of abstraction or detail also requires identifying the right *scale* for the model being developed. Stated differently, thinking about *scaling* means thinking about the magnitude or size of quantities measured with respect to a standard that has the same physical dimensions.

For example, a linear elastic spring can be used to model more than just the relation between force and relative extension of a simple coiled spring, as in an old-fashioned butcher's scale or an automobile spring. We will use a version of Hooke's Law, $F = kx$, to describe the static load-deflection behavior of a ladder's step, but then the spring constant k will reflect the stiffness of the step taken as a whole. This interpretation of k will incorporate more detailed properties of the step, such as the material of which it is made and its dimensions. The validity of using a linear spring to model the ladder's step can be confirmed by measuring and plotting the deflection of a ladder's step as it responds to different loads.

The classic spring equation is also used to model the behavior of tall buildings as they respond to wind loading and to earthquakes. These examples suggest that a simple, highly abstracted model of a building can be developed by aggregating various details within the parameters of that model. That is, the stiffness k for a building would compile or *lump* together a lot of information about how the building is framed, its geometry, its materials, and so on. Thus, for both a ladder step and a tall building, we need detailed expressions that relate their respective stiffness constants to their particular properties.

Similarly, we can use using springs to model atomic bonds if we can develop or show how their spring constants depend on atomic interaction forces, atomic distances, subatomic particle dimensions, and so on. Thus, the spring can be used at both much smaller, *micro* scales to model atomic bonds, as well as at much larger *macro* scales, as for buildings. The notion of scaling includes several ideas, including the effects of geometry on scale, the relationship of function to scale, and the role of size in determining limits — all of which are needed to choose the right scale for a model in relation to the "reality" we want to capture.

Going a step further, we often say that a "real" three-dimensional object behaves like a simple spring. When we say this, we are introducing the notion of a *lumped element model* in which the actual physical properties of a real object or device are aggregated or *lumped* into a less detailed, more abstract expression. For example, we can model an airplane in very different ways, depending on our modeling goals. To lay out a flight plan or trajectory, we can simply consider the airplane as a point mass moving with respect to a spherical coordinate system: The mass of the point is just the total mass of the plane; the effect of the surrounding atmosphere is modeled by introducing a retarding drag force that acts on the mass point in some proportion to the mass' relative speed. To model the more local effects of the movement of air over the plane's wings, a model would have to account for the wing's surface area and be complex enough to incorporate the aerodynamics that occur in different flight regimes. To model and design the flaps used to control the plane's ascent and descent, a model would be developed to include a system for controlling the flaps and to also account for the dynamics of the wing's strength and vibration response. Again, what we incorporate into our lumped elements depends on the scale on which we choose to model, which depends in turn our goals for doing that modeling.

6.2   SOME MATHEMATICAL TOOLS FOR DESIGN MODELING

We now present some tools that we can use to apply the "big picture" principles to develop, validate, apply, and verify such mathematical models. These tools include dimensional analysis, approximations of mathematical functions, linearity, and conservation and balance laws.

6.2.1   Physical dimensions in design (I): Dimensions and units

One powerful idea that is central to mathematical modeling is the following: Every independent term in every equation we use has to be *dimensionally homogeneous* or *dimensionally consistent*; that is, every term has to have the same net physical dimensions. Thus, every term in a balance of mass must have the dimension of mass, and every term in a summation of forces must have the physical dimension of force. We also identify dimensionally consistent equations as *rational* equations. In fact, one important way of validating newly developed mathematical models (or of confirming formulas before using them for calculations) is to ensure that they are rational equations.

The physical quantities used to model objects or systems represent *concepts*, such as time, length, and mass, to which we attach *numerical* measurements or values. If we say a soccer field is 60 meters wide, we are invoking the concept of length or distance, and our numerical measure is 60 meters. The numerical measure implies a comparison with a standard or scale (as we noted in Section 3.2): Common measures provide a frame of reference for making comparisons.

The physical quantities used to model a problem are classed as fundamental or as derived. *Fundamental* or *primary* quantities can be measured on a scale that is independent of those chosen for any other fundamental quantities. In mechanical problems, for example, mass, length, and time are usually taken as the fundamental mechanical *dimensions* or variables. *Derived* quantities generally follow from definitions or physical laws, and they are expressed in terms of the dimensions that were chosen as fundamental. Thus, force is a derived quantity that is defined by Newton's law of motion. If mass, length, and time are chosen as primary quantities, then the dimensions of force are $(\text{mass} \times \text{length})/(\text{time})^2$. We use the notation of brackets [ ] to read as "the dimensions of." If M, L, and T stand for mass, length, and time, respectively, then:

$$[F = \text{force}] = (\text{M} \times \text{L})/(\text{T})^2 \tag{1}$$

Similarly, $[A = \text{area}] = (\text{L})^2$ and $[\rho = \text{density}] = \text{M}/(\text{L})^3$. Also, for any given problem, we need enough fundamental quantities to be able to express each derived quantity in terms of those primary quantities.

The *units* of a quantity are the numerical aspects of a quantity's dimensions expressed in terms of a given physical standard. Thus, a unit is an arbitrary multiple or fraction of a physical standard. The most widely accepted international standard for measuring length is the meter (m), but length can also be measured in units of centimeters (1 cm = 0.01 m) or of feet (0.3048 m). The magnitude or size of the attached number obviously depends on the unit chosen, and this dependence often suggests a choice of units to facilitate calculation or communication. For example, a soccer field width can be said to be 60 m, 6000 cm, or (approximately) 197 ft.

We often want to compute particular numerical measures in different sets of units. Since the physical dimensions of a quantity are the same, there must exist numerical relationships between the different systems of units used to measure the amounts of that quantity (e.g., 1 foot (ft) = 30.48 centimeters (cm), and 1 hour (hr) = 60 minutes (min) = 3600 seconds (sec or s)). This equality of units for a given dimension allows units to be changed or converted with a straightforward calculation, for example,

$$1\frac{\text{lb}}{\text{in}^2} \cong 1\frac{\text{lb}}{\text{in}^2} \times 4.45\,\frac{\text{N}}{\text{lb}} \times \left(\frac{\text{in}}{0.0254\text{m}}\right)^2 \cong 6,897\,\frac{\text{N}}{\text{m}^2} \equiv 6,897\text{Pa} \qquad (2)$$

Each of the multipliers in this conversion equation has an effective value of unity because of the equivalencies of the various units, that is, 1 lb ≅ 4.45 N, and so on. This, in turn, follows from the fact that the numerator and denominator of each of the above multipliers have the same physical dimensions.

We noted earlier that each independent term in a rational equation has the same net dimensions. Thus, we cannot add length to area in the same equation, or mass to time, or charge to stiffness. On the other hand, quantities having the same dimensions but expressed in different units can be added, although with great care, e.g., length in meters and length in feet. The fact that equations must be rational in terms of their dimensions is central to modeling because it is one of the best — and easiest — checks to make to determine whether a model makes sense, has been correctly derived, or even correctly copied!

In a familiar model from mechanics model, the speed of a particle, v, due to the acceleration of gravity, g, when dropped from a height, h, is given by:

$$v = \sqrt{2gh} \qquad (3)$$

Note that both sides of eq. (3) have the same net physical dimensions, that is, L/T on the left-hand side and $[(L/T^2)L]^{1/2}$ on the right. As a result we call eq. (3) *dimensionally homogeneous* because it is totally independent of the system of units being used to measure v, g, and h. However, we often create unit-dependent versions of such equations because they are easier to remember or they make repeated calculations convenient. For example, we may be working entirely in metric units, in which case $g = 9.8\,\text{m/s}^2$, so that

$$v\,(\text{m/s}) = \sqrt{2(9.8)h} \cong 4.43\sqrt{h} \qquad (4)$$

Equation (4) is valid *only* when the particle's height is measured in meters. If we were working with American units only, then $g = 32.17\,\text{ft/sec}^2$ and

$$v\,(\text{ft/sec}) = \sqrt{2(32.17)h} \cong 8.02\sqrt{h} \qquad (5)$$

which is valid *only* when the particle's height is measured in feet. Equations (4) and (5) are *not* dimensionally homogeneous. So, while these formulas may be easier to remember or use, their limited validity must be kept in mind.

There is one other way in which these dimensional considerations come into play that is worth noting. In Section 3.4, while discussing the development of metrics for assessing the achievement of objectives, we introduced the idea of *figures of merit* as sets of *units of interest* for the scales of metrics. When considering the optimization of a design, in the detailed design phase (see Section 6.4.2 below), we will construct objective functions that represent figures of merit as functions whose value is to be optimized. It is very important that we recall and apply the idea that objective functions, like equations, should likewise be rational functions. That is, all of the independent terms in an objective function should have the same net dimensions.

6.2.2   Physical dimensions in design (II): Significant figures

We use numbers a lot in engineering for both design and analysis, but we often need to remind ourselves about the significance of each of those numbers. In particular, people often ask how many decimal places they are expected to keep. But that's the wrong question to ask, because the number of significant figures is *not* determined by the placement of the decimal point. In scientific notation, the *number of significant figures* (NSF) is equal to the number of digits counted starting from the first nonzero digit on the *left* to either (a) the last nonzero digit on the right if there is no decimal point, or (b) the last digit (zero or nonzero) on the right when there is a decimal point. This notation or convention assumes that terminal zeros without decimal points to the right signify only the magnitude or power of ten. In fact, as can be seen from the examples shown in Table 6.1, confusion

TABLE 6.1   Instances of how numbers are generally written

Measurement	Significant Figures	Assessment
5415	Four	Clear
5400	Two (54×10^2) or three (540×10^1) or four (5400)	Not clear
54.0	Three	Clear
54.1	Three	Clear
5.41	Three	Clear
0.00541	Three	Clear
5.41×10^3	Three	Clear
0.054	Two	Clear
0.0540	Two (0.54) or three (0.0540)	Not clear
0.05	One	Clear

Shown are the number of significant figures (NSF) in each number and assessments of the NSF that can be assumed or inferred. Confusion about the NSF arises because of the meaning of the terminal zeros is not stated.

over the NSF arises because of the presence of terminal zeros: We don't know whether those zeroes are intended to signify something, or whether they are placeholders to fill out some arbitrary number of digits.

One way to think about the NSF is to imagine that we are running a test whose outcomes could be A, B or C, and we want to know how often we see the result A. If A occurs four times in a set of ten tests, then we would say that A occurs in 0.4 of the tests. If we got A 400 times in a run of 1,000 tests, we would have found A in 0.400 of the tests—but how do we make it clear that those extra two zeros have meaning? The answer is that we can eliminate any confusion if we write all such numbers, whether from technical calculations and experimental data, in *scientific notation*. In scientific notation we write numbers as products of a "new" number that is normally in the interval 1–10 and a power of 10. Thus, numbers both large and small can be written in one of two equivalent, yet unambiguous forms:

$$514{,}000{,}000 = 0.514 \times 10^9 = 5.14 \times 10^8$$
$$0.000075 = 0.75 \times 10^{-4} = 7.5 \times 10^{-5}$$

Also on the subject of the NSF, we should *always* remember that *the results of any calculation or measurement cannot be any more accurate than the least accurate starting value*. We cannot generate more significant digits or numbers than the smallest number of significant digits in any of our starting data. It is far too easy to become captivated by all of the digits produced by our calculators or computers, but it is really important to remember that *any calculation is only as accurate as the least accurate value we started with*.

6.2.3 Physical dimensions in design (III): Dimensional analysis

We often find it useful to work with or even create *dimensionless* variables or numbers that by design are intended to compare the value of a specific variable with a standard of obvious relevance. For example, hydrologists model some of the behavior of soil in terms of its *porosity*, η, which is defined as the dimensionless ratio $\eta = V_v/V_t$, where V_v is the volume of voids (or interstitial spaces) in the soil and V_t is the total volume of the soil being considered. We also see that this definition of porosity *normalizes* or *scales* the void volume V_v against the total volume V_t. A similar (and more famous) example is Einstein's formula for the relativistic mass of a particle, $m = m_0/\sqrt{1 - (v/c)^2}$, in which the mass m is normalized against the rest mass, m_0, and the particle speed is scaled against the speed of light, c, in the dimensionless ratio v/c. Note that Einstein's formula is dimensionally homogeneous, and that the particle speed is normalized such that $0 \leq v/c \leq 1$ and the mass such that $1 \leq m/m_0 < \infty$.

We can often get a lot of information about some behavior by describing that behavior by a dimensionally correct equation among certain variables. One method for conducting such explorations is embodied in *Buckingham's Pi Theorem*, which can be stated as follows: "A dimensionally homogeneous equation involving n variables in m primary or fundamental dimensions can be reduced to a single relationship among $n–m$ independent dimensionless products." Since a rational equation is one in which every independent, additive term in the equation has the same dimensions, any one term can be defined as a function of all of the others. If we follow Buckingham and use Π_1 to represent a dimensionless term, his famous Pi theorem can be written as:

$$\Pi_1 = \Phi(\Pi_2, \Pi_3 \ldots \Pi_{n-m}). \tag{6a}$$

or, equivalently,

$$\Phi(\Pi_1, \Pi_2, \Pi_3 \ldots \Pi_{n-m}) = 0. \tag{6b}$$

Equations (6) both state that a problem with n derived variables and m primary dimensions or variables requires $n–m$ dimensionless groups to correlate all of its variables.

The Pi theorem is applied by first identifying the n derived variables in a problem: $A_1, A_2, \ldots A_n$. Then m of these derived variables are chosen such that they contain all of the m primary dimensions, say, A_1, A_2, A_3 for $m = 3$. Dimensionless groups are then formed by permuting each of the remaining $n–m$ variables ($A_4, A_5, \ldots A_n$ for $m = 3$) in turn with those ms already chosen:

$$\begin{aligned}
\Pi_1 &= A_1^{a_1} A_2^{b_1} A_3^{c_1} A_4, \\
\Pi_2 &= A_1^{a_2} A_2^{b_2} A_3^{c_2} A_5, \\
&\;\;\vdots \\
\Pi_{n-m} &= A_1^{a_{n-m}} A_2^{b_{n-m}} A_3^{c_{n-m}} A_n.
\end{aligned} \tag{7}$$

The a_i, b_i, and c_i are then chosen to make each of the permuted groups Π_i dimensionless.

As an example, consider the following case. When we are actually detailing the design of a step on a ladder, we will have to model and calculate the deflection of a fixed-ended beam (shown in Figure 6.2) under a centrally applied vertical load P. We will present the appropriate model in Section 6.2.1 when we use it, but let us first see whether we can identify the form we will see by applying the Buckingham Pi theorem. We show in Table 6.2 the five derived variables for this problem (along with their respective dimensions): the deflection δ at the center of the beam, the load P, the beam length L, the modulus of elasticity E of the material of which the beam is made, and the second moment I of the beam's cross-sectional area. The applied load P and the beam length L are chosen as

TABLE 6.2   The five quantities chosen to model the fixed-ended beam that, in turn, will be a model for a step on a ladder. P and L are chosen as fundamental, and δ, E and I are taken as derived

Derived Quantities	Dimensions
Deflection (δ)	L
Load (P)	F
Length (L)	L
Modulus of elasticity (E)	F/L^2
Second moment of area (I)	L^4

the two fundamental quantities for this model. Thus, with $m = 5$ and $n = 2$, the following three dimensionless groups can be formed:

$$\begin{aligned}
\Pi_1 &= P^{a_1} L^{b_1} \delta \\
\Pi_2 &= P^{a_2} L^{b_2} E \\
\Pi_3 &= P^{a_3} L^{b_3} I
\end{aligned} \tag{8}$$

Note first that the two fundamental variables, P and L, appear in each of eqs. (8). Then we apply Buckingham's Pi theorem by rewriting eq. (8) in terms of the physical dimensions of each of the fundamental and derived variables, that is:

$$\begin{aligned}
\Pi_1 &= F^{a_1} L^{b_1} L \\
\Pi_2 &= F^{a_2} L^{b_2} F/L^2 \\
\Pi_3 &= F^{a_3} L^{b_3} L^4
\end{aligned}$$

For each of these numbers to be dimensionless we must set the sums of the exponents *for each physical dimension* to zero,

$$\begin{aligned}
a_1 &= 0 & b_1 + 1 &= 0 \\
a_2 + 1 &= 0 & b_2 - 2 &= 0 \\
a_3 &= 0 & b_3 + 4 &= 0
\end{aligned}$$

which means that

$$\begin{aligned}
a_1 &= 0 & b_1 &= -1 \\
a_2 &= -1 & b_2 &= 0 \\
a_3 &= 0 & b_3 &= -4
\end{aligned}$$

If we substitute these coefficients into eqs. (8) we find the three dimensionless groups needed to model a ladder rung, and they are:

$$\begin{aligned}
\Pi_1 &= \frac{\delta}{L} \\
\Pi_2 &= \frac{L^2 E}{P} \\
\Pi_3 &= \frac{I}{L^4}
\end{aligned} \tag{9}$$

These three groups can be combined to get a single relationship for the deflection we are seeking, in this instance by simply by chain-multiplying the groups,

$$\Pi_1 \Pi_2 \Pi_3 = \left(\frac{\delta}{L}\right)\left(\frac{L^2 E}{P}\right)\left(\frac{I}{L^4}\right) = \frac{\delta E I}{P L^3} \tag{10a}$$

which we can arrange as

$$\delta = \frac{P L^3}{(\Pi_1 \Pi_2 \Pi_3) E I} \tag{10b}$$

and which we can rewrite just one more time as

$$\delta = \frac{PL^3}{C_\delta EI} \tag{10c}$$

where $C_\delta = \Pi_1\Pi_2\Pi_3$ is a dimensionless number—a constant—whose value we will determine later. We can now use eq. (10) to relate the beam variables to one another, as we will do in Section 6.3 when we design our ladder rung.

6.2.4 On physical idealizations and mathematical approximations

We generally *idealize* or approximate situations or objects so that we can model them and apply those models to find behaviors of interest. We make two kinds of idealizations, physical and mathematical, and the order in which we make them is important. Recall an example from basic physics, the pendulum, in which a known mass is hung from a string of prescribed length. *First*, we identify those elements that we believe are important to the problem, so we assume that the string is weightless and acts only in tension, and that the only external force is due to gravity. Further, we assume that any wind resistance is negligible and that the pendulum will swing through only small angles. Our model is (still) verbal, but we have idealized several facets of the pendulum's anticipated behavior by neglecting wind resistance and considering only small angles. Thus, we have a *physical idealization*.

    Second, we translate the physical idealization into a mathematical model. We must ensure consistency here by being careful that our mathematical models replicate exactly what we have assumed in our physical idealization. For example, one common mistake is the approximation of $\cos\theta$ for small angles θ. This requires thought about the meaning of "small," and in relation to what. For small angles we could have either

$$\cos\theta = 1 - \frac{\theta^2}{2!} + \cdots \cong 1 \tag{11a}$$

or

$$(1 - \cos\theta) = \left[1 - \left(1 - \frac{\theta^2}{2!} + \cdots\right)\right] \cong \frac{\theta^2}{2!} \tag{11b}$$

Clearly, eqs. (11a) and (11b) present very different results, either one of which could be a suitable mathematical approximation. The real trick is to properly understand the physical idealization that we are trying to represent.

6.2.5 The role of linearity

Engineers typically try to build models that are, mathematically speaking, *linear* models. We do this because nonlinear problems are almost always much harder to solve, but also because linear models work extraordinarily well for many devices and behaviors of interest. In fact, one of the most often used such approximations is like the one just described in eqs. (11). The more common form of the small-angle approximation is that of $\sin\theta$ for small angles θ. In this case we have a very familiar result:

$$\sin\theta = \theta - \frac{\theta^3}{3!} + \cdots \cong \theta \qquad (12)$$

As we recall from basic physics, it is the small-angle assumption of eq. (12) that enables us to linearize the classical pendulum problem: The basic equation of motion is a linear differential equation that is easily solved and, as one outcome, the tension in the pendulum's string turns out to be constant when the pendulum is restricted to small angles of motion.

Linearity shows up in other contexts. Consider *geometrically similar* objects, that is, objects whose basic geometry is essentially the same. For two right circular cylinders of radius r and respective heights h_1 and h_2, the total volume in the two cylinders is:

$$V_{cy} = \pi r^2 h_1 + \pi r^2 h_2 = \pi r^2 (h_1 + h_2) \qquad (13)$$

Equation (13) demonstrates that the volume is *linearly proportional* to the height of the fluid in the two cylinders. Further, since the total volume can be obtained by adding or *superposing* the two heights, the volume V_{cy} is a *linear function* of the height h. Note, however, that the volume is *not* a linear function of the radius, r. That is, for different radii for the two cylinders, eq. (13) becomes:

$$V_{cy} = \pi h r_1^2 + \pi h r_2^2 \qquad (14)$$

That is, the relationship between volume and radius is nonlinear for the cylinders, so the total volume cannot be calculated just by superposing the two radii. This result, while for a simple, even obvious case, is emblematic of what happens when a linearized model is replaced by its (originating) nonlinear version.

6.2.6 Conservation and balance laws

Many of the mathematical models used in engineering design are statements that some property of an object or system is being conserved. For example, the motion of a body moving on an ideal, frictionless path might be analyzed by stipulating that its *energy is conserved*, that is, energy is neither created nor destroyed. Sometimes, as when modeling the population of an animal colony or the volume of a river flow, *quantities that cross a defined boundary* (whether individual animals or water volumes) *must be balanced*. That is, want to count or measure both what goes in to and what comes out of the boundary of the domain we're watching. Such *balance* or *conservation principles* are applied to assess the effect of maintaining levels of physical attributes. Conservation and balance equations are related — in fact, conservation laws are special cases of balance laws.

The mathematics of balance and conservation laws are in principle straightforward. We start by (conceptually, and sometimes graphically) drawing a boundary around the device or system we are modeling. If we denote the physical attribute or property being monitored as $N(t)$ and the independent variable time as t, a balance law for the *temporal* or time rate of change of that property within the system boundary outlined can be written as:

$$\frac{dN(t)}{dt} = n_{in}(t) + g(t) - n_{out}(t) - c(t) \qquad (15)$$

where $n_{in}(t)$ and $n_{out}(t)$ represent the flow rates of $N(t)$ into (the *influx*) and out of (the *efflux*) the system boundary, $g(t)$ is the rate at which N is generated within the boundary, and $c(t)$ is the rate at which N is consumed within that boundary. Equation (15) is also called a *rate equation* because each term has both the meaning and dimensions of the rate of change with time of the quantity $N(t)$.

In those cases where there is no generation and no consumption within the system boundary (i.e., when $g = c = 0$), the balance law in eq. (15) becomes a *conservation law*:

$$\frac{dN(t)}{dt} = n_{in}(t) - n_{out}(t) \tag{16}$$

Here, then, the rate at which $N(t)$ accumulates within the boundary is equal to the difference between the influx, $n_{in}(t)$, and the efflux, $n_{out}(t)$.

Perhaps the most familiar balance and conservation laws are those associated with Newtonian mechanics. Newton's second law, usually presented as the equation of motion, can be viewed as a balance law because it refers to a balance of forces:

$$\sum \vec{F} = m\vec{a} = \frac{d}{dt}(m\vec{v}) \tag{17}$$

Note, however, that eq. (17) also represents a conservation law because it refers to the conservation of momentum. If there are not net forces acting on the mass m, then $d(m\vec{v})/dt = 0$ and the momentum $m\vec{v}$ is conserved.

The second familiar conservation principle in Newtonian mechanics is the principle of conservation of energy:

$$E(t) = \frac{1}{2}m(\vec{v} \cdot \vec{v} = v^2) + PE = E_0 \tag{18}$$

where PE represents the particular form of potential energy for the system under consideration (e.g., mgh for gravitational potential and $kx^2/2$ for a linear spring) and E_0 is the constant total (kinetic plus potential) energy. For a nonideal system, energy is not conserved and the result is the work-energy principle:

$$\left[\frac{1}{2}mv_2^2 - \frac{1}{2}mv_1^2\right] + (PE)_2 - (PE)_1 = \int_1^2 \vec{F} \cdot d\vec{s} \tag{19}$$

Thus, the difference in the total (kinetic and potential) energy between states 1 and 2 is equal to the work done by the forces acting on the system as they traverse the path from state 1 to state 2.

6.3   DESIGN MODELING OF A LADDER RUNG

In this section and the next we want to demonstrate the processes of modeling and of preliminary and detailed design. Specifically, we will model and design the step or rung of a ladder. As we model that step, we will post signs [*in brackets*] to indicate which modeling principles (as represented by the questions in Section 6.1.1) we are applying. In order to design the ladder's step we need a model that predicts its behavior [**Why**], which means that we want to be able to understand how the step's attributes (e.g., size, shape, material, connections to the ladder's frame, etc.) affect its ability to support given loads [***Find***]. We are told that the ladder must support a person of specified weight W_p carrying a specified weight W_w [***Given***]. We will model the behavior of the step (and the ladder) using a standard model of linear elastic beam behavior (described below in Section 6.3.1) and standard models of the elastic behavior of materials [***Assume***]. We will develop and apply the model of the step using basic principles of mechanics [***How***] and show just how the total weight that it can support depends on its geometric and material properties [***Predict***]. Note how we are limiting our modeling effort: At this point we are not analyzing the size, shape or materials of the side frames, any cross bracing, or any footpads on a rung. We are also precluding any linkages among the various ladder supports. One or more mathematical models would be needed to develop these parts of the ladder, but now we will just model an individual rung or step.

Let us now apply some basic mechanics principles. In Figure 6.1 we show three sketches of a person on a ladder, the first of which is a *free-body diagram* (FBD) of the

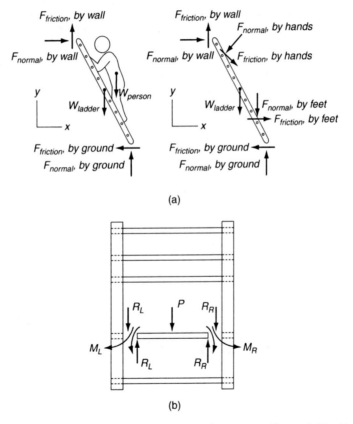

(a)

(b)

FIGURE 6.1 *Free-body diagrams* (FBDs) of various aspects of a person standing on a ladder: (a) side elevation-FBDs of person and ladder taken as a system (Courtesy of Sheppard and Tongue.); (b) a front-elevation FBD of the ladder showing vectors of the force $F_{normal} = P = W_p + W_w$ due to the person standing on a rung, as well as the vertical forces (R_L, R_R) and *moments* (M_L, M_R) by which the ladder's side rails support the rung.

person and the ladder taken as a system. (A free-body diagram is a visual tool that mechanical and civil engineers use to sketch or portray the forces acting on the system under consideration. A comparable tool in electrical engineering is the circuit diagram.) The second sketch shows an FBD of the entire ladder. The third drawing shows elevations of FBDs of the rung, on which are shown vectors representing (a) the force exerted by the load-carrying person and (b) the vertical forces and *moments* provided by the ladder's frame to support the step. In order to fully understand the import of these FBDs and the modeling to follow, we need to import some results about structures called *beams* that are typically introduced in courses called *strength of materials* ("strengths") or *mechanics of materials* ("mechomat").

6.3.1   Modeling a ladder rung as an elementary beam

Imagine the two scenarios shown in Figure 6.2, both of which show a vertical load P supported by a transverse (i.e., normal or, here, horizontal) element. Figure 6.2(a) shows a *cable* or rope, along with FBDs of two sections of the cable. We see that the load seems to make the cable kink, and that a vertical load can be supported by a tensile force T in the cable or rope. In Figure 6.2(b) we show a *beam*, along with two FBDs of the beam divided into two sections. In the first FBD we see that the external vertical force in each section is supported by reactions R_A and R_B at each beam support and an internally developed *shear force, V*. However, there is nothing to prevent either section from rotating or spinning because each has an unbalanced couple or *moment* in the configuration shown. In the second beam FBD we have included *bending moments, M*, that are internally developed couples (or moments) that maintain moment equilibrium and thus prevent each section from spinning out of

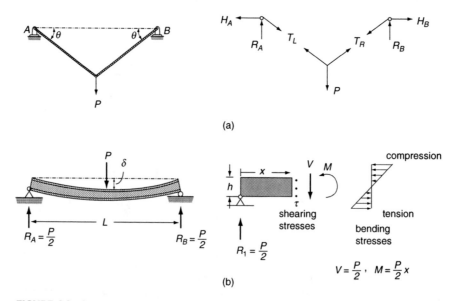

FIGURE 6.2   Supporting a vertical force with a transverse (horizontal) structure: (a) *cable* and FBDs of two sections of the cable; (b) a *beam* and a FBD of one section of the beam that also shows how a moment (couple) is developed by axially directed normal stresses on the beam's cross-sectional area.

control. These bending moments are developed by in-plane stresses along the axis of the beam, so that it is a set of *horizontal* stresses that support a *vertical* load in a beam!

For our purposes, the important aspects of elementary beam theory are that beams behave like linear springs and the stiffness of a beam depends on several of the beam's parameters. In fact, for a simple beam, if a load P is applied at the midway point of a beam of length L, the deflection of the beam under that point, δ, is given by

$$\delta = \frac{PL^3}{C_\delta EI} \tag{20a}$$

which is the result derived as eq. (10c) using dimensional analysis. We can also rewrite eq. (20a) as an analog of the classical spring formula, $F = kx$, that is,

$$P = \left(\frac{C_\delta EI}{L^3}\right)\delta \tag{20b}$$

In eqs. (20) E is the *modulus of elasticity* of the material of which the beam is made, I is the *second moment of the cross-section of the area* of the beam (see Figure 6.2(b)), and C_δ is a dimensionless constant that depends on the boundary conditions at the beam's ends. The modulus of elasticity, which is often called *Young's modulus*, is a measure of the stiffness of the material and has the same physical dimensions as stress, that is, $[E] = \mathsf{F}/\mathsf{L}^2$. Young's modulus typically has values measured in (metric) units of gigapascals, $1\,\text{GPa} = 10^9\,\text{Pa}$, where the pascal (Pa) is the SI unit of stress defined as $1\,\text{Pa} = 1\,\text{N}/\text{m}^2$. The second moment of the area I is a geometrical property of the beam's cross-section and has physical dimensions $[I] = \mathsf{L}^4$. By substituting these dimensions into eqs. (20) we can verify that the beam's deflection has the appropriate physical dimensions for a measure of movement, $[\delta] = \mathsf{L}$, and confirm that eqs. (20) agree exactly with the result derived using dimensional analysis and presented in eq. (10c) — which also confirms the usefulness of that modeling tool! Further, the effective spring constant of the beam has the same dimensions as the classical spring constant: $[C_\delta EI/L^3] = [k] = \mathsf{F}/\mathsf{L}$.

The other physical quantity of great interest in beam theory is the bending stress along the axis of the beam. As may be visualized from Figure 6.2(b), it is the bending stress that creates the bending moment and its consequent shear force that enable a long thin structure — the beam — to support a load that acts in a direction normal to the (long) axis of that beam. Again, from the "strengths" (or "mechomat") results that are a staple of most engineering curricula, the maximum stress in the loaded beam is:

$$\sigma = \frac{PLh}{2C_\sigma I} \tag{21}$$

where h is the height of the beam's cross-section (see Figure 6.2). Stress has the same physical dimensions as pressure, that is, $[\sigma] = \mathsf{F}/\mathsf{L}^2$.

While we have several times stressed the analog with elementary spring characteristics, there is a crucial difference — especially for the designer — between the classic spring formula and eqs. (20) and (21) for the beam. The spring formula has only one constant or design variable, k, that can be chosen or manipulated. Thus, there is not much design freedom for a spring. On the other hand, for a beam that has to span a given length L, there are three variables that can be varied: E, I and h. (To the extent we can choose how the beam is supported at its ends, we can also choose between appropriate sets of constants, C_δ and C_σ.) The increased number of variables means that we can design to achieve objectives expressed in terms of the beam's deflection (eq. (20a)) and its maximum stress (eq. (21)). Thus, we shall later talk about *designing the beam for stiffness*, when the deflection is our focus, or *designing the beam for strength*, when the maximum stress is our focus.

What about the supports at the beam's (or, mindful of why we're doing this, our rung's) ends? There are several kinds of supports that can be stipulated or modeled. The two limiting cases that are of most relevance are pictured in Figures 6.3. The first such

(a)

(b)

(c)

(d)

FIGURE 6.3 Connecting ladder rungs to ladder frames. (a, b) Top and bottom views of how metal rungs are attached to a ladder's fiberglass frame; note the gap between the top surface of the rung and the frame, so the support is neither simple or fixed. (c) On this metal ladder the connections go through the frame into the hollow (curved) box that forms the rung; it is also an intermediate support. (d) This very old wooden stepladder has solid rungs that are bolted to the frame, but there are extra (partially visible) supports that bring the connection closer to fixed or clamped.

case is a *simple* (or *pinned* or *hinged*) support that provides a vertical reaction force and prevents any vertical deflection; as shown in Figure 6.3(b), the ends of the beam are free to rotate. Simple supports can clearly provide vertical reactions that will support any vertical load on the rung. The other limiting case, shown in Figure 6.3(c), is the *fixed* (or *rigid* or *clamped*) support: It provides both a vertical reaction force that prevents vertical deflection and a moment that forces the slope of the deflection of the beam to vanish; that is, the moment prevents any rotation at the beam's end. These two limiting cases, informed by our actual experience, suggest that we will have to make another modeling-design assumption when we design the rung.

With eqs. (20) and (21) in hand and a tentative decision taken about the kinds of beam supports we will consider, we have specified the equations we will use, the calculations we can make and the types of answers we might expect [*Predict*]. Thus, we have established a principled model that we can now use for the preliminary and detailed design of a ladder rung.

6.3.2 Design criteria

What are our design criteria, that is, against what requirements do we assess the performance of our designs? In part, that depends on both our objectives and our constraints. There are no doubt some top-level objectives that were identified during conceptual design:

- minimize the mass of material used in order to achieve a ladder that is light in weight; and
- minimize the cost in order to achieve an inexpensive ladder.

But there are two other design aspects that must be considered and that can be categorized as objectives or constraints. Those two issues derive from not wanting the rung to break or fail when someone stands on it, and not wanting the rung to deflect too much lest that someone feel uncomfortable. We need to specify what it means to require that the rung "not break or fail" and "not deflect too much." And we need to specify whether these are both (possibly competing) objectives, constraints, or some combination.

A material breaks or fails when any one of three *failure strengths* is exceeded. Determining values of the design variables such that the rung's bending stress does not exceed specified failure strengths is what we mean by *design for strength*. The three failure strengths are values of the stresses at which a material fails under, respectively, a *tensile* stress, a *bending* test, or a *tensile test that produces permanent deformation*. These three failure strengths are materials properties that have been measured and tabulated for most materials. As a result, and as we will soon see, our design problem will become a *materials selection* problem. We can generally lump the three modes of failure together and refer to the minimum of the three for a given material as the failure strength of interest, σ_f. Since material properties are largely established through laboratory testing and experience, our degree of confidence varies with the material. We reflect that variation of confidence by stating that the failure strength should be divided by a safety factor S, with S being as low as 1.2 for well understood materials and as high as 5 for materials for which our experience is not as widespread. (Of course, other uncertainties can be incorporated into S.) Then the strength requirement would be expressed in terms of the bending stress as $\sigma \leq \sigma_f/S$. Failure strengths typically have values on the order of a megapascal, $1\,\text{MPa} = 10^6\,\text{Pa}$, where the pascal (Pa) is the SI unit of stress defined as $1\,\text{Pa} = 1\,\text{N/m}^2$.

A rung deflects too much when a specified maximum deflection is exceeded. Determining values of the design variables such that the rung's midpoint deflection does not exceed specified deflection limits is what we mean by *design for stiffness*. The specified upper limit generally follows from ergonomic considerations because we don't want a ladder step to feel wobbly when we are standing on a rung. Thus, codes or standards often specify a maximum deflection δ_{max}, and that maximum deflection is in turn expressed as a fraction of the rung's length L. The deflection requirement would then be expressed in terms of the deflection as $\delta \leq \delta_{max} = C_f L$, where C_f is a very small number, say $C_f = 0.01$.

We need to address one more (design) loose end: Do we choose pairs of constants, C_δ and C_σ, that correspond to beams with simple supports or to beams with fixed supports? Experience suggests that the ends of a step would be most accurately modeled as fixed or clamped. However, since the side rails of the frame are not truly rigid, there will always be a (very) small amount of rotation at the rung's ends (see Figure 6.4). Thus, we will model the rung as a beam on simple supports, knowing it will be a more flexible and more conservative model that will over-predict both the stress in and the deflection of the step [*Assume*]. As a result, our final design will bend less and carry bigger loads than our model predicts. The values of the constants for simple supports, $C_\delta = 48$ and $C_\sigma = 4$, were found from exact solutions for the deflection and the bending stress of such a beam that is loaded by a vertical load P at its midpoint.

6.3.3 Design optimization

Optimization is the mathematical technique used to determine the best or the *optimum* solution — the one most likely to achieve a specified goal — from a set of candidate solutions. When seeking an optimum solution we speak of maximizing or minimizing an *objective function*, while satisfying certain *constraints*. Not surprisingly, optimization is a complex subject that takes up a large number of books. However, we will not need the full range of available optimization tools in order to do our design tasks of selecting the dimensions of and a material for a ladder step. We have four objectives we want to achieve: We want to minimize the mass and the cost subject to the strength constraint, and the mass and cost subject to the stiffness or deflection constraint. There are several ways to proceed. We could simply look at how the mass and the cost vary with different materials and then intuitively rank both cost and mass for both strength and stiffness. If we were experienced structural designers, or if our intuition was sufficiently well developed, we might take note of the fact that the stiffness constraint is (usually) much more severe than the strength constraint. Thus, it is rather unlikely that the strength constraint ($\sigma \leq \sigma_f/S$) will be violated if the stiffness constraint ($\delta \leq \delta_{max} = C_f L$) is met. If such a case emerged, we might have to revise our goal for cost and our constraint on stiffness, and then check again whether the strength is sufficient.

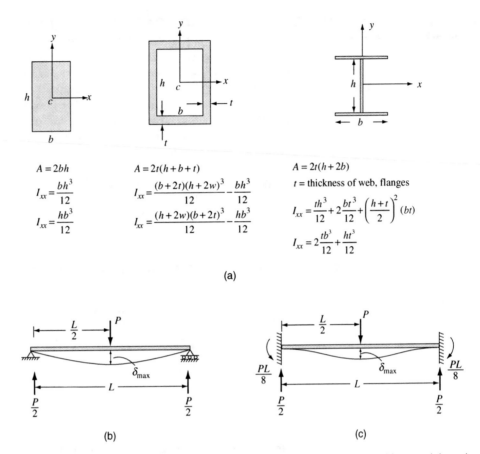

$$A = 2bh$$

$$I_{xx} = \frac{bh^3}{12}$$

$$I_{xx} = \frac{hb^3}{12}$$

$$A = 2t(h+b+t)$$

$$I_{xx} = \frac{(b+2t)(h+2w)^3}{12} - \frac{bh^3}{12}$$

$$I_{xx} = \frac{(h+2w)(b+2t)^3}{12} - \frac{hb^3}{12}$$

$$A = 2t(h+2b)$$

t = thickness of web, flanges

$$I_{xx} = \frac{th^3}{12} + 2\frac{bt^3}{12} + \left(\frac{h+t}{2}\right)^2 (bt)$$

$$I_{xx} = 2\frac{tb^3}{12} + \frac{ht^3}{12}$$

(a)

(b)

(c)

FIGURE 6.4 Some aspects of elementary beam models: (a) rectangular, hollow box, I-beam and channel cross-sections, including thicknesses (h) and second moments (I); (b) a beam with *simple supports* at both ends; (c) a beam with *fixed supports* at both ends.

We also want to ensure that we get a reasonable result for the step thickness. For example, polymer foams may be superior in both cost and stiffness, but the final step thickness may be 0.5 m, which is clearly impractical for a ladder. (Such a large thickness would also violate the assumptions underlying the beam model whose results are given in eqs. (20) and (21) [**Validate**]).

It is also important to keep in mind the intended application. If our ladder is headed to a moon base, the mass would be critically important but the cost much less so. If the ladder is to be sold by a "big box" retailer, then cost may be of paramount importance.

Finally, then, the basic optimization process is to compare values of the mass and the cost for different designs and then choose those designs that maximize the performance (e.g., strength or stiffness) and minimize mass or cost. In Section 6.4 we will do just that, after which we will choose materials that maximize performance and minimize mass or cost.

6.4   PRELIMINARY AND DETAILED DESIGN OF A LADDER RUNG

In this section we undertake elements of preliminary design and of detailed design. In the "real" preliminary design of a ladder we would consider beams of various cross-sections and likely make some estimates of which shapes are likely to be more efficient when made of different materials. We might then choose one shape for further development, along with a range or set of materials. Then, in "real" detailed design we would refine that design by working to optimize it, making it as light and cheap as possible. We would also decide how to attach the rungs to the ladder frame (e.g., with rivets, welds, or bolts) and then "size" and fix the locations of those attachments. In our case, we will use preliminary design to illustrate and contrast materials selection when designing for strength and when designing for deflection (or stiffness). In our detailed design we will optimize rung designs to achieve minimum mass and minimum cost.

6.4.1   Preliminary design considerations for a ladder rung

With both failure and deflection criteria defined, and both loose ends tied up, we can see that our design problem is such that we generally want the rung's midpoint deflection δ and its maximum bending stress σ to be such that

$$\delta = \frac{PL^3}{48EI} \leq \delta_{max} = C_f L \tag{22a}$$

and

$$\sigma = \frac{PL}{8hI} \leq \frac{\sigma_f}{S} \tag{22b}$$

where P represents the combined weight of someone standing on the ladder and of the package that person is carrying; that is, $P = W_p + W_w$. An important question is, Do we treat these as inequalities (objectives) or do we adopt the equal signs (constraints) for both the deflection and the stress? The answer is that we cannot treat both as constraints. While there are nominally three design variables (E, I, h), the fact is that I and h are so strongly related that they are effectively a single variable. Still more important, E and I (and h) are not truly independent variables. In fact, as we pointed out in the discussion after eqs. (20), the material property (E) and the geometric properties (I or h and L) are incorporated into a single effective stiffness, namely, $k_{eff} = 48EI/L^3$. The implication of this is that we can design for strength or we can design for stiffness, but we cannot do both at once. Thus, we can choose to minimize the deflection, in which case we are designing for stiffness and we must then check to ensure that the corresponding bending stress is below the failure criterion. To design for stiffness, we start by equating the bending stress to the failure strength, after which we calculate the corresponding deflection and assess whether we can (or cannot) accept that value.

As noted, the rung's effective stiffness depends on both material and geometric properties. In many structural design problems, the material is chosen or specified in advance. In that case, the design variable that remains is the cross-sectional area of the rung as represented by its second moment I. As shown in Figure 6.4(a), the second moment I and the rung's thickness h can be used to model a wide variety of shapes, including rectangular cross sections, I-beams and channel sections. The reasons that I-beams, channels and similar such sections are widely used are that they are more efficient than rectangular cross-sections in that they sustain higher stresses per unit weight, and that modern material processing capabilities make it easy to manufacture such shapes in great volume. In fact, if we look again at Figure 6.3, we see that only the wooden ladder rung has a complete rectangular cross section. However, knowing that our results will perhaps be unrealistic, we will limit our exploration of this aspect of the design space by assuming that the step has a rectangular cross section, with width b [*Assume*]. In this case, then, $I = bh^3/12$, and our current set of design variables has changed from E, I, and h to E, b, and h.

Finally, we will also assume that the rung width b is constrained, as in fact it is. For example, the *American National Standard for Ladders — Wood Safety Requirements* (published by The American National Standards Institute) stipulates fixed rung widths for a variety of wooden stepladders. Thus, even though we are not constraining our design to wooden rungs, we will assume that the width b is a specified quantity [*Assume*]. As a result, we now actually have only two design variables, E and h.

6.4.2 Preliminary design of a ladder rung for stiffness

We now formulate the first of two different preliminary design problems for rungs of rectangular cross sections and fixed widths, namely, we *design for stiffness* by constraining the deflection. We do that because we don't know in advance what stiffness is required to achieve a specified deflection, although we do know (and specify) the limiting value that we place on that deflection. We might also think of design for stiffness as *design for deflection*. From the first of eqs. (22) we note that $PL^3/48EI = C_f L$, which then yields:

$$\frac{PL^3}{4Ebh^3} = C_f L \tag{23}$$

Equation (23) can be solved for the thickness design variable, h:

$$h = \left(\frac{PL^2}{4EbC_f}\right)^{1/3} \tag{24}$$

Equation (24) determines the rung thickness and its value clearly depends on the given values of P, L, b, and C_f, as well as on the value of the modulus E, which is as yet unspecified. Typically we would have a range of materials in mind (e.g., aluminum, steel, wood or a composite for a ladder), and we would calculate the corresponding thicknesses of our rung accordingly. However, we do have to ensure that the step does not fail, and thus by substituting eq. (24) into the second of eqs. (22) we would check that

$$\sigma = \frac{PL}{8hI} = \frac{3PL}{2bh^2} = \left(\frac{54PC_f^2}{bL}\right)^{1/3} E^{2/3} \le \frac{\sigma_f}{S} \tag{25}$$

TABLE 6.3 Possible designs for stiffness of the rung of a ladder

Material	E (GPa)	h (mm)	$\sigma/(\sigma_f/S)$
Aluminum	70	9.2	1.51
Steel	212	6.4	0.63
Wood	9	18.3	1.06
CFRP	110	7.9	0.90

The design load (i.e, the weight supported) is $P = 1350$ N; the rung length $L = 350$ mm; the rung width $b = 75$ mm; the safety factor $S = 1.5$; and the constraint constant $C_f = 0.01$. Note that two of the materials, aluminum and wood, violate our strength constraint that $\sigma/(\sigma_f/S) \leq 1$.

or, with the stress cast in a dimensionless ratio,

$$\frac{\sigma}{\sigma_f/S} = \left(\frac{54PC_f^2}{bL}\right)^{1/3} \frac{E^{2/3}}{\sigma_f/S} \leq 1 \tag{26}$$

In Table 6.3 we show some thickness and modulus values of possible rungs designed to meet the stiffness constraint. (We will provide far more extensive data and results in Section 6.4.5 where we conclude our detailed design.) We note that the thicknesses all seem reasonably small, but that the ratio of the stress to the failure stress, $\sigma/(\sigma_f/S)$, is not always less than 1. Our intuition that the stiffness constraint would be more severe than the strength constraint turned out to be false for two of the materials, although the wood rung is a very near miss, even with a small safety factor [*Verify*]. A different kind of wood might have produced a more satisfying result. However, these results confirm one of the basic reasons for making a model, that is, to (numerically) check on our intuition.

Anticipating the detailed design work to be done below to minimize both the weight and the cost of the ladder step, we also note that eq. (25) is cast as a product of two factors. The first factor, $(54PC_f^2/bL)^{1/3}$, incorporates the function and geometry of the rung, as well as the design's constraint. Note that every element in the first factor, which is called a *structural index*, is either a number or a known quantity. The second factor, $E^{2/3}/(\sigma_f/S)$, reflects the material property that must be chosen to ensure that failure is avoided. The factor $E^{2/3}/(\sigma_f/S)$ can be considered as the *material index* (MI) for this design:

$$\mathrm{MI}_\delta \equiv \frac{E^{2/3}}{\sigma_f/S} \tag{27}$$

Note, finally, that the physical dimensions of the structural index ensure that its product with the material index has the proper physical dimensions for calculating the requisite bending stress.

6.4.3 Preliminary design of a ladder rung for strength

We *design for strength* by constraining the stress to ensure that failure does not occur. We then have to calculate the corresponding deflection and decide whether we can accept that deflection. Thus, the statement of designing for strength begins with the second of eqs. (22), $\sigma = PL/8hI = \sigma_f/S$, which means constraining the stress such that:

$$\sigma = \frac{PL}{8hI} = \frac{3PL}{2bh^2} = \frac{\sigma_f}{S} \tag{28}$$

We now solve for the thickness design variable, h, using the constraint (28):

$$h = \left(\frac{3SPL}{2b\sigma_f}\right)^{1/2} \tag{29}$$

Then the deflection that corresponds to this strength design is formulated by substituting eq. (29) into the first of eqs. (22):

$$\frac{\delta}{L} = \left(\frac{bL}{54P}\right)^{1/2} \frac{(\sigma_f/S)^{3/2}}{E} \tag{30}$$

Note, too, that here we have cast the deflection in a dimensionless ratio.

Again, eq. (30) is dimensionally correct and its individual factors can also be identified: $(bL/54P)^{1/2}$ is the structural index and $MI_\sigma = (\sigma_f/S)^{3/2}/E$ is the material index. It is interesting to observe that MI_σ is strongly (and almost inversely) related to MI_δ: $MI_\sigma = (MI_\delta)^{-2/3}$. This reinforces our earlier comment that we cannot design for both deflection (stiffness) and strength at the same time because, to a significant degree, these two objectives or paradigms are at odds with each other. In Table 6.4 we show some results for strength design for the same four materials listed in Table 6.3. Here, too, the thicknesses all seem reasonably small, but the ratio of the midpoint deflection to the rung length, δ/L, is sometimes — but not always — larger than the 0.01 limit prescribed in the stiffness design. The two materials that failed the strength constraint in the stiffness design, aluminum and wood, here pass the stiffness constraint in the strength design, and vice versa for the other two, steel and CFRP. We were right only one-half of the time with our assumption about stiffness being a more severe constraint than strength [*Verify*].

6.4.4 Detailed design of a ladder rung (I): Minimizing the rung mass

Now, and somewhat artificially, we enter the phase of detailed design. We focus on refining the preliminary design of the rung to address the objectives listed at the beginning of

TABLE 6.4 Possible designs for strength design of the rung of a ladder

Material	E (GPa)	σ_f (MPa)	h (mm)	δ/L
Aluminum	70	110	11.4	0.005
Steel	212	550	5.1	0.020
Wood	9	40	18.8	0.009
CFRP	110	250	7.5	0.012

The design load (i.e. the weight supported) is $P = 1350$ N; the safety factor is $S = 1.5$; the rung length is $L = 350$ mm; and the rung width $b = 7575$m. Note that two of the materials, steel and CFRP, violate our stiffness constraint that $\delta/L \leq 0.01$.

Section 6.3.2. We now want to optimize the rung's behavior by endeavoring to minimize the mass of material used, leaving cost optimization until later (Section 6.4.5). At the same time, we will continue to distinguish between objectives and constraints with regard to the strength and the deflection (stiffness) of the rung.

The mass of the rectangular rung that we want to minimize is given by:

$$m = \rho b h L \tag{31}$$

where ρ is the mass density, another materials property. Following a path that departs from that taken when we designed, respectively, for stiffness (deflection) and for strength, we combine our design equations into one *objective function* that reflects our constraints and has a single design variable. That new, compiled design variable will be a material index that depends only on material properties. We find that MI for strength design first, by eliminating h between eqs. (29) and (31) and solving for m. The result is the following optimization problem:

$$\text{Minimize } m_\sigma^h : m_\sigma^h = \left(\frac{27SPbL^3}{8}\right)^{1/2} \left(\frac{\rho}{\sigma_f^{1/2}}\right) \tag{32}$$

We have introduced some new notation in eq. (32): The superscript h on the optimized mass identifies the design variable here (the thickness h) while the subscript σ indicates that we are designing for strength. Equation (32) is dimensionally homogeneous, and its factors are once again identifiable: $(27SPbL^3/8)^{1/2}$ is the rung's structural factor and $\rho/\sigma_f^{1/2}$ is the material index, here a function of two material properties, the density ρ and the failure strength σ_f. Since the other factor in eq. (32) is comprised of numbers and known quantities, we minimize the mass m by finding a material that minimizes the material index $\text{MI}_{m\sigma}^h = \rho/\sigma_f^{1/2}$, where the subscript shows that we are minimizing the mass (m) with a strength (σ) constraint, and the superscript shows that the thickness (h) is the design variable.

Identifying appropriate and acceptable materials is a harder problem than the one we faced in the previous preliminary stiffness and strength designs. Fortunately there are computer-based tools that we can successfully apply to screen large numbers of materials. In Figure 6.5 we show a *materials selection chart* generated in the software package C. E. S. Selector. That charts plots the logarithm of the density ρ (ordinate, or y-axis) against the logarithm of the failure strength σ_f (abscissa, or x-axis). We use logarithms of the material constants because of the range of magnitudes of the material properties. And we placed ρ and σ_f on their particular axes so as to make our next step more intuitive.

Figure 6.5 also includes a design guideline along which the $\text{MI}_{m\sigma}^h$ is a constant, that is, $\text{MI}_{m\sigma}^h = \rho/\sigma_f^{1/2} = K_{m\sigma}^h$. The constant $K_{m\sigma}^h$ is called the *utility* as it represents how well a material is being utilized: Here smaller values of $K_{m\sigma}^h$ mean smaller values of the material index, and thus smaller values of the mass m. This utility guideline is a linear equation when expressed in logarithmic form:

$$\log \rho = \frac{1}{2}\log \sigma_f + \log K_{m\sigma}^h \tag{33}$$

where $\log \equiv \log_{10}$ is the standard logarithm to base 10.

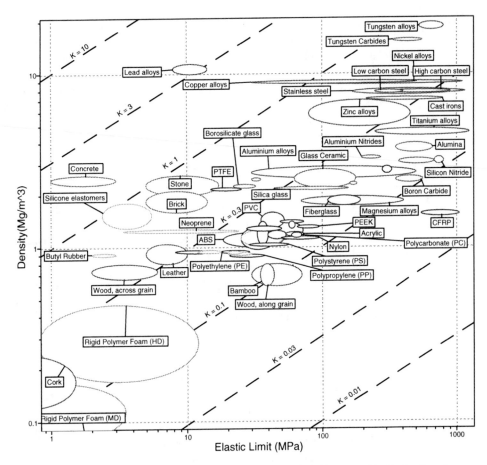

FIGURE 6.5 A materials selection chart for minimizing the mass of a ladder rung subject to the strength constraint $(\sigma = \sigma_f)$. Our material index is $MI^h_{m\sigma} = \rho/\sigma_f^{1/2} = K^h_{m\sigma}$, where the lower the value of $K^h_{m\sigma}$, the more desirable the material. The chart was generated in the software package C.E.S. Selector, produced by Granta Design Limited.

The materials data is shown in clumps or groupings of various kinds of materials, and the design guideline is a demarcation line that helps us identify materials that work to achieve our objective. In particular, the smaller the value of $K^h_{m\sigma}$, the more optimal the material, so that the region of highest strength and lowest density is the lower right corner of Figure 6.5. There are several features of the plot that are worth noting. First, carbon-fiber-reinforced polymer (CFRP), which is also known as carbon composite or graphite epoxy, is by far the best material from the standpoint of mass or weight. Second, note that wood and bamboo are almost a desirable as CFRP, and that rigid polymer foams are also almost as good. But a quick calculation of the resulting thickness of a polymer foam step will place it out of the running. Nonetheless, while other factors dictated that we dismiss polymer foam as a rung material, it provided a very good answer to the question we asked of our design model. Magnesium and aluminum alloys are the best metals in Figure 6.5.

Commercial ladders are often made from aluminum, and from fiberglass as well because it is roughly comparable to aluminum. There are also a number of other materials that appear desirable, namely titanium alloys, and ceramics such as silicon carbide and boron nitride. However, that will change as we later look to minimize cost.

In order to avoid having the ladder step feel too floppy or saggy to its user, we design for stiffness by once again or prescribing the maximum deflection. We want to minimize the mass (eq. (31)) with the thickness h being the design variable that we eliminate between eqs. (24) and (31). We then find the following optimization problem:

$$\text{Minimize } m_\delta^h : m_\delta^h = \left(\frac{Pb^2L^5}{4C_f}\right)^{1/3}\left(\frac{\rho}{E^{1/3}}\right) \tag{34}$$

Equation (34) is dimensionally homogeneous, and its factors are once again identifiable: $(Pb^2L^5/4C_f)^{1/3}$ is the step's structural factor and $\rho/E^{1/3}$ is the material index, now a function of the density ρ and the elastic modulus E. Since the structural factor in eq. (34) is entirely known, we minimize the mass m by finding a material that minimizes the material index $\text{MI}_{m\delta}^h = \rho/E^{1/3}$.

Figure 6.6 shows a log-log plot of density ρ (y-axis) against Young's modulus E (x-axis). A line of constant utility for the material index is $\text{MI}_{m\delta}^h = \rho/E^{1/3} = K_{m\delta}^h$. In logarithmic form that line of constant utility is:

$$\log \rho = \frac{1}{3}\log E + \log K_{m\delta}^h \tag{35}$$

where, again, $K_{m\delta}^h$ is the utility constant for minimizing mass under a constraint on the maximum deflection. The smaller the value of $K_{m\delta}^h$, the more optimal the material. The region of highest stiffness and lowest density is the lower right corner of the plot. In terms of stiffness, medium-density (MD) polymer foam is best, and wood, bamboo, CFRP, and high-density (HD) polymer foam are about of equal optimality and rank next. Although the thickness of foam rungs would be impractically large, our model did not rule that out because it did not set any limits on h. Magnesium alloys are slightly superior to aluminum alloys and fiberglass, and all three are slightly behind wood from a stiffness/mass standpoint. There are still a number of ceramic materials (e.g., boron carbide) that are still in contention at this point.

6.4.5   Detailed design of a ladder rung (II): Minimizing the rung cost

We now turn our attention to the second of the objectives listed at the beginning of Section 6.3.2, namely, we want to minimize the cost of the materials used to make the ladder step. At the same time, we will continue to distinguish between objectives and constraints with regard to the strength and the deflection (stiffness) of the rung.

For the purposes of ranking designs, it is usually sufficient to approximate the cost of a part such as a ladder step as the product of the published cost per unit mass of the unformed material, C_m, and the mass of the part. For detailed design, one would typically get quotes on actual material prices in the quantities sought and the cost of fabrication and assembly. Values of C_m can be found in a number of sources, including on the Internet, in materials books and journals, and using software packages. However, it is very important

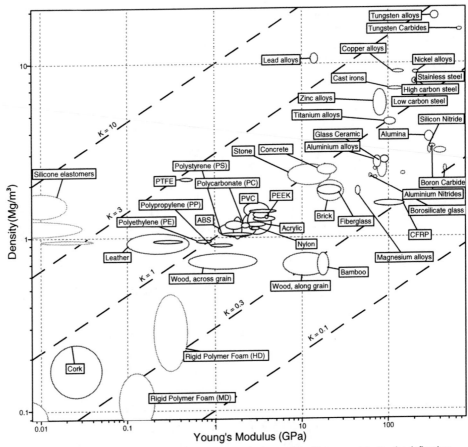

FIGURE 6.6 A materials selection chart for minimizing the mass of a ladder rung subject to the deflection constraint ($\delta = \delta_{max} = C_f L$). Our material index is $MI^h_{m\delta} = \rho/E^{1/3} = K^h_{m\delta}$, where the lower the value of $K^h_{m\delta}$, the more desirable the material. The chart was generated in the software package C.E.S. Selector, produced by Granta Design Limited.

that all of the materials costs used for a given design project come from the same source to ensure consistency in how such values are derived and tabulated.

Since we already know the mass of the step for both strength and stiffness criteria, the cost is simply $C_m m$. Then the problems of cost optimization for strength and stiffness design are, respectively,

$$\text{Minimize } \$^h_\sigma : \$^h_\sigma = \left(\frac{27SPbL^3}{8}\right)^{1/2}\left(\frac{C_m\rho}{\sigma_f^{1/2}}\right) \tag{36}$$

and

$$\text{Minimize } \$^h_\delta : \$^h_\delta = \left(\frac{Pb^2L^5}{4C_f}\right)^{1/3}\left(\frac{C_m\rho}{E^{1/3}}\right) \tag{37}$$

where we use the $ sign to note cost in any currency (i.e., not just US dollars) and where we have grouped the cost per unit mass C_m together with the material indices (MIs) because that cost is specific to each material. Then, instead of seeking the minima of the material indices in each case, we now seek the minima of *cost indices* (CIs), respectively defined as:

$$CI^h_\sigma = \left(\frac{C_m \rho}{\sigma_f^{1/2}} \right) \tag{38}$$

and

$$CI^h_\delta = \left(\frac{C_m \rho}{E^{1/3}} \right) \tag{39}$$

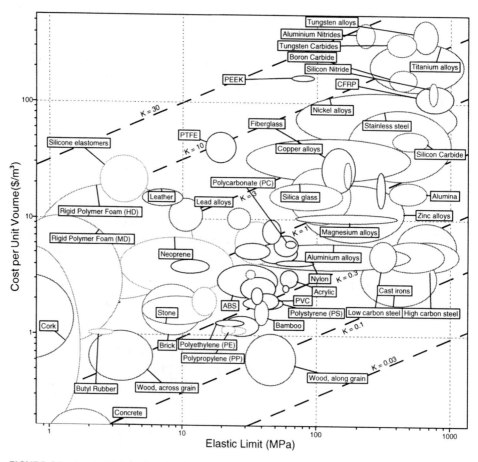

FIGURE 6.7 A materials selection chart for minimizing the cost of a ladder step subject to the strength constraint $(\sigma = \sigma_f)$. Our cost index is $CI^h_\sigma = C_m \rho / \sigma_f^{1/2} = K^h_{S\sigma}$, where the lower the value of $K^h_{S\sigma}$, the more desirable the material. The chart was generated in the software package C.E.S. Selector, produced by Granta Design Limited.

Figure 6.7 shows a log-log plot of cost per unit volume $C_m\rho$ (y-axis) against failure strength σ_f (x-axis). The plot is set up to reflect the cost index that is now set equal to a utility constant $K^h_{S\sigma}$, that is, $CI^h_\sigma = C_m\rho/\sigma^{1/2}_f = K^h_{S\sigma}$. Expressed logarithmically, our utility guideline for minimizing cost without exceeding the failure strength is then:

$$\log C_m\rho = \frac{1}{2}\log \sigma_f + \log K^h_{S\sigma} \tag{40}$$

The lower the value of $K^h_{S\sigma}$, the more optimal the material. The region of highest strength and lowest cost is once again the lower right corner of the plot. There are noteworthy features in Figure 6.7, the most striking of which is that CFRP, which was so desirable from a mass standpoint, is near the bottom from a cost standpoint. CFRP is used extensively in aerospace and military applications but much less so — and for good reason — for consumer products. In terms of cost, wood, concrete, and cast iron are the three best materials. However, concrete and cast iron tend to make very heavy ladders and the concrete steps would be thicker than desirable. Magnesium alloys do not do as well as the aluminum alloys. Fiberglass is even worse than magnesium, and the ceramics are all significantly less desirable because of their high cost. An interesting choice is the non-stainless steels (which have risen because of their low cost), but our model does not account for corrosion, and unpainted steel is prone to rust.

In Figure 6.8 we plot the cost per unit volume $C_m\rho$ (y-axis) against Young's modulus E (x-axis). The material index here is $CI^h_\delta = C_m\rho/E^{1/3} = K^h_{S\delta}$, where $K^h_{S\delta}$ is the utility constant. The logarithmic form of our utility guideline is:

$$\log c_m\rho = \frac{1}{3}\log E + \log K^h_{S\delta} \tag{41}$$

As usual, smaller values of $K^h_{S\delta}$ signify more optimal material choices, with the region being the lower right corner of the plot. Here, concrete is much, much better than wood, the next best material. Perhaps there is a reason that the stairs inside and outside commercial construction are usually made of concrete? In terms of cost, CFRP is even worse for stiffness than it was for strength, while aluminum is not bad. The steels are still desirable, but fiberglass has fallen somewhat, and the ceramics (which are very stiff) have risen.

At this point we have identified four different rankings of materials that depend on whether we deem stiffness or strength as more important, and whether we assess mass or cost as more important. What material should we choose, and why should we choose it?

6.4.6 Detailed design of a ladder rung (III): Real materials results

With the principles of selecting materials for detailed design now established, we turn to choosing some materials to identify the specific physical dimensions, masses, and costs our models actually produce. The data in the C.E.S. Selector charts show a range of values for certain classes of materials. In order to do numerical comparisons, we have to select specific materials. The following eight materials were chosen from the C.E.S.

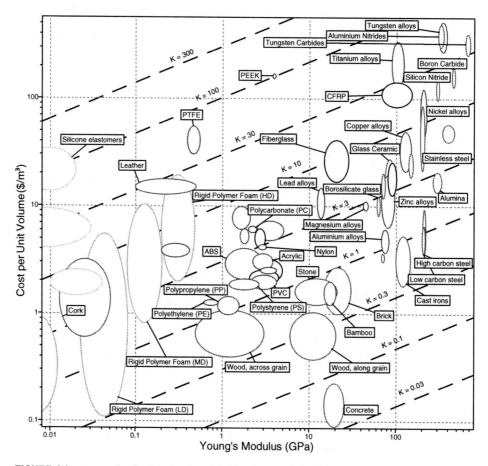

FIGURE 6.8 A materials selection chart for minimizing the cost of a ladder rung subject to the deflection constraint ($\delta = \delta_{max} = C_f L$). Our cost index is $CI_\delta^h = \rho/E^{1/3} = K_{\$\delta}^h$, where the smaller the value of $K_{\$\delta}^h$, the more desirable the material. The chart was generated in the software package C.E.S. Selector, produced by Granta Design Limited.

database as specific examples of materials in the classes that appeared interesting in Figures 6.5–6.8:

1. pine (Pinus spp.) in the longitudinal direction to represent wood along the grain;
2. wrought aluminum alloy, 6061, T451 to represent aluminum alloys;
3. epoxy SMC (glass fibre) to represent fiberglass;
4. epoxy SMC (carbon fibre) to represent CFRP;
5. carbon steel AISI 1040, oil-quenched and tempered at 425°C, to represent carbon steel (it is a medium-carbon steel);
6. low-alloy white cast iron (BS grade 1B) as cast iron;

TABLE 6.5 The eight materials sorted from minimum to maximum mass using the strength criterion as determined by eqs. (26) and (29)

Material	$\rho(kg/m^3)$	σ_f (MPa)	m (kg (lb$_m$))	h (mm (in))
Styrofoam	20	0.12	0.185 (0.41)	279 (10.98)
Pine	500	40	0.254 (0.56)	15 (0.60)
CFRP	1550	250	0.314 (0.69)	6 (0.24)
Fiberglass	1650	150	0.432 (0.95)	8 (0.31)
Aluminum	2700	110	0.826 (1.82)	9 (0.36)
Steel	7850	550	1.074 (2.37)	4 (0.16)
Cast iron	7700	325	1.370 (3.02)	5 (0.21)
Concrete	2400	7	2.910 (6.41)	37 (1.44)

7. high-performance concrete as concrete; and

8. polystyrene foam closed cell (0.020) as a polymer foam (it is an ultra-low-density styrofoam).

The mass, cost, and thickness were calculated for each of the materials for both the strength and stiffness constraints by using eqs. (24) and (29) to establish the thickness, eqs. (32) and (34) to minimize mass, and eqs. (36) and (37) to minimize cost. The chosen values for the design variables were the load, $P = 1330$ N (300 lb$_f$), the step length, $L = 356$ mm (14 in), the step width, $b = 76.2$ mm (3 in), and the safety factor, $S = 1.5$. These values are typical of the values specified in the ANSI standards for wood ladders.

For each of the four combinations (mass/strength, cost/strength, mass/stiffness and cost/stiffness), we sorted the materials from best to worst. Finally, we examined each material to determine whether strength or stiffness was the limiting constraint in the design. Then we chose the limiting case and sorted the results to minimize first mass and then cost. The results are presented in Tables 6.5–6.10.

The data in Table 6.5 show that Styrofoam provides the best — the lowest — mass for design for strength at 185 grams per step, but as mentioned earlier, the step thickness of 279 mm (11 in) is impractical. Next comes pine, which is used extensively in commercial ladders, then CFRP, which is not (because it is expensive). Then come the two other

TABLE 6.6 The eight materials sorted from minimum to maximum cost using the strength criterion as determined by eqs. (26) and (33)

Material	Unit Cost (US$/kg)	m (kg (lb$_m$))	Cost (US$)	h (mm (in))
Pine	$1.00	0.254 (0.56)	$0.25	15 (0.60)
Concrete	$0.10	2.910 (6.41)	$0.29	37 (1.44)
Styrofoam	$2.50	0.185 (0.41)	$0.46	279 (10.98)
Steel	$0.64	1.074 (2.37)	$0.69	4 (0.16)
Cast iron	$0.55	1.370 (3.02)	$0.75	5 (0.21)
Aluminum	$2.00	0.826 (1.82)	$1.65	9 (0.36)
Fiberglass	$11.00	0.432 (0.95)	$4.75	8 (0.31)
CFRP	$81.00	0.314 (0.69)	$25.47	6 (0.24)

TABLE 6.7 The eight materials sorted from minimum to maximum mass using the stiffness criterion as determined by eqs. (21) and (31)

Material	$\rho(kg/m^3)$	E (GPa)	m (kg (lb_m))	h (mm (in))
Styrofoam	20	0.005	0.135 (0.30)	249 (9.82)
Pine	500	9	0.278 (0.61)	20 (0.81)
CFRP	1550	110	0.374 (0.82)	9 (0.35)
Fiberglass	1650	20	0.702 (1.55)	16 (0.62)
Aluminum	2700	70	0.757 (1.67)	10 (0.41)
Concrete	2400	40	0.811 (1.79)	12 (0.49)
Steel	7850	212	1.521 (3.35)	7 (0.28)
Cast iron	7700	200	1.521 (3.35)	7 (0.29)

TABLE 6.8 The eight materials sorted from minimum to maximum cost using the stiffness criterion as determined by eqs. (21) and (34)

Material	Unit Cost/ (US\$/kg)	m (kg (lbm))	Cost (US\$)	h (mm (in))
Concrete	\$0.10	0.811 (1.79)	\$0.08	12 (0.49)
Pine	\$1.00	0.278 (0.61)	\$0.28	20 (0.81)
Styrofoam	\$2.50	0.135 (0.30)	\$0.34	249 (9.82)
Cast iron	\$0.55	1.521 (3.35)	\$0.84	7 (0.29)
Steel	\$0.64	1.521 (3.35)	\$0.97	7 (0.28)
Aluminum	\$2.00	0.757 (1.67)	\$1.51	10 (0.41)
Fiberglass	\$11.00	0.702 (1.55)	\$7.72	16 (0.62)
CFRP	\$81.00	0.374 (0.82)	\$30.27	9 (0.35)

TABLE 6.9 The eight materials sorted from minimum to maximum mass using the strength or stiffness criterion (from Tables 6.5 and 6.7)

Material	Limiting Constraint	m (kg (lb_m))	h (mm (in))	Cost (US\$)
Styrofoam	Strength	0.19 (0.41)	279 (10.98)	\$0.46
Pine	Stiffness	0.28 (0.61)	20 (0.81)	\$0.28
CFRP	Stiffness	0.37 (0.82)	9 (0.35)	\$30.27
Fiberglass	Stiffness	0.70 (1.55)	16 (0.62)	\$7.72
Aluminum	Strength	0.83 (1.82)	10 (0.41)	\$1.65
Steel	Stiffness	1.52 (3.35)	7 (0.28)	\$0.97
Cast iron	Stiffness	1.52 (3.35)	7 (0.29)	\$0.84
Concrete	Strength	2.91 (6.41)	37 (1.44)	\$0.29

TABLE 6.10 The eight materials sorted from minimum to maximum cost using the strength or stiffness criterion (from Tables 6.6 and 6.8)

Material	Limiting Constraint	Cost (US$)	h (mm (in))	m (kg (lb$_m$))
Pine	Stiffness	$0.28	20 (0.81)	0.28 (0.61)
Concrete	Strength	$0.29	37 (1.44)	2.91 (6.41)
Styrofoam	Strength	$0.46	279 (10.98)	0.19 (0.41)
Cast iron	Stiffness	$0.84	7 (0.29)	1.52 (3.35)
Steel	Stiffness	$0.97	7 (0.28)	1.52 (3.35)
Aluminum	Strength	$1.65	10 (0.41)	0.83 (1.82)
Fiberglass	Stiffness	$7.72	16 (0.62)	0.70 (1.55)
CFRP	Stiffness	$30.27	9 (0.35)	0.37 (0.82)

common ladder materials, fiberglass and aluminum. Concrete is dead last at 2.9 kg (6.5 lbm) and 37 mm (1.4 in).

The numbers shown given in Table 6.6 show that pine is the cheapest (i.e., has the lowest cost) material for design for strength at $0.25 per step. Next come concrete and styrofoam, which are eliminated by either their weight (concrete) or thickness (styrofoam). Steel and cast iron, which are both inexpensive materials, are in the middle. Near the bottom come the two other commercial materials, aluminum and fiberglass, at almost $2 and almost $5 per step, respectively, and CFRP is dead last at over $25 per step.

We show minimized masses for design for stiffness in Table 6.7. Styrofoam is still best at 135 grams per step, but as we found before, its step thickness of 249 mm (10 in) is impractical. Next come pine and CFRP. The two other common ladder materials, fiberglass and aluminum, are in the middle of the pack. Concrete is no longer dead last because of its high stiffness. Steel and cast iron are simultaneously the heaviest and thinnest materials at 1.5 kg (3.4 lb$_m$) and 7 mm (0.3 in), respectively, but their densities are their downfall here.

The results for minimized cost for design for stiffness case are displayed in Table 6.8. Concrete is by far the best at $0.08 per step, more than three times cheaper than second place pine at $0.28 per step. Near the bottom are aluminum and fiberglass at almost $2 and almost $8 per step, respectively. CFRP is dead last at over $30 per step.

It seems clear from the data given in Tables 6.5–6.8 that we need to look at both stiffness and strength for these eight materials, because the limiting case changes from material to material. In Tables 6.9 and 6.10 we take the limiting constraint case for a *given material*. When we look at the minimum mass for the limiting cases, styrofoam rises to the top. Styrofoam would be a wonderful material — were it not for the thickness issue! Three of the next four materials (i.e., pine, fiberglass and aluminum) are used in commercial ladders. The one that isn't, CFRP, is too expensive at more than 18 times the cost of the next most expensive. Steel, cast iron, and concrete are too heavy to be practical if there are better choices. The steps alone on a 2 m (6 foot) concrete ladder would weigh almost 18 kg (40 lb$_m$)!

On a cost basis, pine is the best at $0.28 per step. As before, aluminum and fiberglass are near the bottom because of their high costs of almost $2 and almost $8 per step, respectively. Because of their *low* cost, concrete, cast iron and steel are often used for stairs or ladders that don't need to be moved, but they are clearly too heavy for portable ladders. CFRP is dead last at over $30 per step.

The examination of specific materials did yield a few surprises — the very high cost of CFRP and the thickness of Styrofoam steps — but overall the results agreed very well with what we learned from the materials selection charts.

6.4.7   Remarks on material selection and detailed design

The design modeling exercise we have just completed is enlightening for several reasons. One is that each of the models answered a very specific question, and the answer to that question didn't always agree with our intuition. We would not have initially considered rigid foam, concrete or cast iron for steps on a ladder, but they proved to be optimal in the context of the specific design question posed. We also see why certain materials are chosen for certain applications, for example, why military (and increasingly commercial) aircraft use composite materials extensively. These are contexts where weight is a driving force in their design. However, cost remains a strong driver for many, if not most, consumer products.

We are also beginning to realize the limitations of our model. Steels would be eliminated in many applications because of corrosion, which we have not included in our thinking, and our model also doesn't account for what happens under an impact load. For examples, if we dropped a hammer on a polymer foam step, it would probably embed itself in the foam rung. However, the high-tech ceramics such as boron nitride and silicon carbide might provide an opportunity for an entrepreneur who is willing to take some risks because they are resistant to such fracturing.

It's become fairly clear why wood and aluminum are used extensively for commercial ladders. They are at or near the top in all four categories. And there is an additional trick that can be used with aluminum but not so easily with wood. It is easy and inexpensive to extrude beam shapes out of aluminum (see Figure 6.3 again) that have a much higher stiffness than a rectangular cross-section, whereas we would waste a lot of material if we tried to extrude wood. One other question remains (see Exercise 6.12), namely, why is fiberglass also used extensively for commercial ladders?

6.4.8   Remarks on the formulation of design problems

Another interesting aspect of our design and optimization efforts is that we always chose the beam's thickness h as our design variable. Would our results be different if we fixed the rung thickness h and left the width b free to vary? From a structural point of view it wouldn't make much sense, but an interesting point does emerge. For strength design, for example, we would eliminate the width b from the constraint (28), which yields:

$$b = \frac{3SPL}{2h^2\sigma_f} \tag{42}$$

so that the mass to be optimized becomes:

$$\text{Minimize } m_\sigma^b : \ m_\sigma^b = \left(\frac{3SPL^2}{2h}\right)\left(\frac{\rho}{\sigma_f}\right) \tag{43}$$

The factors in eq. (43) can once again be identified as a changed structural index, $3SPL^2/2h$, and the material index, ρ/σ_f. Since the structural factor is known, we can once again minimize the mass m by choosing a material with a minimum material index, ρ/σ_f. This MI differs noticeably from the result obtained when the thickness was optimized, as can be seen from eq. (32). Interestingly enough, the situation changes again if we look at steps with square cross sections, $b = h = \sqrt{A}$. Eliminating this design variable from the constraint (28) means that

$$b = h = \left(\frac{3SPL}{2\sigma_f}\right)^{1/3} \tag{44}$$

Since the volume of the rung is now $V = AL$, where $A = h^2 = b^2$ is the beam's cross-sectional area, the mass to be optimized becomes:

$$\text{Minimize } m_\sigma^A : m_\sigma^A = \left(\frac{3SPL^{5/2}}{2}\right)^{2/3}\left(\frac{\rho}{\sigma_f^{2/3}}\right) \tag{45}$$

The structural factor has changed once again, as has the material index, which is now $\rho/\sigma_f^{2/3}$.

Comparing eqs. (32), (43) and (45), we see that while they all vary linearly with the density ρ (as they well should!), their structural indices and their material indices (MIs) all change with a change of the free design variable. Note in particular that when our design variable h is measured in the direction of the load, we find a minimum mass when the material index $\rho/\sqrt{\sigma_f}$ is a minimum. When the geometry such that our design dimension is in a direction normal to the load, it will have a minimum mass at the minimum ρ/σ_f material index. Finally, the beam will have a minimum mass at the minimum material index $\rho/(\sigma_f)^{2/3}$ when the design variable is effectively the square root of the area of the cross-section. These variations in the MIs occur because when we choose a different design variable, we are effectively changing the functional relationship between the stress and that free variable.

Similarly, for stiffness or deflection design, if b is free to vary, then the mass to be minimized is:

$$m_\delta^b = \left(\frac{PL^3}{4C_f h^2}\right)\frac{\rho}{E} \tag{46}$$

If the cross section is square and $A = \sqrt{b} = \sqrt{h}$ is free to vary, then the relevant mass expression is:

$$m_\delta^A = \left(\frac{PL^4}{4}\right)^{1/2}\frac{\rho}{E^{1/2}} \tag{47}$$

Thus, as before, if we compare eqs. (34), (46) and (47), we see changes in the structural factors and in the material indices because, as with the strength results just given, we are changing the functional dependence of the stress distribution on the geometric design variable. Thus, when the design variable h is measured in the direction of the load, we find a minimum mass at when the material index $\rho/E^{1/3}$ is a minimum. When the geometry such that our design dimension is in a direction normal to the load, it will have a minimum mass when ρ/E is a minimum index. Finally, when the design variable is effectively the square root of the area of the cross section, the beam will have a minimum mass at the minimum material index $\rho/E^{1/2}$. It is a useful exercise to sketch lines of different slopes on the material selection charts to see how the desirability of various materials increases or decreases.

We clearly want to minimize mass and cost indices, but we may also choose instead to maximize performance indices that are expressed as reciprocals of the materials or cost indices. That is, while some people prefer to minimize material indices because then they are consistently minimizing, others prefer performance indices because it feels more intuitive to seek maxima. Either approach is fine, as they lead to exactly the same results (and the materials selection software makes it easy to assign axes differently), as long as it is very clear which index is being used and in which direction it is being optimized.

In this vein, we could have cast our two mass minimization problems in terms of performance indices that are reciprocals of materials indices, that is, $\mathrm{PI}_{m(\sigma,\delta)} = \left(\mathrm{MI}_{m(\sigma,\delta)}\right)^{-1}$. Then we could summarize the variation of the performance indices with the (free) design variable (i.e., h, b, or A). For strength optimization, the variation of PIs can be ordered according to the (decreasing) exponent of the failure strength σ_f:

$$\mathrm{PI}_\sigma^b = \sigma_f/\rho, \ \mathrm{PI}_\sigma^A = \sigma_f^{2/3}/\rho, \ \mathrm{PI}_\sigma^h = \sigma_f^{1/2}/\rho \tag{48}$$

Similarly, the PIs for the corresponding stiffness or deflection optimization problems are ordered by the (decreasing) exponent of Young's modulus E:

$$\mathrm{PI}_\delta^b = E/\rho, \ \mathrm{PI}_\delta^A = E^{1/2}/\rho, \ \mathrm{PI}_\delta^h = E^{1/3}/\rho \tag{49}$$

Note that in each of eqs. (48) and (49) the density ρ appears in the denominator in the same way. The variation that does occur with respect to different design variables is either in the failure strength σ_f for strength design or in the elastic modulus E for the stiffness (deflection) design.

6.4.9 Closing remarks on mathematics, physics, and design

To sum up more generally, we have seen that preliminary and detailed design require both careful mathematical modeling and appropriately focused research to obtain relevant data. We have examined only one (small) case of mathematical modeling and its results, but designing beams and doing research to identify appropriate materials are both eminently practical skills in their own right. Further, the lessons regarding dimensions, scaling, simplifying assumptions, and how a model answers only the questions asked of it are lessons that can be applied to almost all modeling (and design) efforts.

It is also worth noting that one of the reasons that engineering programs emphasize engineering science content is a reflection of the extent to which such modeling and related research activities must be performed to do good engineering. (In traditional engineering programs the balance is noticeably tilted toward engineering science courses, but there are programs that place more emphasis on design content and experiences.) In any event, an engineer who does these tasks well and investigates design alternatives thoroughly can make the difference between a company that is a leader in its field and one that is not.

6.5 NOTES

Section 6.1: The discussion of mathematical modeling has its roots in (Dym 2004) and (Dym 2007).

Section 6.2: The modeling of beams is taken from (Dym 1997), while the approach to material selection is due to (Ashby 1999).

Section 6.3: Again, the approach to material selection, especially the use of materials selection charts, is based on (Ashby 1999).

6.6 EXERCISES

6.1 What is the value of the gravitational acceleration g when expressed in the dimensions of furlongs and fortnights? (*Hints*: The *furlong* is a unit of length used at race tracks, and a *fortnight* is an old British term for a certain unit of time.)

6.2 Demonstrate that eq. (43) is correct given eqs. (31) and (42).

6.3 Demonstrate that eq. (45) is correct given eqs. (31) and (44).

6.4 Derive eq. (46).

6.5 Derive eq. (47).

6.6 Plot lines of equal utility on Figure 6.5 assuming that eq. (43) is the objective function. How does the relative ranking of the materials change?

6.7 Plot lines of equal utility on Figure 6.5 assuming that eq. (45) is the objective function. How does the relative ranking of the materials change?

6.8 Do the reasons for selecting among steels, aluminum alloys, and magnesium alloys for structural members change as the sorts of loads applied and the geometric constraints change?

6.9 Plot lines of equal utility on Figure 6.6 assuming that eq. (46) is the objective function. How does the relative ranking of the materials change?

6.10 Plot lines of equal utility on Figure 6.6 assuming that eq. (47) is the objective function. How does the relative ranking of the materials change?

6.11 Do the reasons for selecting among wood, fiberglass, and CFRP for structural members change as the sorts of loads applied and the geometric constraints change?

6.12 Why is fiberglass used so extensively to make commercial ladders?

CHAPTER 7

COMMUNICATING THE DESIGN OUTCOME (I): BUILDING MODELS AND PROTOTYPES

Here's my design; can we build it?

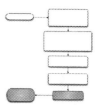

As we will explore in this and the two following chapters, design results can be communicated and reported in several ways. In this chapter we focus on building models and prototypes, and we do so in two ways. First we discuss prototypes, models and proofs of concept in general, even philosophical terms, emphasizing the "why" of such building. Then we "drill down" to the practical aspects of actually building a model or prototype in a shop, and typically a school shop. Here we talk about what is actually involved in making a simple model or prototype made of wood, plastic (polymers) or metal, and using hand tools. In our particular context, when prototypes and models are built partly to do basic tests, but generally even more to demonstrate to clients (and faculty advisors and student colleagues) how spiffy our design truly is, we are touching upon design tasks 12 and 15 of Figure 2.3.

7.1 PROTOTYPES, MODELS, AND PROOFS OF CONCEPT

We now discuss three-dimensional, physical realizations of concepts for designed artifacts; that is, we will talk about objects made to strongly resemble the object being designed, if not actually mimic "the real thing." There are several versions of physical things that could be made, including prototypes, models, and proofs of concept, and they are often made by the designer.

Prototypes are "original models on which something is patterned." They are also defined as the "first full-scale and usually functional forms of a new type or design of a construction (such as an airplane)." In this context, prototypes are working models of designed artifacts. They are tested in the same operating environments in which they're expected to function as final products. It is interesting that aircraft companies routinely build prototypes, while rarely, if ever, does anyone build a prototype of a building.

A *model* is "a miniature representation of something," or a "pattern of something to be made," or "an example for imitation or emulation." We use models to *represent* some devices or processes. They may be paper models or computer models or physical models. We use them to illustrate certain behaviors or phenomena as we try to verify the validity of an underlying (predictive) theory. Models are usually smaller and made of different materials than are the original artifacts they represent, and they are typically tested in a laboratory or in some other controlled environment to validate their expected behavior.

A *proof of concept*, in this context, refers to a model of some part of a design that is used specifically to test whether a particular concept will actually work as proposed. As we will describe further in Section 7.1.3, doing proof-of-concept tests means doing controlled experiments to prove or disprove a concept.

7.1.1 Prototypes and models are not the same thing

The definitions of prototypes and models sound enough alike that it prompts a question: Are prototypes and models the same thing? The answer is, "Not exactly." The distinctions between prototypes and models may have more to do with the intent behind their making and the environments in which they are tested than with any clear dictionary-type differences. Prototypes are intended to demonstrate that a product will function as designed, so they are tested in their actual operating environments or in similar, uncontrolled environments that are as close to their relevant "real worlds" as possible. Models are intentionally tested in controlled environments that allow the model builder (and the designer, if they are not the same person) to understand the particular behavior or phenomenon that is being modeled. An airplane prototype is made of the same materials and has the same size, shape, and configuration as those intended to fly in that series (i.e., Boeing 747s or Airbus 310s). A model airplane would likely be much smaller. It might be "flown" in a wind tunnel or for sheer enjoyment, but it is not a prototype.

A prototype is the first of its kind; a model represents a device or a process.

Engineers often build models of buildings, for example, to do wind-tunnel testing of proposed skyscrapers, but these models are not prototypes. Rather, building models used in a wind-tunnel simulation of a cityscape with a new high-rise are essentially toy building blocks that are meant to imitate the skyline. They are not buildings that work in the sense of aircraft prototypes that actually fly. So, why do aeronautical engineers build prototype airplanes, while civil engineers do not build prototype buildings? What do they do in other fields?

7.1.2 Testing prototypes, models, and concepts

We introduced testing in discussing both models and prototypes. In design the type of testing that is often most important is *proof of concept* testing in which a new concept, or a particular device or configuration, can be shown to work in the manner in which it was designed. When Alexander Graham Bell successfully summoned his assistant from another room with his new-fangled gadget, Bell had proven the concept of the telephone. Similarly, when John Bardeen, Walter Houser Brattain, and William Bradford Shockley successfully controlled the flow of electrons through crystals, they proved the concept of the solid-state electronic valve, know as the transistor, that replaced vacuum tubes. Laboratory demonstrations of wing structures and building connections can also be considered as proof-of-concept tests when they are used to validate a new wing structure configuration or a new kind of connection. In fact, even market surveys of new products — where samples are mailed out or stuffed into sacks in the Sunday papers — can be conceived of as proof-of-concept tests that test the receptivity of a target market to a new product.

Proof-of-concept tests are scientific endeavors. We set out reasoned and supported hypotheses that are tested and then validated or disproved. Turning on a new artifact and seeing whether or not it "works" is not a proper proof-of-concept demonstration. An experiment must be designed, with hypotheses to be disproved if certain outcomes result. Remember that prototypes and models differ in their underlying "reasons for being" and in their testing environments. While models are tested in controlled or laboratory environments, and prototypes are tested in uncontrolled or "real-world" environments, the tests are *controlled* tests in both cases. Similarly, when we are doing proof-of-concept tests, we are doing controlled experiments in which the failure to disprove a concept may be key.

For example, suppose we had chosen Mylar containers as our new beverage product and we're designing them to withstand shipping and handling, both in the manufacturing plant and in the store. If we think of all of the things that could go awry (e.g., stacks of shipping pallets that could topple) and analyze the mechanics of what happens in such incidents, then we might conclude that the principal design criterion is that the Mylar containers should withstand a force of X Newtons. We would then set up an experiment in which we apply a force of X N, perhaps by dropping the containers from a properly calculated height. If the bags survived that drop, we could say that they'd likely survive shipping and handling. However, we could not absolutely guarantee survival because there is no way we could completely anticipate every conceivable thing that might happen to a beverage-filled Mylar container. On the other hand, if the Mylar container fails a properly designed drop test, we can then be certain that it will not survive shipping and handling, and so our concept is disproved. The National Aeronautics and Space Administration (NASA) conducted a similar proof-of-concept test for gas-filled shock absorbers for the Mars lander. There are potential issues of legal liability involved in product testing — for example, for how much non-standard use of a product can a manufacturer be held responsible? — but they are beyond the scope of this text.

Prototypes, models, and proof-of-concept testing have different roles in engineering design because of their intents and test environments. These distinctions must be borne in mind while planning them for the design process.

7.1.3 When do we build a prototype?

The answer is, "It depends." The decision to build a prototype depends on a number of things, including: the size and type of the design space, the costs of building a prototype, the ease of building that prototype, the role that a full-size prototype might play in ensuring the widespread acceptance of a new design, and the number of copies of the final artifact that are expected to be made or built. Aircraft and buildings provide interesting illustrations because of ample commonalities and sharp differences. The design spaces of both aircraft and high-rises are large and complex. There are, literally, millions of parts in each, so many, many design choices are made along the way. The costs of building both airplanes and tall buildings are also quite high. In addition, at this point in time, we have ample experience with both aeronautical and structural technologies, so that we generally have a pretty good idea of what we're about in these two domains. So, again, why prototype aircraft and not prototype buildings? In fact, don't the complexity and expense of building even a prototype aircraft argue directly against the idea of building such prototypes?

Notwithstanding all of our past experience with successful aircraft, we build prototypes of airplanes because, in large part, the chances of a catastrophic failure of a "paper design" are still unacceptably high, especially for the highly regulated and very competitive commercial airline industry that is the customer for new civilian aircraft. That is, we are simply not willing to pay the price of having a brand new airplane take off for the first time with a full load of passengers, only to watch hundreds of lives being lost — as well as the concomitant loss of investment and of confidence in future variants of that particular plane. It is in part an ethical issue, because we do bear responsibilities for technical decisions when they impinge upon our fellow humans. It is also in part an economic issue, because the cost of a prototype is economically justifiable when weighed against potential losses. Also, we build prototypes of airplanes because those particular planes are not simply thrown away as "losses" after testing; they are retained and used as the first in the series of the many full-size designs that are the rest of the fleet of that kind of airplane.

Buildings do fail catastrophically, during and after construction. However, this occurs so rarely that there is little perceived value in requiring prototype testing of buildings before occupancy. Building failures are rare in part because high-rises can be tested, inspected, and experienced gradually, as they are being built, floor by floor. The continuous inspection that takes place during the construction of a building, from the foundation on up, has its counterpart in the numerous inspections and certifications that accompany the manufacture and assembly of a commercial airliner. But the maiden flight of an airplane is a binary issue, that is, the plane either flies or it doesn't, and a failure is not likely to be a graceful degradation!

Another interesting aspect of comparing the design and testing of airplanes to that of buildings has to do with the number of copies being made. We have already noted that prototype aircraft are not discarded after their initial test flights; they are flown and used. In fact, airframe manufacturers are in business to build and sell as many copies of their prototype aircraft as they can, so engineering economics plays a role in the decision to build a prototype. The economics are complicated because the manufacturing cost of the first plane in a series is very high. Technical decisions are made about the kinds of tooling and the numbers of machines needed to make an airplane, and economic tradeoffs between the anticipated revenue from the sale of the airplane and the cost the manufacturing process are evaluated. We will address some manufacturing-cost issues in Chapter 11.

Another lesson we can learn from thinking about buildings and airplanes is that there is no obvious correlation between the size and cost of prototyping — or the decision to build a prototype — and the size and type of the design space. And while it might seem that the decision to build a prototype might be strongly influenced by the relative ease of building it, the aircraft case shows that there are times when even costly, complicated prototypes must be built. On the other hand, if it is cheap and easy to do, then it generally would seem a good idea to build a prototype. There certainly are instances where prototypes are commonplace, for example, in the software business. Long before a new program is shrink-wrapped and shipped, it is alpha- and beta-tested as early versions are prototyped, tested, evaluated, and, hopefully, fixed.

If there is a single lesson about prototypes, beyond that it is generally good to build them, it is that the project schedule and budget should reflect plans for doing so. More often than not a prototype is required, although there may be instances in which resources or time are not available. In weapons development contracts, for example, the U.S. Department of Defense virtually always requires that design concepts be demonstrated so that their performance can be evaluated before costly procurements. At the same time, it is interesting that aircraft companies (and others) are demonstrating that advances in computer-aided design and analysis allow them to replace some elements of prototype development with sophisticated simulation.

Sometimes we build prototypes of parts of large, complex systems to use as models to check how well those parts behave or function. For example, structural engineers build full-size connections, say, at a point where several columns and beams intersect in a geometrically complicated way, and test them in the laboratory. Similarly, aeronautical engineers build full-size airplane wings and load them with sandbags to validate analytical models of how these wing structures behave when loaded. A prototype of a part of a larger artifact is built in both instances, and then used to model behavior that needed to be understood as part of completing the overall design. Thus, again, we use prototypes to demonstrate functionality in the real world of the object being designed, and we use models in the laboratory to investigate and validate behavior of a miniature or of part of a large system.

7.2 BUILDING MODELS AND PROTOTYPES

We now present some of the principles, heuristics, and guidelines for designing, building, and testing prototypes and models. The important questions we ask here are: What do we want to learn from the model or prototype? Who is going to make it? What parts or components can be bought? How, and from what, is it going to be made? And, how much will it cost? We have already answered the first question in Section 7.1, but we should keep our answers in mind as we turn to the implementation details of (actually) building a model or a prototype.

7.2.1 Who is going to make it?

We have two basic choices when we want a model or prototype: we make and/or assemble it in house or we outsource it. With sufficient time and money, we can make just about anything we want, from an application-specific integrated circuit, to a computational-fluid-dynamics (CFD) model of a pressure relief valve, to a pilot-scale oil refinery. Thus it is clear that three major factors enter into the decision about who makes our model: expertise, expense and time.

Many companies and schools keep machinists, electronics technicians, and programmers on staff for building models. Many companies and engineering schools also have facilities that individuals can use to build prototypes, and some schools even require students to learn how to use such facilities. However, it is rather rare to have facilities or expertise for making intricate or tight-tolerance items. Thus, we first want to ask if anyone on our design team has the necessary expertise, or is willing to learn. We should identify expertise and facilities that are available in-house. If the needed expertise is missing in-house, we should plan on having parts or components outsourced.

Time and cost are usually intertwined. If we need a part "yesterday," it's not likely that we can outsource it without a spending a lot of money, but we may be able to go to our own machine shop and machine it in an hour. However, engineers aren't always allowed to use their company's machine shop, so we might have to convince a machinist to do it. Then the likelihood of getting it done right away will depend on the machinist's workload and her willingness to oblige. We can see that it is a good idea to cultivate good relationships with a company's (or a school's!) machinists and technicians. It is always a good idea to try to give them meaningful lead times, to ask their advice often, and to not ask for things that are silly or impossible. Treat technicians and machinists — and indeed, all staff — as professionals and as equals. This will make it much more likely that we can get their help when something is needed immediately.

Sometimes, especially for specialized items, it may still be cheaper and/or faster to have specific items outsourced. For example, gears and printed circuit boards are likely to be gotten cheaper and faster when outsourced.

If we do something outsourced, it will go much more smoothly if we prepare detailed specifications for what we are outsourcing. That might include properly drawn, toleranced and specified mechanical drawings for machined or manufactured parts (see Chapter 8); accurate and double-checked Gerber files for printed circuit boards; or complete and correct part numbers for parts or components.

7.2.2 What parts or components can be bought?

There are a lot of parts and components that are best bought from suppliers, unless we happen to be in the business of designing and making those particular items. For example, it is rarely worth the time, equipment, and expense to make screws or transistors: Common mass-produced items should always be bought — although it also a good idea to check our institution's stockroom(s) to see whether the parts are already available on campus! Fasteners, such as nails, nuts, bolts, screws, and retaining rings should almost always be bought, as should common mechanical parts or devices such as pulleys, wheels, gears, casters, transmissions, and hinges. Similarly, electronic, electromechanical and optical components such as resistors, capacitors, integrated circuits, electric motors, solenoids, light-emitting diodes (LED's), lenses, and photodiodes can be bought.

The Internet and the widespread use of *just-in-time* manufacturing have dramatically changed the ease with which we can find suppliers and parts. We show a list of some widely available suppliers and their URLs in Table 7.1. Many companies allow on-line ordering and provide fast (i.e., overnight) shipping of in-stock items. Some companies will ship both small lots and large quantities, and some web sites provide excellent search

TABLE 7.1   A list of suppliers and their URLs for a variety of products that might be useful in building models and prototypes

What is being sought?	Supplier	URL
Materials, mechanical items	McMaster-Carr	<http://www.mcmaster.com/>
	Grainger	<http://www.grainger.com/>
Electronic supplies	Digi-Key	<http://www.digikey.com/>
	Newark	<http://www.newark.com/>
	Mouser	<http://www.mouser.com/>
Optical, opto-mechanical, electro-optical components	Thorlabs Newport	<http://www.thorlabs.com/> <http://www.newport.com/>
Suppliers (i.e., catalogs of other suppliers)	Thomas Register Global Spec	<http://www.thomasnet.com/> <http://www.globalspec.com/>

capabilities (assuming we know the proper term for the part we want) and display real-time inventories.

It really does pay to spend some time searching for parts and components. An experienced engineer or a librarian can often be very helpful as we search. We should note the prices or cost when we find what we want because that will come in handy for our budget. We can likely find many, or even all, of our prototype components already available on line or in a store — and it's far better to learn that before we start building, instead of just as we finish.

7.2.3   Building a model safely

Safety in the workshop is critically important. Power tools can easily cause dismemberment and death. A moment's inattention can lead to a permanent change in lifestyle and career. *Take safety warnings seriously*:

- Do not use equipment or machinery for which you have not been trained.
- Use protective gear and dress properly.
- Do not use power or machine tools when you are tired or intoxicated.
- Always have a buddy with you in the shop.

Almost all machine tools come with brochures that detail their safe use. *Read that documentation*. The time you spend may save you a finger or an eye. Most facilities have training programs or videos available, so be sure to ask for them and use them. In fact, you are often not allowed to use machinery or shops until you have passed the safety training. *Do not try to bypass or cheat on the training*. There are many training videos and resources available on-line. For example, one set of videos on the use of metal lathes, mills, and sheet-metal and woodworking tools is available at <http://www.eng.hmc.edu/ E8/Videos.htm>. If other resources are not available, please view these videos before going into a shop. The main rule is: *Keep your body parts away from sharp moving objects*. There is almost always a safe way to do something. Learn what that safe way is.

Individual shops will have their own safety requirements for dress and protective gear. Consider the following to be a minimal set of requirements:

- *Always wear eye protection, safety goggles or safety glasses, while using tools or when you are near someone who is.* Drills, lathes, mills, and saws produce shavings that often become airborne. Hammers do shatter, and objects being struck by hammers often break or go flying off. Wrenches and screwdrivers tend to be less hazardous to eyes, but freak accidents do occur. *Keep your eyes safe.*

- *Keep your hair short or pulled back and out of the way.* Spinning drills, mills, and lathes seem to have a magnetic attraction for long hair, and it can easily get trapped by a rapidly spinning drill chuck or a speeding saw band. *Keep you hair intact.*

- *Always wear full-coverage shoes with a sturdy sole.* Sandals or flip-flops will not protect you from dropped tools or hot metal shavings. Thin soles will not protect you from sharp objects found on the floor. *Protect your feet.*

- *Wear long, non-baggy pants.* As with shoes, long pants will protect your legs against hot flying metal shavings and other hazards. Baggy pants can get caught in moving or rotating machinery. Pants don't have to be tight, but the closer they fit, the less chance you have of getting them caught on something.

- *Wear short-sleeved shirts or blouses.* Many serious injuries and fatalities have been caused by loose clothing getting caught in moving machinery. Short sleeves represent a tradeoff between protecting an arm and getting a sleeve caught in a drill press or a mill. Rolled-up long sleeves don't count because they can unroll and get tangled in the machinery. Any loose fabric that is within a foot of your hands is likely to get caught in a machine tool that you are using.

- *Do not wear jewelry around machine tools.* Take the jewelry off and store it somewhere safe. Necklaces and bracelets are the most dangerous items, but rings, earrings, and other piercings can get caught and do damage.

Check that there is proper protection against fumes or particulates. Make sure ventilation is adequate, and that you meet requirements regarding particulates.

Fatigue and intoxication cause many industrial accidents. It is far better to miss a deadline (or risk a lower grade!) and keep all of your fingers than it is to miss the deadline (or get the lower grade) anyway because you had to make an unplanned trip to the emergency room or spend time in the hospital!

7.2.4 How, and from what, will my model be made?

Now we turn to basic principles and best practices for making, machining, and assembling mechanical parts. Having said that, note that this section will not detail the myriads of manufacturing processes because we assume here that you are going to be making, or at least assembling, your prototype yourself. And look at Figure 7.1 for an example of a spiffy model of the design ideas that the second Danbury arm support team set forth in Figure 5.10.

7.2.4.1 Plan!   The carpenter's maxim is: *measure twice, cut once.* Time spent creating detailed plans and annotations *before* cutting or machining will pay huge dividends

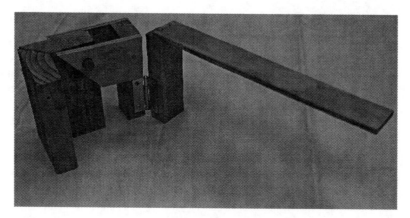

FIGURE 7.1 A wooden model of the design put forth by the second team working on the Danbury arm support. The team's ideas were displayed in Figure 5.10 and the model was constructed to help demonstrate and clarify the nature of the arm motions that were to be supported.

in ensuring that things fit together the very first time they are assembled, and this will surely minimize reworking the problem or remaking the parts.

Computer-aided design and drawing (CADD) packages are very useful for creating solid models and preparing engineering drawings (see Chapter 8). They should be used to (formally) draw actual shaped or machined parts after conceptual sketches have been drawn by hand. Many manufacturers have solid models on their websites for parts and components. These pre-made models can save a lot of time because we can use them to do a virtual assembly in the CADD package to ensure that clearances and contacts are correct. It also helps us think about how we will do the final assembly. After checking that everything will fit together properly, we should create fully dimensioned mechanical drawings from the solid models for each part. It may not be necessary to geometrically tolerance the dimensions on the print for a prototype (again, see Chapter 8!), but it helps to identify which dimensions are critical and to have a mental model of how close they have to be. It is also useful to annotate the printouts. It may seem excessive to create engineering drawings for a simple prototype, but with practice and today's CADD tools it takes very little time to create the model and drawing of a reasonably complicated part — much less time and much less stress than reworking the model while making or machining the parts.

Detailed plans also include a *bill of materials*. It is easier to construct a bill of materials during a virtual or paper assembly to identify all of the necessary parts *before* everything has been done. Then, checking the availability of the parts can help greatly in scheduling. It may also be useful to construct a *process router*, which is a list of instructions on how the prototype is to be fabricated and assembled. Note that in accord with the discussion of fabrication specifications in Section 8.1.2, a bill of materials can be viewed as a prescriptive specification, and a process router as a procedural specification.

7.2.4.2 *Materials* True prototypes are typically made from the same materials that are intended for the final design. Of course, those materials may change for the final design as a response to what was learned from the prototype. A model, on the other hand, can be constructed from whatever material will assist in answering the questions for which the model was designed to pose. The most common materials for model construction are paper, cardboard, wood, plywood, polymers (such as PVC, ABS, polystyrene, and acrylic), aluminum, and mild steel.

Paper and cardboard are usually measured and marked with rulers, pencils, compasses, and templates or stencils. They are usually cut with scissors, paper cutters, or knives such as box cutters or hobby knives. They are usually folded or rolled into their final shapes, and fastened with glue, paste, tape, staples or roundhead fasteners. Paper and cardboard are suitable for inexpensive models that do not have to support large loads. They are also good stand-ins when a properly equipped workshop is not available.

Wood is usually purchased as lumber. Lumber comes from a number of different trees, each with its own properties, and may come green or dried. Green wood is lumber that has not been allowed to dry after being cut to shape and will change in dimension quite drastically with time. Dried lumber, especially kiln-dried, has had much of the moisture removed from it and will remain much more dimensionally stable. Lumber is also classified as softwood (such as Douglas fir, pine, or redwood), or hardwood (such as oak, cherry, or walnut). The nominal sizes of softwood lumber range from $1'' \times 2''$ to $8'' \times 8''$. The most common nominal size is $2'' \times 4''$. Standard lengths range from 4 to 16 ft. The true dimensions are usually less than the nominal and vary with water content. A dried softwood $2'' \times 4''$ is actually closer to $1\frac{1}{2}'' \times 3\frac{1}{2}''$. The nominal length is usually close to the true length. Hardwoods are also available in fractional sizes from $3/8''$ on up. The nominal sizes of hardwoods are usually a little closer to the true dimensions. Softwoods are much less expensive than hardwoods, while hardwoods tend to be stronger and more wear resistant.

Wood is an anisotropic material; that is, the properties along the grain differ greatly from those across the grain. The tensile strength is much greater along the grain than across. Thus we have to think carefully about stress directions when using wood. For example, as noted in Chapter 6, beams carry their loads by developing stresses in the direction normal to their given applied loads.

Wood can be cut with assorted hand saws and power saws, including band saws and scroll saws. It can be shaped with a wood lathe, a router, or a power sander. It can be drilled with either a hand drill or a drill press.

Wood dimensions and flatness will vary with humidity and exposure to water or other absorbable fluids. It is very difficult to maintain tight dimensional tolerances in wood. Expansion spaces have to be designed into closed wood structures (such as boxes or drawers) so that undue stress is not placed in the wood as it expands and contracts in response to the weather.

Plywood is a composite material that is made from thin layers of wood glued together. There are a large number of different grades and thicknesses. The standard size sheet is $4' \times 8'$. Plywood is much more dimensionally stable than wood, and much more uniform in properties. However, it is also anisotropic, being much stronger in the plane of the layers than it is in a direction normal to the layers. It can be shaped with the same tools used for wood, but will cause faster wear on the cutting tools.

Polymers or *plastics* such as PVC, ABS, polystyrene, and acrylic are available in a number of pre-formed shapes, including sheets, bars, strips, film, rods, disks, tubes and pipes, and U-channels. Most of the polymers can be shaped with the same tools as wood. They can also be cut or shaped on the same machine tools used for metals if the tool's speed is suitably adjusted. Polymers can hold tight tolerances quite well and, depending on the polymer, can be machined into quite intricate shapes. There are usually solvent-based adhesives for joining one piece of a polymer to another piece of the same polymer. If properly done, the joint has the same strength as the bulk material. Polymers are not as strong or stiff as aluminum or steel, but they can be quite strong in some applications.

Aluminum is available in a large number of grades and shapes. Bulk aluminum is available as sheets, bars, strips, film, rods, disks, tubes, and U-channels, among other shapes. Aluminum is quite strong and light. It is not as strong as steel, but it is easier to shape and machine. It holds dimensional tolerances very well. It can be cut with a band saw or hacksaw and machined with metal lathes or mills, and holes can be drilled with a mill, drill press or hand drill. Aluminum has very high thermal and electrical conductivities. If not brought into contact with iron or steel, aluminum is quite resistant to corrosion at room temperature. Aluminum melts at a fairly low temperature and is not suitable for parts that will be exposed to high temperatures. Certain aluminum alloys can be welded, but it requires specialized tools. Aluminum is best joined with fasteners such as machine screws or rivets.

Mild steel is denser and stronger than aluminum. Bulk steel is also available as sheets, bars, strips, film, rods, disks, tubes, U-channels, and other shapes. It can also be cut with a band saw or hacksaw, machined with metal lathes or mills, and holes can be drilled with a mill, drill press, or hand drill. However, the cutting tools will experience increased wear with steel and the cutting speeds will be lower. Steel does require protection against corrosion. Perhaps the major use of steel in hand-built prototypes is in sheet form, and sheet steel is easily spot welded (see Section 7.5).

Two final details on materials. First, the choice of materials will be governed by cost, model performance requirements (e.g., are we building a true prototype or a model?), and access to cutting and shaping tools. Second, we will also need fasteners made of appropriate materials in order to connect the parts of our model or prototype together. Fasteners include things such as nails, wood screws, machine screws, sheet metal screws, bolts, nuts, washers and pins. We will detail the process of selecting fasteners in Section 7.3; the techniques for properly using and installing a fastener are described immediately below.

7.2.4.3 *Building techniques*   This section describes some of the basic techniques for shaping and joining materials. There are many sites on the web and in reference libraries that provide further information.

Straight edges in wood and polymers are usually best cut with a table saw or a band saw that has a guide rail. The rail is set at the required distance from the blade, the piece is held firmly against the guide rail, and the piece is then pushed through the saw. Be sure to use a push stick or rod if necessary to keep your fingers a safe distance away from the blade.

Straight edges in metal are usually rough cut with a band saw or cutting wheel, and then finished or faced with a mill.

Curved edges in wood or plastic are usually cut with a band saw or scroll saw. The desired profile is drawn in pencil on the wood and then the pencil line is used to guide the saw. Care must be taken not to bind the saw blade (put lateral forces on it) when cutting curves. The curve can be hand-or power-sanded to smooth off the saw cut.

Curved edges in metal may be cut with a proper band saw. The blade should be appropriate for the metal being cut and a great deal of patience will be required because it will take much longer to cut metal than it would a similar thickness of wood. The curve can be filed or hand-or power-sanded to a final shape.

Cylindrically symmetric profiles can be formed in wood on a wood lathe, and in metal and polymers on a metal lathe. Since the operation of lathes and similar tools is beyond our scope, be sure to obtain the proper training if you feel that use of such tools would be appropriate to your task.

Holes are used to allow passage for a fastener or for something like an axle, a cable, or a tube. In mass-produced items, the locations of such *joiner holes* are specified with geometric tolerances that determine whether parts are easily interchangeable. In models and one-of's, it is usually quicker and easier to clamp two parts together and drill through both at once to obtain both of the required holes. The parts may then not be interchangeable, depending on the care with which the holes were drilled, but the holes will be aligned and the fastener or cable will pass through properly. Similarly, it may be better to drill a passageway hole after parts are assembled. Then the hole is guaranteed to pass smoothly through all of the required parts.

Pieces of wood may be joined together in several different ways, the quickest being staples or nails. Both nails and staples can produce strong joints, but they can also easily split the wood if they are improperly sized or used. Wood pieces can also be glued together using white glue or carpenter's glue, but the glue joints must be made along the grain on both pieces. Pieces of wood cannot be successfully glued together if their glue is applied across end grains. The need to glue along the grain is one of the reasons that the *mortise-and-tenon* joint is used to make wood cabinets and furniture. The strongest wood joint is formed when gluing is combined with a fastener, such as a nail.

A wood screw is used to join two pieces of wood together or to fasten another material, such as plastic, to a wooden surface. The top piece of wood should have a clearance hole drilled so that the screw can slide through without touching the hole walls. The bottom piece should have a pilot hole drilled to keep the wood from splitting. Table 7.2 gives approximate dimensions for both clearance and pilot holes for different screw sizes. The actual size will depend on the hardness and moisture content of the wood.

Rub a screw with soap or wax before screwing it into hardwood. If brass screws are wanted, thread their holes with a steel screw of the same size before screwing in the brass screw. Brass is much softer than steel and the screw might be damaged (especially in hardwood) if it is used both to thread the hole and to hold the pieces together.

TABLE 7.2 Dimensions of screws and clearance and pilot holes for steel screws used to join two pieces of wood

		inch				mm	
screw size	diam	clearance hole	pilot hole (softwood)	pilot hole (hardwood)	clearance hole	pilot hole (softwood)	pilot hole (hardwood)
0	0.060	1/16	none	1/32	1.6	none	0.8
1	0.073	5/64	none	1/32	2	none	0.8
2	0.086	3/32	none	3/64	2.4	none	1.2
3	0.099	7/64	none	1/16	2.8	none	1.6
4	0.112	7/64	none	1/16	2.8	none	1.6
5	0.125	1/8	none	5/64	3.2	none	2
6	0.138	9/64	1/16	5/64	3.6	1.6	2
7	0.151	5/32	1/16	3/32	4	1.6	2.4
8	0.164	11/64	5/64	3/32	4.5	2	2.4
9	0.177	3/16	5/64	7/64	5	2	2.8
10	0.190	3/16	3/32	7/64	5	2.4	2.8
11	0.203	13/64	3/32	1/8	5.5	2.4	3.2
12	0.216	7/32	7/64	1/8	5.5	2.8	3.2
14	0.242	1/4	7/64	9/64	6.5	2.8	3.6
16	0.268	17/64	9/64	5/32	7	3.6	4
18	0.294	19/64	9/64	3/16	7.2	3.6	5
20	0.320	21/64	11/64	13/64	8.5	4.5	5.5
24	0.372	3/8	3/16	7/32	9	5	5.5

The heads of wood screws are flat, oval, or round. Both oval and round heads will protrude above the wood's surface of the wood. The clearance hole for either a flat head or an oval head screw should be countersunk using a *countersink* to make that small conical depression. The good news is that a properly done wood screw joint will be much stronger than a nail joint and run almost no risk of splitting the wood. The bad news is that wood screws are more expensive than nails and drilling clearance and pilot holes takes some time.

Pieces of metal may be joined by several methods. One is by drilling clearance holes through both pieces, passing a bolt through the holes and fastening it with a nut on the far end. The clearance hole on the top piece should be countersunk if either a flat- or oval-head screw is used. Another approach would be to drill a clearance hole in the first piece and then drill and tap a hole in the second piece. A round-head machine screw is then passed through the first piece and screwed into the second. The clearance hole on the top piece can be *counter-bored* (drilled with a flat-bottomed hole slightly larger than the screw head) if we don't want the screw to protrude above the top of the piece. It is bad practice to countersink the top piece if the bottom piece is threaded, because the screw head may snap off as it is tightened. In the parlance of GD&T (geometric dimensioning and tolerancing — see Chapter 8), we have a fixed-fixed fastener and any error in position will result in huge shear stresses on the fastener.

We *assemble* our prototype once we have fabricated or bought all of the parts we need. Our choice of assembly tools will depend on our choices of materials and fasteners. A hammer is useful for nails and pins, and for shaping malleable metals. It is also useful for tapping close-fitting parts together. Screwdrivers should match the types of slots on your screws and bolts. Wrenches should be used for nuts and bolts, and life is much easier if at least one adjustable end wrench is at hand. Pliers are used to hold and squeeze things, but should never be used to hold a nut or a bolt. Instead, find a properly fitting wrench and use it. If the model is held together with a lot of screws, a power screwdriver can sharply reduce hand fatigue and aching.

Finally, try to find a large work surface and keep it clean and orderly as assembly progresses. If a process router or assembly sequence was developed during planning, try to follow it as far as possible. Also, it is best to dry-fit parts in place before gluing or fastening them to make sure that they actually do go together properly. One of the most common mistakes in assembling prototypes is to forget that you need access to a fastener to put it in and tighten it, so put parts on the inside of an enclosed space before enclosing that space.

7.2.5 How much will it cost?

An important step in model building is the identification of the costs of materials and assembly. Many a project has been hindered by not adequately planning for the cost of building a model or prototype.

The first step in estimating the cost is to develop the bill of materials as mentioned above. Be sure to include fasteners and any tools that are not in hand. The parts can be priced once the bill of materials has been established. Those prices might have been taken from manufacturers' or suppliers' websites as materials and parts were selected, or those suppliers and manufacturers must be called to get the prices. The budget will also include the cost of outsourced assembly or machining. In most companies, engineers also have to charge their time against the specific project they are working on, so it is a good habit to keep track of the time spent (and its estimated cost), even if it is not necessary for the current budget. (It makes for a good report to state that equipment and supplies were kept within a $125 budget, but that the assembly labor was worth $10,000!)

It is wise to plan for miscalculations of the prices and the amounts of parts and materials that are used. Prices always seem to go up between the time items are priced and the time they are bought. And people often forget to include sales tax and shipping in their costs. So it's good to leave a margin for error, say 10% to 15%, especially if this is your first model-building project. Be sure, too, that any big-budget items meet your needs *before* you order them. A 10% reserve won't help if the $100 item in the $125 budget is the wrong one.

Computer spreadsheets were developed to replace paper spreadsheets or ledger sheets, which have been used for budgets and expenses since the early days of business and accounting. Make liberal use of computer spreadsheets for budgets and for bills of materials. If bills of materials and budgets are integrated, changes in the budget driven by changes in materials can be handled quickly and painlessly.

7.3   SELECTING A FASTENER

A crucial aspect of almost all objects or devices that have more than one part is the nature of *fasteners* that are used to join the device's parts to each other. Fasteners and fastening methods are categorized as *permanent*, meaning that the fastener cannot be undone, and *temporary*, meaning that the fastener can be undone in a nondestructive manner. Welds, rivets and some adhesives are instances of permanent fasteners. Screws, nuts and bolts, and paper clips are examples of temporary fasteners. There are tens of thousands of different fasteners. For example, a quick search of one distributor's web site showed that we could order 78 different sizes of zinc-plated, steel Phillips flat-head wood screw. Since it would be impossible to cover all existing fasteners — why would we select a Truss Opsit[®] Self Tapping Left-Handed Thread Screw? — we will describe just the most common fasteners and the reasons for selecting them.

Fastener selection is typically done during both the preliminary and detailed design stages. It is worth noting that each fastener is designed to meet some objective(s), satisfy some constraint(s) and serve some function(s). Thus, in addition to being of practical importance in design and model fabrication, fastener selection represents an implementation of basic design concepts. We will organize our discussion of fasteners first by material (e.g., wood, plastics and metals) and secondarily along the distinction between permanent and temporary fasteners.

7.3.1   Fastening wood

Wood fastening or joining is usually performed with adhesives such as white glue, impact fasteners such as nails and staples, wood screws, or craft joints such as dovetails or dowel pins. Most wood adhesives and impact fasteners are permanent fasteners. Wood screws are normally temporary fasteners. Craft joints are usually permanent, but can be temporary. Proper craft joints usually involve a fair amount of expertise in woodworking, so we will not cover them here. There is a lot of information available on the web about woodworking generally and about dovetail joints, mortise and tenon joints, and much, much more.

7.3.1.1   *Permanent wood fasteners*   We will limit our discussion to the most common adhesives for wood joining: white glue, carpenter's glue, hot-melt glue, contact cement, and nails.

White glue is inexpensive and strong if used properly. It is not moisture or heat resistant, so is not appropriate for use outdoors or in high-temperature environments. The fumes are not hazardous.

Carpenter's glue has close to the same consistency as white glue. It is strong if used properly, and has moderate moisture and heat resistance. It fills gaps well. The fumes are not hazardous.

Hot-melt glue, often just called *hot glue*, melts at high temperature and solidifies at room temperature. It is applied with a glue gun that heats the glue to its melting point. Its strength is moderate to low, and it is moisture resistant but not heat resistant. It is excellent for quick assembly and short-lived prototypes. It is easy to burn yourself with the hot glue, so use caution with it.

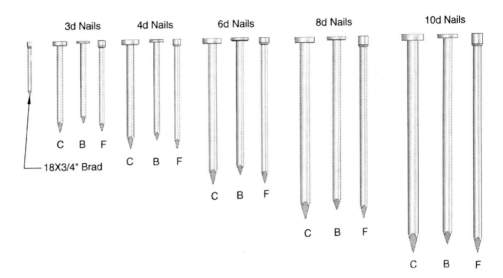

FIGURE 7.2 A collection of four styles of standard nails of varying size: a brad, common nails (C), box nails (B), and finishing nails (F).

Contact cement is most often used to bond veneers or plastic laminates to wood. It is extremely permanent, strong, and heat and moisture resistant. It must be applied to both surfaces to be joined and allowed to dry until tacky. Then the two surfaces are joined. The fumes are hazardous, as is contact with the uncured cement, so protective gloves, goggles, and adequate ventilation are needed.

Nails are considered permanent fasteners even though they can sometimes be removed without permanent damage — although we should not always count on being able to remove a nail. A nail holds two boards together by friction and by the nail head, if there is one. Nails are not normally considered precision fasteners and will display a fairly wide range of dimensions as manufactured. When practical, a nail should be sized so that approximately 2/3 of its length is in the bottom board.

There are many different kinds of nails, of which *common nails* are the most common (see Figure 7.2). They are sized in *pennies* (the approximate weight in pounds of 1000 nails), which is abbreviated as "d." They range from 2d, which are 1 in long and made from 15-gage wire, to 60d, which are 6 in long and made from 2-gage wire. A common nail is for general-purpose joining of boards.

Finishing nails have a small, almost non-existent head (Figure 7.2), and are slightly smaller in diameter than common nails. Finishing nails are countersunk with a nailset so their heads are below the wood's surface. They are used for cabinetry and in other circumstances where the nail head should not show.

Box nails (Figure 7.2) are used to join thin pieces of dry wood. They have a blunt tip to avoid splitting the wood. Box nails also have slightly smaller diameters than common nails, and they often have a coating that heats and melts as they are driven: The coating then solidifies and glues the nail in place.

Brads are small wire nails that resemble small finishing nails (Figure 7.2). They are typically used for attaching molding to walls or other places where small inconspicuous nails are required.

7.3.1.2 *Temporary wood fasteners*

Wood screws are the most common temporary fasteners in wood. Wood screws come with flat heads, oval heads or round heads. They are typically made of three materials: brass, galvanized steel or stainless steel. Brass is generally used for decorative applications as it is soft and easily damaged. Stainless steel is the most expensive, but it is the most resistant to rust and corrosion. Galvanized steel is the most common. Table 7.2 (in Section 7.2.4.2) listed standard wood screw sizes, their corresponding diameters, and their clearance and pilot hole sizes. Wood screws range in length from ½ in to 3½ in.

Round-head screws (Figure 7.3) protrude above the surface of the wood (typically for cosmetic reasons) and the screw head rests flush against the top surface of the wood. They are most often used for mounting hardware such as hinges or knobs onto wood.

Oval-head screws (Figure 7.3) resemble a cross between a flat-head and a round-head screw. The head is designed to protrude above the surface of the wood (again for cosmetic reasons), but the hole should be countersunk. Oval head screws are most often used to attach pre-countersunk hardware, such as hinges, to wood.

Flat-head screws (Figure 7.3) are used where the screw cannot protrude above the surface of the finished wood. The hole for the screw should be countersunk unless the wood is particularly soft, in which case the screw can simply be driven so that its head is below the surface of the wood.

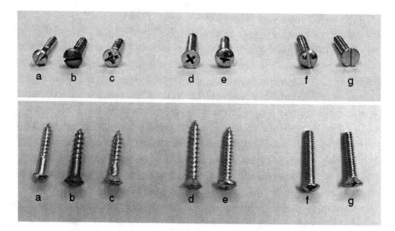

FIGURE 7.3 Assorted screws (temporary fasteners), each a No. 10 size, and all but (d) 1 in long: (a) steel slotted round-head wood screw; (b) brass slotted oval-head wood screw; (c) steel Philips flat-head wood screw; (d) steel Philips flat-head sheet metal screw (1 1/4 in long); (e) steel Philips pan-head sheet metal screw; (f) steel slotted round-head machine screw; and (g) steel slotted flat-head machine screw.

Screw slots come in slotted, Philips, and specialty varieties. The specialty slot, such as a Torx, requires a special head to drive it, and is used in such applications as closing the clam shell of the DeWalt D21008K corded power drill pictured in Figure 4.3. Fasteners with Philips heads can support greater driving forces than those with slotted heads; they are preferred if the screws are to experience high torque while being tightened. Some of the specialty slots can support even greater driving force than the Philips type.

7.3.2 Polymer fastening

Permanent polymer joining is most commonly done with adhesives that generally fall into two classes, solvent-based cements that are polymer specific and general adhesives such as epoxy. The general adhesives must generally be used when joining different polymers or a plastic to wood or metal. In rare cases, polymers are joined using friction welding. Temporary fasteners are usually threaded fasteners such as machine screws and nuts and bolts, very much like those used for metal, except that the fasteners may be made from a polymeric material such as nylon or acetyl resin. Accordingly, our discussion of threaded fasteners will be deferred to the section on fastening metals.

7.3.2.1 *Permanent polymer fastening* When they can be used, *solvent cements* are the preferred means of joining two pieces of a polymer together. If done properly, the joint will have the same strength and characteristics as the bulk material. Typically, the solvent will dissolve some of the original material and then evaporate, permitting the material to re-solidify. Solvent cements work best when the two surfaces being joined have a nearly perfect physical joint with no gaps or holes. (Some thicker solvent cements can fill gaps.) If too much solvent cement is used, the material may be weakened. Specific solvent cements include plastic model cement for polystyrene, primer and cement for PVC piping, and acrylic solvent cement for joining acrylic. Each of these solvent cements typically come with their own instructions that should be carefully read and followed exactly.

General adhesives should be chosen after it has been determined whether or not a particular adhesive is recommended for joining the chosen materials. The first to be examined should be epoxies and cyanoacrylates (superglues). Epoxies do very well in bonding porous materials and a good-to-poor job of bonding nonporous material, depending on the specific material. Cyanoacrylates work very well on smooth nonporous materials but do poorly on porous materials. Contact cements should be examined next if epoxies or cyanoacrylates prove unacceptable.

7.3.2.2 *Temporary polymer fastening* Our discussion of threaded plastic fasteners is deferred to the corresponding section on metal fastening because they are similar to temporary metal fasteners.

7.3.3 Metal fasteners

The principal permanent means of joining metals are soldering/brazing, welding and riveting. Threaded fasteners are the principal temporary means of joining metals.

FIGURE 7.4 A Miller LMSW–52 spot welder.

7.3.3.1 *Permanent metal joining or fastening*   *Welding* involves melting portions of the two pieces to be joined and (usually) adding some additional metal. The joint is formed when the metal re-solidifies. Welding is most often used to join two pieces of ferrous metals (steels and cast iron), but it can be performed on aluminum and other metals by expert welders under the correct circumstances. Arc welding involves specialized training and equipment and is well beyond the level of this text.

Spot welding is done to join two pieces of (usually ferrous) sheet metal, usually with a relatively low-cost and safe *spot welder* that consists of two long arms that end in electrode tips (see Figure 7.4). The two pieces of sheet steel are squeezed between the two electrode tips and a brief-but-large current is passed through the electrodes and the sheet steel. The current resistively melts a small spot between the sheets which then solidifies and forms the joint. (Like other power tools, the spot welder has its own procedural and safety instructions that should be attentively followed.)

Soldering and *brazing* use a piece of low-melting-temperature metal to join two pieces of higher-temperature-melting metal. The difference between the two is the temperature at which the joining metal melts. By convention, using a joining metal that melts below 800°F (425°C) or 450°C (840°F) is soldering, and using a joining metal that melts above that is brazing. Depending on the size of the joint and the temperature involved, a soldering iron, a soldering gun or a butane torch can be used for the heating and melting. It is important to have good mechanical contact between the pieces to be welded or soldered before heating. The molten metal will be drawn into the gap between the pieces by capillary action. The joint is not as strong as a weld, but can be done with much less training or specialized equipment.

Rivets are the final common permanent fastener used to join metal parts. There are two principal types: *Solid rivets* are used when there is access to both sides of the joint and maximum strength is required. *Blind rivets* — often called pop rivets, although

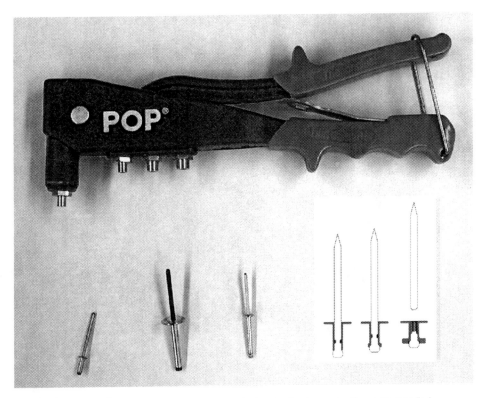

FIGURE 7.5 A Pop[R] pop rivet gun (top) and a graphic showing just how a rivet is actually installed (bottom).

POP[R] is a registered brand name for a blind rivet — can be installed when there is access only to one side of the joint. Solid rivets require specialized training and equipment and are mentioned only in passing.

Blind rivets are installed with a rivet gun. A hole is drilled through the two pieces to be joined, in accord with the manufacturer's recommendations. A blind rivet is placed in the rivet gun, and then inserted into the hole. The handle on the rivet gun is squeezed until the mandrel snaps off. Figure 7.5 shows such a rivet gun and the procedure for installing a blind rivet.

Rivets can support tensile loads but are more often used for shear loads. The machine screws discussed in the next section are used for tensile loads.

7.3.3.2 *Temporary metal fasteners*
The principal temporary fasteners for metal are sheet-metal screws, machine screws, cap screws, and bolts and nuts. There is no universal definition that differentiates a screw from a bolt. Some maintain that screws are threaded fasteners that come to a point and bolts are threaded fasteners with a constant thread diameter and a square end. By this definition, machine screws are actually bolts. Others argue that screws are meant to be turned or rotated while being attached, and bolts

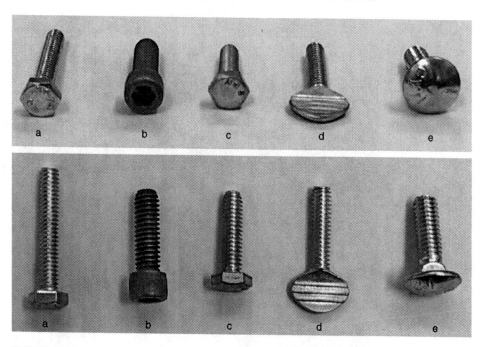

FIGURE 7.6 Assorted screw and bolt (temporary) fasteners: (a) steel hex bolt $1/4-20 \times 2$ in; (b) black oxide steel socket head cap screw $5/16-18 \times 1$ in; (c) steel hex-cap screw $1/4-20 \times 1$ in; (d) steel thumb screw $1/4-20 \times 1$ in; and (e) steel carriage screw $5/16-18 \times 1$ in.

are designed not to rotate during or after attachment and that they (usually) have smooth non-grippable tops. Under this definition, a hex-head bolt is a screw. The lesson for us is that we must exercise care when we refer to screws or bolts; perhaps it's best to follow local custom. An assortment of threaded fasteners is shown in Figures 7.3 and 7.6.

Machine screws come in a wide variety of head types, slot types, materials, diameters and lengths. The most common head types are pan, round, cheese, flat, and oval. There are many variants on these basic types. The choice of head type depends on whether the resultant joined surface must be flush and whether the fastener can be fixed or floating. Fixed and floating fasteners will be discussed in Section 8.3.2.6, but briefly: The location of a fixed fastener cannot be adjusted in position as it is tightened, and the location of a floating fastener can be adjusted slightly as it is tightened. Flat heads require countersinking of the surface and result in a flush surface but a fixed fastener (Figure 7.3(g)). Oval heads require countersinking and result in a rounded but protruding surface and a fixed fastener. Pan, round, and cheese heads (and no, not the Green Bay type) all result in protruding heads, but floating fasteners. The difference is the degree to which they protrude. However, if there is room, the holes for all three may be counter-bored, resulting in a flush surface, but with a noticeable gap around the screw head (Figure 7.3(f)).

Cap screws are sometimes considered machine screws and sometimes considered as a separate category. Cap screws have either hex heads or socket heads. The hex heads are designed to be tightened with a wrench. Hex-head cap screws are almost never used in

TABLE 7.3 Common *inch* screw sizes and dimensions and clearance hole dimensions for machine screws

Screw Size	Major Diameter	Pitch Diameter	Minor Diameter	Normal Clearance Hole	Close Clearance Hole
0–80	0.060	0.052	0.044	0.073	0.067
1–64	0.073	0.063	0.053	0.089	0.081
2–56	0.086	0.074	0.064	0.106	0.094
3–48	0.099	0.086	0.073	0.120	0.106
4–40	0.112	0.096	0.081	0.136	0.125
5–40	0.125	0.109	0.094	0.154	0.140
6–32	0.138	0.118	0.099	0.169	0.154
8–32	0.164	0.144	0.125	0.193	0.180
10–24	0.190	0.163	0.138	0.221	0.205
1/4–20	0.250	0.218	0.188	0.281	0.266
5/16–18	0.313	0.276	0.243	0.344	0.328
3/8–16	0.375	0.334	0.297	0.406	0.390
7/16–14	0.438	0.391	0.349	0.469	0.453
1/2–13	0.500	0.450	0.404	0.531	0.516

counter-bored holes due to the difficulty of getting a wrench in the hole to tighten the screw (Figure 7.6(c)). Socket head cap screws are designed to be tightened with a hex key or Allen wrench and are frequently used in counter-bored holes to leave a flush surface (Figure 7.6(b)).

Machine screws are most commonly made of steel, stainless steel, aluminum, brass, or nylon. There are others designed for specific applications. The material chosen is a function of cost, strength needed, and compatibility with the metals being joined.

The dimensions of machine screws are governed by standards (see Section 7.4). *Inch fasteners* are specified with a thread diameter and the number of threads per inch (TPI). Diameters smaller than 1/4-inch are specified with a gauge number. A 1/4-20 is an inch fastener with a 1/4-inch thread diameter and 20 TPI. *Metric fasteners* are specified with a thread diameter and a thread pitch (the distance between adjacent threads). An M6 × 1 is a metric thread fastener with a 6-mm thread diameter and a thread pitch of 1 mm. Common *inch* screw sizes and clearance hole dimensions are listed in Table 7.3, and common *metric* thread sizes and clearance hole dimensions are listed in Table 7.4.

When we specify clearance holes for threaded fasteners, we must take into account the skill of the machinist and the cost of precision machining. The normal clearance holes are for reasonably competent yet inexpensive machining. The close clearance holes are for precise and more expensive machining. The available tolerance that we can specify in geometric dimensioning and tolerancing (GD&T) is the difference between the clearance hole and the major diameter. For example, as we will discuss in Section 8.2.3.6, a 1/4-20 machine screw with a close clearance hole will have only $0.266 - 0.250 = 0.016$ in available for tolerancing.

TABLE 7.4　Common *metric* thread sizes and dimensions and clearance hole dimensions for machine screws

Screw Size	Major Diameter	Pitch Diameter	Minor Diameter	Normal Clearance Hole	Close Clearance Hole
M1.6 × 0.35	1.60	1.37	1.17	1.9	1.75
M2 × 0.4	2.00	1.74	1.51	2.5	2.25
M2.5 × 0.45	2.50	2.21	1.95	3.0	2.75
M3 × 0.5	3.00	2.68	2.39	3.7	3.3
M3.5 × 0.6	3.50	3.11	2.76	4.3	3.9
M4 × 0.7	4.00	3.55	3.14	4.8	4.4
M5 × 0.8	5.00	4.48	4.02	5.8	5.4
M6 × 1	6.00	5.35	4.77	6.8	6.4
M8 × 1.25	8.00	7.19	6.47	8.8	8.4
M10 × 1.5	10.00	9.03	8.16	11.0	10.5
M12 × 1.75	12.00	10.86	9.85	13.0	12.5
M14 × 2	14.00	12.70	11.55	15.0	14.5

7.3.4　What size temporary fastener should I choose?

It is a sad-but-true fact that the majority of threaded fasteners are chosen because they "look right" to the experienced designer. But the proper way to choose the diameter of fastener is to:

- calculate the force that the fastener is expected to endure,
- include a reasonable safety factor, and
- choose a fastener that exceeds the strength required.

The two forces that a screw is likely to experience are a tensile force (along the axis of the screw) and a shear force (across the axis of the screw). The calculation of these forces in a complex piece of machinery is beyond our scope of this book (although it can be found in a typical strengths or mechomat text). The manufacturer's specification that is of interest to us is the *proof load*, which is the load the fastener must withstand without undergoing permanent plastic deformation. We would typically choose a fastener with a proof load four times the expected maximum load (this corresponds to a safety factor $S = 4$). Also, we would normally tighten the fastener to have a *pre-load* of 90% of the proof load. The torque required to pre-load the bolt may be estimated as:

$$T = 0.2F_l d \tag{1}$$

where T is the torque, F_l is the proof load, and d is the nominal diameter of the fastener. For example, if the maximum tensile load is expected to be 1550 N, the proof load should be $4 \times 1550\,\text{N} = 6200\,\text{N}$. After a search we found a manufacturer that has a steel Philips pan-head machine screw M6 × 1 with a proof load of 6230 N, so we would choose this screw. From Table 7.4 we note that the pitch diameter is 5.35 mm = 0.00535 m. The pre-load is then $0.9 \times 6200\,\text{N} = 5580\,\text{N}$. We would then use a torque wrench to tighten this screw to a torque of $0.2 \times 5580\,\text{N} \times 0.00535\,\text{m} = 29.9\,\text{Nm}$.

In closing we would note that the subject of fastener selection has filled many a volume and more than a few manufacturer's catalogs. The suggestions made above should be viewed as a starting point, not as the final word on fastener selection. Having said that, it is still the case that our guidelines will be adequate for designing and building most common models or prototypes. If the design is critical, or if the designer is not widely experienced with fasteners, we would seek expert guidance from a mentor, a machinist, or a reference book.

7.4 NOTES

Section 7.2: Anyone with any interest at all in woodworking should definitely read (Abram 1996)! There are also many sources for data about common construction techniques and fasteners on the Internet. Particularly helpful (and consulted during the writing of both Sections 7.2 and 7.3) are: The *Industrial Screw* website at <http://www.industrialscrew.com/index.cfm?page=tech>;
Lowe's How To Library at the Lowe's website <http://www.lowes.com>;
Bob Vila's How To Library <http://www.bobvila.com/HowTo_Library/>; and
eHow <http://www.ehow.com/>.

Section 7.3: The ANSI B18 series covers rivets, bolts, nuts, machine, and cap screws, and washers in American engineering units. The thread size is governed by the Unified Thread Standard, ANSI B1.1, ANSI B1.10M, and ANSI B1.15. Metric screw threads are governed by ISO 68-1, ISO 261, ISO 262, and ISO 965-1. The *This-To-That* website, <http://www.thistothat.com/>, gives specific guidance on selecting adhesives for joining two materials together, i.e., this to that.

7.5 EXERCISES

7.1 Create sketches or CAD drawings of a hollow $6'' \times 6'' \times 6''$ cube with a frame of $1/4''$ square 6061 aluminum stock and side panels of $1/16''$ thick polystyrene sheets.

7.2 Select the means of fastening together the frame and attaching the side panels for the hollow cube of Exercise 7.1.

7.3 Develop a bill of materials and a budget for the hollow cube of Exercise 7.1. (*Hint*: Visit a supply website such as *McMaster-Carr*, <http://www.mcmaster.com/>.)

7.4 Develop a process router for the hollow cube of Exercise 7.1.

7.5 Create sketches or CAD drawings of a hollow $3' \times 3' \times 3'$ cube with a frame of $1'' \times 2''$ (nominal) furring strips and side panels of $1/4''$ thick BC plywood.

7.6 Select the means of fastening together the frame and attaching the side panels of the hollow cube of Exercise 7.5.

7.7 Develop a bill of materials and a budget for the hollow cube of Exercise 7.5. (*Hint*: Try websites such as *Lowe's*, <http://www.lowes.com/>, or *Home Depot*, <http://www.homedepot.com/>.)

7.8 Develop a process router for the hollow cube of Exercise 7.5.

7.9 Two sheets of $1/4''$-thick aluminum are to be joined by a $1/4''$–20 socket-head cap screw with through holes and a nut. Determine the drill sizes required to drill the holes in the top and bottom sheets. Also fully specify the screw and the nut. (*Hint*: Check fastener tolerances as described in Section 8.3.)

7.10 Two sheets of $1/4''$-thick aluminum are to be joined by a $1/4''$–20 flat-head machine screw with through holes and a nut. Determine the drill sizes required to drill the holes in the top and bottom sheets. Also fully specify the screw and the nut. (*Hint*: Check fastener tolerances as described in Section 8.3.)

7.11 Two sheets of $1/4''$-thick aluminum are to be joined by a $1/4''$–20 socket head cap screw with a through hole on the top plate and a blind threaded hole on the bottom sheet. Determine the drill sizes required to drill the holes in the top and bottom sheets. Specify the drilling depth and the tap for the bottom hole. Also fully specify the screw. (*Hint*: Check fastener tolerances as described in Section 8.3.)

7.12 A member of your design team wants to join two sheets of $1/4''$-thick aluminum with a $1/4$–20 flat head machine screw with a through hole in the top sheet and a threaded through hole in the bottom sheet. Specify fully your recommendation to your team member. (*Hint*: Check fastener tolerances as described in Section 8.3.)

7.13 For Exercise 7.9 and a safety factor of 4, determine the maximum tensile load that can be applied between the top and bottom sheet, and the pre-load and the tightening torque on the screw. (*Hint*: You did remember to look up the proof load on the screw when you fully specified it?)

CHAPTER 8

COMMUNICATING THE DESIGN OUTCOME (II): ENGINEERING DRAWINGS

Here's my design; can you make it?

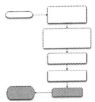

REPORTING IS an essential part of a design project: The project is not complete if its results have not been communicated to the client and to other stakeholders designated by the client. Final design results can be communicated in several ways, including oral presentations, final reports (that may include design drawings and/or fabrication specifications), and prototypes and models such as those discussed in Chapter 7. In this chapter we consider some common guidelines for producing engineering design drawings as the first part of our discussion of design task 15 as identified in Figure 2.3.

8.1 ENGINEERING DESIGN DRAWINGS SPEAK TO MANY AUDIENCES

Being able to communicate effectively is an essential, critical skill for engineers. We communicate in teams, through oral presentations, in written documents, and through technical drawings of our designs. We communicate with a client when we frame the design problem, as we work through the design process, and as we near the end and we create standardized, detailed drawings that portray our design. As noted in Chapter 7, we communicate when we build models to demonstrate or evaluate our design's effectiveness. And perhaps as important as anything else, we also communicate when we take our ideas from our heads and explain them to others. We devote this chapter to the creation of design drawings — an essential modality for effective communication.

In addition to communicating with client(s) about a project, a design team must also communicate, even if only indirectly and through the client, with the maker or manufacturer of the designed artifact. This is where "the design rubber really meets the road" because the builder or fabricator of that design may never meet the team that created the design. Generally, the only "instructions" that the fabricator sees are those representations or descriptions of the designed object that are included in the final design report. This means that these representations and descriptions must be complete, unambiguous, clear, and readily understood. The relevant question is, then, What can we designers do to ensure that our product descriptions will result in the built object being just what we designed?

The answer is deceptively simple: When we communicate design results to a manufacturer, we must think very carefully about the fabrication specifications we are writing. This means paying particular attention to the various kinds of drawings we do during a design project and to the different standards that we associate with final design drawings.

8.1.1 Design drawings

We first turn to design drawings, which can include sketches, freehand drawings, and computer-aided-design-and-drafting (CADD) models that extend from simple wire-frame drawings (e.g., something very much like stick figures) through elaborate solid models (e.g., elaborate "paintings" that include color and three-dimensional perspective). Drawing is very important in design, especially in mechanical design, because a great deal of information is created and transmitted in the drawing process.

In historical terms, we are talking about the process of putting "marks on paper." These marks include both sketches and drawing and *marginalia*, that is, notes written in the margins. The sketches are of objects and their associated functions, as well as related plots and graphs. The marginalia include notes in text form, lists, dimensions, and calculations. Thus, the drawings enable a parallel display of information as they can be surrounded with adjacent notes, smaller pictures, formulas, and other pointers to ideas related to the object being drawn and designed. We see, here, that putting notes next to a sketch is a powerful way to organize information, certainly more powerful than the linear, sequential arrangement imposed by the structure of sentences and paragraphs. We show an example that illustrates some of these features in Figure 8.1. This is a sketch made by a designer working on the packaging for a battery-powered computer clock. The packaging consists of a plastic envelope and the electrical contacts. The designer has written down some manufacturing notes adjacent to the drawing of the spring contact. Further, it would not be unusual for the designer to have scribbled modeling notes (e.g., "model the spring as a cantilever of stiffness. . . "), or calculations (e.g., calculating the spring stiffness from the cantilever beam model), or other information relating to the unfolding design. Note that some of this information also readily translates into aspects of fabrication specifications.

Marginalia of all sorts are familiar sights to anyone who has worked in an engineering environment. We often draw pictures and surround them with text and equations. We

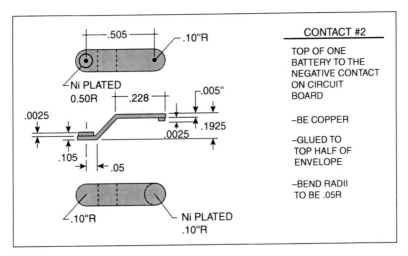

FIGURE 8.1 Design information adjacent to a sketch of the designed object (after (Ullman, Wood, and Craig 1990)). Note how clear and neat this sketch is, and that the notes are in easy-to-read block letters (recall our discussion of such sketches in Section 5.2.3.2).

also draw sketches in the margins of documents to elaborate a verbal description, to fortify understanding, to indicate more emphatically a coordinate system or sign convention. Thus, it should come as no surprise that sketches and drawings are essential to engineering design. (It is interesting that while some classic engineering design textbooks stress the importance of graphical communication, drawing and graphics seem to have vanished from engineering curricula!) In some fields — for example, architecture — sketching, geometry, perspective, and visualization are acknowledged as the very underpinnings of the field.

Of particular importance to the designer is the fact that graphic images are used to communicate with other designers, the client, and the manufacturing organization. Drawings are used in the design process in several different ways, including to:

- serve as a launching pad for a brand-new design;
- support the analysis of a design as it evolves;
- simulate the behavior or performance of a design;
- provide a record of the shape or geometry of a design;
- facilitate the communication of design ideas among designers;
- ensure that a design is complete (as a drawing and its associated marginalia may remind us of still-undone parts of that design); and
- communicate the final design to the manufacturing specialists.

As a result of the many uses for sketches and drawings, there are several different kinds of drawings that are formally identified in the design process. One list of the kinds of design drawings is strongly evocative of mechanical product design:

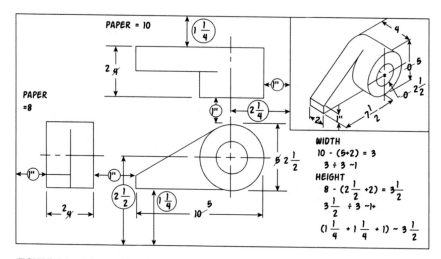

FIGURE 8.2   A *layout drawing* that has been drawn to scale, does not show tolerances, and is certainly subject to change as the design process continues. Adapted from (Boyer et al. 1991).

- *Layout drawings* are working drawings that show the major parts or components of a device and their relationship (see Figure 8.2). They are usually drawn to scale, do not show tolerances (see below), and are subject to change as the design process continues.

- *Detail drawings* show the individual parts or components of a device and their relationship (see Figure 8.3). These drawings must show tolerances, and must indicate materials and any special processing requirements. Detail drawings are drawn in conformance with existing standards (discussed below), and are changed only when a formal "change order" provides authorization.

- *Assembly drawings* show how the individual parts or components of a device fit together. An *exploded view* is commonly used to show such "fit" relationships (see Figure 8.4). Components are identified by a part number or an entry on an attached *bill of materials*, and they may include detail drawings if the major views cannot show all of the required information.

In describing the three principal kinds of mechanical design drawings, we have used some technical terms that need definition. First, drawings show *tolerances* when they define the permissible ranges of variation in critical or sensitive dimensions. As a practical matter, it is literally impossible to make any two objects to be *exactly* the same. They may appear to be the same because of the limits of our ability to distinguish differences at extremely small or fine resolution. However, when we are producing many copies of the same thing, we want them to function pretty much the same way, so we must limit as best we can any variation from their ideally designed form. That's why we impose tolerances that prescribe limits on the manufacturer and what he produces.

We have also noted the existence of drawing standards. *Standards* explicitly articulate the best current engineering practices in routine or common design situations. Thus,

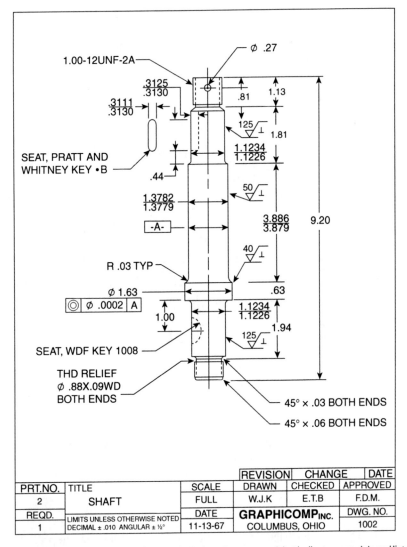

FIGURE 8.3   A *detail drawing* that includes tolerances and that indicates materials and lists special processing requirements. It was drawn in conformance with ASME drawing standards. Adapted from (Boyer et al. 1991).

standards indicate performance bars that must be met for drawings (e.g, ASME Y14.5M–1994 *Dimensions and Tolerancing*), for the fire safety of buildings built within the United States (e.g., the *Life Safety Code* of the National Fire Protection Association), for boilers (e.g., the ASME *Pressure Vessel Code*), and so on. The American National Standards Institute (ANSI) serves as a clearinghouse for the individual standards written by professional societies (e.g., ASME, IEEE) and associations (e.g., NFPA, AISC) that govern various phases of design. ANSI also serves as the national spokesman for the United States in working with other countries and groups of countries (e.g., the European Union) to ensure

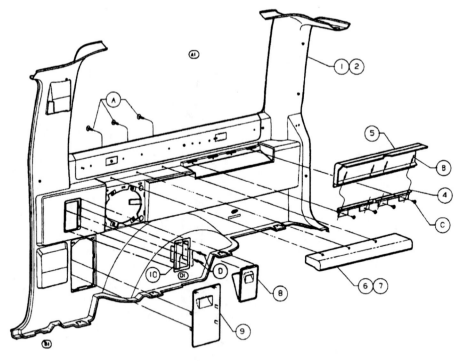

FIGURE 8.4   This *assembly drawing* uses an *exploded view* (remember Figure 4.3?) to show how some of the individual parts of an automobile fit together. Components are identified by a part number or an entry on an attached bill of materials (not shown here). Adapted from (Boyer et al. 1991).

compatibility and consistency wherever possible. A complete listing of U.S. product standards can be found in the *Product Standards Index*. The drawing standards specified in ASME Y14.5M–1994 will be described in some depth in Section 8.2.

8.1.2   Fabrication specifications

As we noted in Chapter 1, the endpoint of a successful design project is the set of plans that form the basis on which the designed artifact will be built. It is not enough to say that this set of plans, which we have identified as the fabrication specifications and which includes the final design drawings, must be clear, well organized, neat, and orderly. There are some very specific properties we want the fabrication specifications to have: namely, they should be *unambiguous* (i.e., the role and place of each and every component and part must be unmistakable), *complete* (i.e., comprehensive and entire in their scope), and *transparent* (i.e., readily understood by the manufacturer or fabricator).

We require fabrication specifications to have these characteristics because we want to make it possible for the designed artifact to be built by someone totally unconnected to the designer or the design process. Further, that artifact must perform just as the designer intended because the designers may not be around to catch errors or to make suggestions, and the maker cannot turn around to seek clarification or ask on-the-spot questions.

We have long passed the time when designers were also craftsmen who made what they designed. As a result, we can no longer allow designers much latitude or shorthand in specifying their design work because they are unlikely to be involved in the actual manufacture of the design result.

Fabrication specifications are normally proposed and written in the detailed design stage (cf. Chart 4 in Chapter 2). Since our primary focus is conceptual design, we will not discuss fabrication specifications in depth. However, there are some aspects that are worth anticipating even early in the design process. One is that many of the components and parts that will be specified are likely to be purchased from vendors, such as automobile springs, O-rings, DRAM chips, and so on. This means that a great deal of detailed, disciplinary knowledge comes into play. This detailed knowledge is often critically important to the lives of a design and its users. For example, many well known catastrophic failures have resulted from inappropriate parts being specified, including the Hyatt Regency walkway connections, the Challenger O-rings, and the roof bracing of the Hartford Coliseum. The devil really is in the details!

Of course, it sometimes is the manufacture or use of a device that exposes deficiencies that were not anticipated in the original design. That is, the way that designed objects are used and maintained produces results that were not foreseen. The F–104 fighter plane, for example, was called "the widowmaker" because test pilots found that they could do flight maneuvers that the plane's designers did not anticipate (and that they also didn't think were appropriate, when they finally learned of these pilot maneuvers!). An American Airlines DC–10 crashed in 1979 because its owners did a maintenance procedure in a way that undermined the design of the engine's supporting structures and their connections to the plane's wings. How much fortune-telling ability must a designer have? How far into the future, and how well, must a designer foresee the uses and misuses to which her work will be put? There are clearly ethical and legal issues here, but for the moment our intent is only to convey the fact that design details, such as fabrication specifications, are really important.

Given that many parts and components can be bought from catalogs, while others are made anew, what sort of information must the designer include in a fabrication specification? Briefly, there are many kinds of requirements that can be specified in a fabrication specification, some of which are:

- the physical dimensions
- the kinds of materials to be used
- unusual assembly conditions (e.g, bridge construction scaffolding)
- operating conditions (in the anticipated use environment)
- operating parameters (defining the artifact's response and behavior)
- maintenance and lifecycle requirements
- reliability requirements
- packaging requirements
- shipping requirements
- external markings, especially usage and warning labels
- unusual or special needs (e.g., must use synthetic motor oil)

This relatively short list of the different kinds of issues that must be addressed in a fabrication specification makes the point about our requirements for the properties of such a specification. The specification of the kind of spring action we see in a nail clipper might not seem a big deal, but the springs in the landing structure of a commercial aircraft had better be specified very carefully!

One final note here. In the same way that there are different ways to write design requirements (cf. Section 1.2 and Chapter 5), we can anticipate different ways of writing fabrication specifications. When we specify a particular part and its number in a vendor's catalog, we are writing a *prescriptive* fabrication specification; when we specify a class of devices that do certain things, we are presenting a *procedural* fabrication specification; and when we leave it up to a supplier or the fabricator to insert something that achieves a certain function to a specified level, we are setting forth a *performance* fabrication specification.

8.1.3   Philosophical notes on specifications and drawings and pictures

Since there are so many standards that define practices in so many different engineering disciplines and domains, and since these are less likely to come into play in conceptual design, we close our discussion of design drawings and fabrication specifications with a few philosophical notes.

First, different engineering disciplines use different approaches that arise in large part because of the different ways in which the disciplines have grown and evolved, and they continue because of the various needs of each discipline. In mechanical design, for example, in order to make a complex piece that has a large number of components that fit together under extremely tight tolerances, there is no way to complete that design other than by constructing the sequence of drawings described earlier. There is no matching topological equivalent, that is, we cannot usually specify a mechanical device well enough to make it by drawing springs, masses, dampers (or dashpots), pistons, etc. We have to draw explicit depictions of the actual devices. In circuit design, on the other hand, both practice and technology have merged to the point that a circuit designer may be finished when she has drawn a circuit diagram, the analogy of the spring-mass-damper sketch. We won't discuss the many reasons that practices differ so much among the engineering disciplines, or the various practices themselves. Nevertheless, it is important that designers be aware that while there are habits and styles of thought that are common to the design enterprise, there are practices and standards that are unique to each discipline, and it is the designer's responsibility to learn and use them wisely.

We also want to reinforce the theme that some external pictorial representation, in whatever medium, is absolutely essential for the successful completion of all but the most trivial of designs. Think of how often we pick up a pencil or a piece of chalk to sketch something as we explain it, whether to other designers, students, teachers, and so on. Perhaps this occurs more often in mechanical or structural design because the corresponding artifacts quite often have forms and topologies that make their functions rather evident. Think, for example, of such mechanical devices as gears, levers, and pulleys. Think, too, of beams, columns, arches, and dams. This evocation of function through form is not always clear. Sometimes we use more abstract drawings to show functional verisimilitude without the detail of sketches that are based on physical forms. Three examples of this sort of drawing abstraction reflect the kinds of discipline-dependent differences we discussed above: (1) the use of circuit diagrams to represent electronic devices; (2) the use of flowcharts to represent chemical-engineering-process plant designs, and (3) the use of block diagrams (and their corresponding algebras) to represent control systems. These pictures and charts and diagrams, with all of the different levels of abstraction we have seen, only serve to extend our limited abilities, as humans, to flesh out complicated pictures that exist solely within our minds.

Perhaps this is no more than a reflection of a more accurate translation of a favorite Chinese proverb, "One showing is worth a hundred sayings." It may also reflect a German proverb, "The eyes believe themselves; the ears believe other people." In fact, a good sketch or rendering can be very persuasive, especially when a design concept is new or controversial. Drawings serve as excellent means of grouping information because their nature allows us (on a pad, at the board, and soon in CADD programs) to put additional information about an object in an area adjacent to its "home" in a drawing. This can be done for the design of a complex object as a whole, or on a more localized, part-by-part basis. Again, drawings and diagrams are very effective at making geometrical and topological information very explicit. However, we do have to remember that drawings and pictures are limited in their ability to express the ordering of information, either in a chain of logic or in time.

Our final observation in this regard is that we have not made any reference to photographic images. Certainly, photos have much of the content and impact that we ascribed to other graphical descriptions, but they are not widely used in engineering design. One possible exception is the use of optical lithographic techniques to lay out very large scale integrated (VLSI) circuits, wherein a photography-like process is used. It is also the case that we are increasingly collecting data by photographic means (e.g., geographic data obtained from satellites). With computer-based scanning and enhancement techniques, we should expect that design information will be represented and used in this way. One sign of this trend is the increasing interest in geographic information systems (GIS), which are highly specialized database systems designed to manage and display information referenced to global geographic coordinates. It is easy to envision that satellite photos will be used together with GIS and other computer-based design tools in design projects involving large distances and spaces (e.g., hazardous-waste disposal sites and urban transportation systems). Thus, we should not forget photography as a form of graphical representation of design knowledge, alongside sketches, drawings, and diagrams.

8.2 GEOMETRIC DIMENSIONING AND TOLERANCING

We will now further discuss the requirements for a specific type of technical drawing described in the previous section — the detail drawing. This type of drawing is used to communicate the details of your design to the manufacturer or machinist. As such, it must contain as much information as possible while still maintaining clarity in meaning (and not becoming too cluttered!). Engineers have developed a system of standard symbols and conventions to meet this goal — these standards and conventions are described in detail in the following sections.

We begin and motivate our discussion of geometric dimensioning and tolerancing (GD&T) by imagining the following. Suppose you were designing a device, say a three-hole punch (see Exercise 8.1) or a simple desk lamp or a hammer, with the goal of having someone fabricate or make your design. What would you need to put down on paper, in both words and pictures, for you to be sure that the maker would fabricate exactly what you want? If you set down that description on paper and handed it to a friend or colleague, would she know exactly what you intend and want? This imaginary exercise is far more difficult than it sounds. In fact, imagine even a simpler version, akin to the way we introduced design problems. Supposed you had said to someone, "Please join this piece of metal to that piece of wood." Is that enough of a description of what you mean, for example, if you're attaching steel rails to wooden railroad ties, or if you're designing a clock to be housed in an elegant piece of grained maple?

The point is that as engineers we require common standards by which we can communicate our designs to those makers or machinists or fabricators who will actually make or build them. There are certain essential components that every drawing must have to ensure that it is interpreted as it is intended. These components include:

- standard symbols to indicate particular items;
- clear lettering;
- standard drawing views;
- clear, steady lines;
- appropriate notes, including specifications of materials;
- a title on the drawing;
- the designer's initials and the date it was drawn;
- dimensions and units; and
- permissible variations, or tolerances.

Geometric dimensioning and tolerancing (GD&T) is a standard approach that was designed to meet these requirements. GD&T refers to the ASME Y14.5M–1994 *Dimensions and Tolerancing* standards that establish a common language in which we create engineering drawings. It is important for us as engineers to learn this language so that we can communicate our designs appropriately. Figure 8.5 displays a technical drawing that conforms to the GD&T system and ASME standards. This drawing shows a screwdriver handle that all Harvey Mudd engineering students are required to manufacture as part of a second-year course, E8: *Design Representation and Realization*. In this chapter we will address and define the various symbols used in this drawing. Of particular note at this point are the descriptive title, the date on the drawing, and the designer's initials. In addition, the engineer has included a note to specify the material as cast acrylic and that the finish should be polished. Dimensions and tolerances, specified in accord with GD&T standards and rules, are carefully detailed on the drawing. These GD&T rules and guidelines are described below.

8.2.1 Dimensioning

In order to understand the geometric dimensioning and tolerancing system, we must first understand the appropriate method for *dimensioning* or putting dimensions on a drawing.

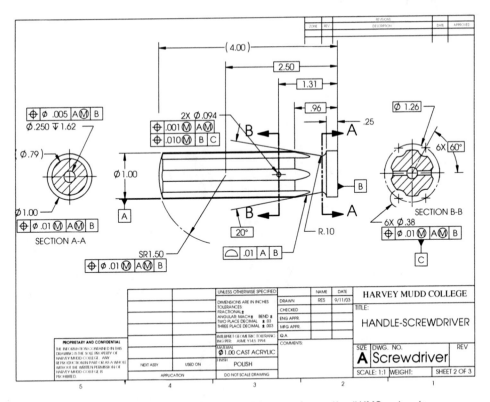

FIGURE 8.5 A detail drawing of the handle of a screwdriver manufactured by all HMC engineering students. This drawing uses a set of symbols and the particular placement of these symbols conveys information about the size and location of certain features of the screwdriver handle. In addition, the drawing contains information about the materials to be used, the finish of the part, the person who created it, and the date it was created.

Dimension placement, symbols, and conventions are all important to the common language for engineers and machinists. The following concepts are essential for understanding dimensioning of technical drawings.

8.2.1.1 *Orthographic views* Most technical drawings show orthographic views of the object being represented. Orthographic views are drawings based on the projection of the object onto a plane. The best way to visualize an orthographic drawing is to imagine a box around the object, with a projection of the object onto each surface of the box. The box is then unfolded to give rise to the six primary views of the orthographic drawing: top, front and bottom views; and right side, left side and rear views (Figure 8.6). It should be noted that this particular type of orthographic drawing uses *third-angle projection* in which the drawing is derived from an image projected *onto a plane between the observer and the object*, so that the order is observer, projected view, object. Or, said somewhat differently, in third-angle projection the image is projected onto a plane *in front of* the object.

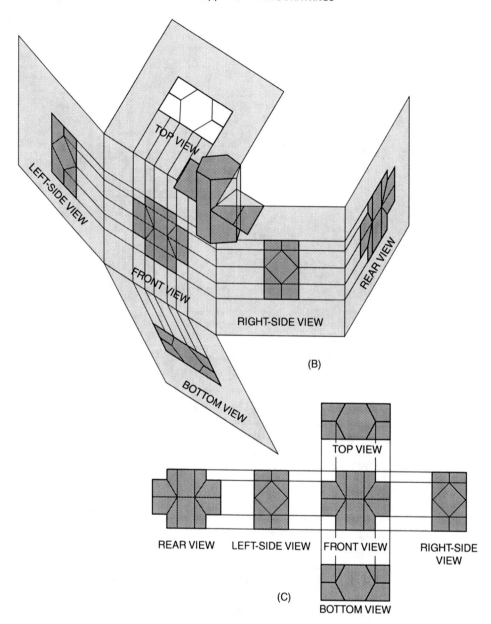

FIGURE 8.6 The six *orthographic* views of an object. The orthographic views are created by projecting the object onto a plane. This can be visualized by imagining a box around an object with a projection of the object onto each face; the unfolding of the box leads to the six views: front, top, bottom, right side, left side and back views. Note that in practice, we often need to use only the front, top, and right side views to fully describe an object, as the others are redundant. (From (Goetsch, Nelson and Chalk, 2000)).

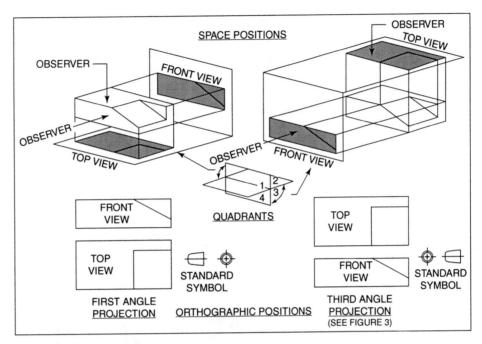

FIGURE 8.7   First- and third-angle projections. The difference between these two orthographic projections lies in the location of the plane on which the object is projected. In first-angle projection, the object is projected onto a plane *behind* it. In third-angle projection, the object is projected onto a plane *in front of* it. Note the different symbols used to represent each drawing type. (Reprinted from ASME Y14.3–1975 and ASME Y14.5–1994 (R2004), by permission of the American Society of Mechanical Engineers. All rights reserved.)

In Japan and some European countries, a different type of orthographic view is used: In *first-angle projection* the drawing is derived from an image projected *onto a plane behind the object*, so that the order is observer, object, projected view. That is, in first-angle projection the image is projected onto a plane *behind* the object. The two types of orthographic views can lead to very different drawings and it is important to know which system is being used. Figure 8.7 shows the views in first- and third-angle projections, as well as the symbols used to denote which system is depicted. It is important to note that all six views of the orthographic projection are not always required. We can often fully define an object with front, top, and right side views (in third-angle projection), or front, bottom, and right side (in first-angle projection). In some cases we need only the front and top views. It is important to note that the orthographic views are to be laid out as one drawing, that is, the three (or two) views must line up as they are laid out in the projection, with features aligning in the views.

Choosing an appropriate front view for an orthographic drawing is essential for ensuring its correct interpretation. It is much easier to figure out what is being represented given the right front view because that front view is seen first and it represents the most basic and characteristic profile of the object being drawn. In addition, the front view

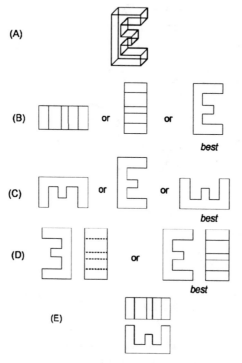

FIGURE 8.8 Choosing a front view: (A) isometric view of object to be drawn; (B) front view should be chosen to show the most informative profile of the object; (C) the front view should be chosen so that the view shows the most stable version of the object; (D) a front view should be chosen to minimize the hidden lines of an object in other views; and (E) the best choice of views for this object: front and top views.

should be stable (i.e., heavy on the bottom), and should have as few hidden lines as possible. Consider a block letter "E", as in Figure 8.8. Several poor choices for front views of this object are shown, as well as several "best" front views.

8.2.1.2 Metric vs. inch dimensioning
Metric and inch dimensions (and that *is* what they're called) are specified differently in the ASME standard, which enables us to tell at first glance which system of units is used on a drawing. American (inch or in) dimensions are specified to have no zero before the decimal point (e.g., .5 in). Further, the dimension must contain the same number of decimal places as the tolerance for that dimension. For example, if the tolerance for a given dimension is .01 in, the dimension must be .50 in. Metric dimensions include a zero before the decimal point (e.g., 0.5 mm). A metric dimension does not need to match the number of decimal places with the tolerance, and no decimal point or zero is included if the dimension is a whole number.

8.2.1.3 Line types
Technical drawings use several different types of lines. The weight and style of these lines, as well as their placement on the drawings, are specified in the ASME standard. Most CADD packages include settings for the ASME standard, so that they will automatically place the lines correctly if they are initially set correctly. *Extension lines* come out from a part and leave a visible gap between the part and the line. Examples of extension lines can be seen in Figure 8.5; these are the vertical lines up from the screwdriver handle. *Dimension lines* are typically broken for the numbers and are placed at least 10 mm apart from the object on the drawing. Subsequent dimension lines

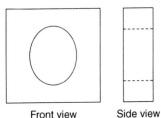

Front view   Side view

FIGURE 8.9 Hidden lines are indicated by dashed lines. They are used to represent a feature in a view where that feature is not explicitly seen. For example, the hole shown in the front view is located by hidden lines in the side view.

are at least 6 mm apart. Examples of these lines can also be seen in Figure 8.5. *Leader lines* are used to indicate surfaces and hole diameters. They should: be at an angle between 30 and 60 degrees, point toward the center of a hole, and include only one dimension per leader line. The left side view of the screwdriver print shows a leader line pointing to the outer diameter of the piece; note that the arrow points directly at the center of the diameter. *Hidden lines* are dashed lines ($- - - - - - - -$) that indicate the presence of a feature that is seen in another view. Hidden lines indicate the presence of a hole in the object shown in Figure 8.9. The front view shows the hole, the side view uses hidden lines to indicate where the hole is located. *Center lines* are used to indicate a cylinder and are represented by the type of dashed line shown in Figure 8.10. The presence of this type of line alone indicates a cylindrical feature; the side view showing the part is a cylinder is not needed.

8.2.1.4 *Orientation, spacing and placement of dimensions* The most common practice is to orient all dimensions so that they can be read when the drawing is held horizontally. It is also acceptable to use an aligned system in which dimensions are oriented either vertically so that they can be read from the right or horizontally so that they can be read from the bottom. As mentioned before, the minimum spacing between adjacent dimensions is specified as 6 mm. The placement of dimensions is also important. Dimensions should be stacked with the shortest dimensions placed closest to the object and longer dimensions beyond them. This system avoids crossing extension lines, thereby minimizing confusion. Dimensions should also be staggered to make it easier to read them. In addition, in orthographic drawing views, the dimensions should be placed *between* the drawing views, that is, between the front and top views, and between the front and right side views. The overall size, length, height and depth should be specified.

8.2.1.5 *Size dimensions and location dimensions* It is important to distinguish between size dimensions and location dimensions. *Size dimensions* define the

Front view   Side view

FIGURE 8.10 Center lines are indicated by a different type of dashed line and describe cylindrical features.

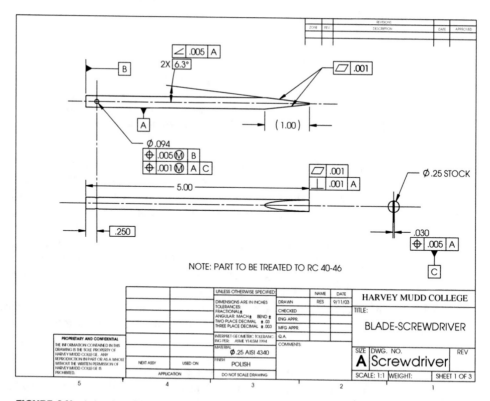

FIGURE 8.11 A drawing of a screwdriver blade indicates all the types of dimensions: *basic* (indicated by boxing of the numbers); *reference* (indicated by parentheses); *stock* (indicated by STOCK following the dimension); and *size/location dimensions* (for example, the overall blade length, 5.00 in).

size of features: overall height, length, thickness, diameter of a hole, size of a slot, etc. *Location dimensions* specify where a feature is located with respect to other features or the edge of an object. Location dimensions define the center of a hole or the location of a slot, for example, with respect to the edge of a part or with respect to another feature. The general rule of thumb is to dimension size dimensions first, then do location dimensions. Remember that both size and location dimensions will have tolerances associated with them, a concept that we will revisit later in this chapter.

In addition to size and location dimensions, there are three other dimension types that are important: basic dimensions, reference dimensions, and stock dimensions. Figure 8.11 is a drawing of a screwdriver blade, from the same screwdriver whose handle appears in Figure 8.5, and it illustrates all of these types of dimensions. (And by the way, are the dimensions in this drawing in millimeters or inches?) First, the overall blade length (5.00 in) in Figure 8.11 is an example of the *size* dimension just above. The boxed numbers in the drawing are *basic dimensions* (e.g., the .250 dimension from the blade end to the hole on the left side of the drawing). Basic dimensions define the basis for permissible variation, or tolerance, in the geometric dimensioning and tolerancing system. In other words, they define the theoretically exact point from the end of the blade from which to measure the variation in the location of the hole. We will revisit this concept in the next section on tolerancing. *Reference dimensions* are indicated in parentheses, for example, the (1.00) dimension on the screwdriver blade in the top view. A reference dimension is a point of information for the machinist and is not a requirement. It means that if the part has been produced correctly, the blade length should be around one inch. The last type of dimension is a *stock dimension*, and it is indicated by writing .25 STOCK on the drawing. This indicates that the material used for the part comes from the manufacturer with a specified size and associated tolerance; no further tolerance specification is required.

Every specified dimension requires an associated tolerance in a technical drawing except basic, reference, and stock dimensions. This makes sense when we realize the purpose of these types of dimensions. One more important note should be made: Any number on a technical drawing that does not have a tolerance directly associated with it still has a specified tolerance on the drawing. These tolerances are specified in the title block of a technical drawing and are called *block tolerances*. In Figure 8.11 the block tolerances can be seen to be .XX $\pm$ .03. The tolerance is determined by the number of decimal places in the dimension. This means that the 5.00 overall blade length has a tolerance of $\pm$ .03 inches.

8.2.2 Some best practices of dimensioning

Each of the concepts and dimension types outlined in Section 8.2.1 can be integrated into a set of guidelines for dimensioning a technical drawing. Three of the most important rules for dimensioning are:

- Dimension the *size dimensions first*, then do the location dimensions.
- *Do not double dimension.* In an orthographic drawing, it is not necessary to specify the same dimension twice. For example, specifying the depth of an object on the right and top views leads to unnecessary clutter in the drawing. It is not good drawing practice.
- *Do not dimension to hidden lines.* Dimension a feature where it is visible. A hole, for example, should be dimensioned in the view where it is visible. This is good practice that leads to clearer technical drawings.

8.2.3 Geometric tolerancing

Now that we have covered some basic dimensioning symbols and rules, we turn to geometric tolerancing. A *tolerance* is the permissible variation of a part. Tolerances are applied to all size and location dimensions. Tolerances are required because we, as engineers, need to know how much a part can vary from its specifications before it no longer functions as intended. Defining tolerances requires that we know and understand the function of a given part. It is good practice to specify tolerances only as tightly as we need because parts become much more expensive to manufacture as their tolerances become smaller. Figure 8.12 shows the relative cost of increased tolerances. Here the y-axis tracks the percentage of increase in cost of making the hole, and the x-axis shows both the size and location tolerances on the hole. (Location tolerances will be described below.) This figure not only gives us an idea of the increased cost as the tolerances become tighter, but

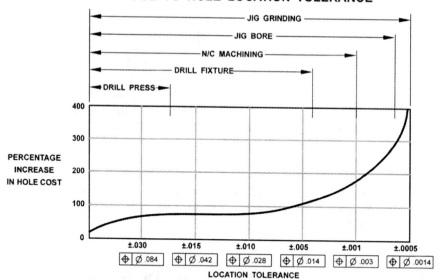

FIGURE 8.12 Relative cost of manufacturing as hole location tolerance gets smaller. The cost goes up significantly as smaller tolerances are prescribed. In addition, special equipment is required to meet tight tolerances. (Reprinted by permission of Technical Documentation Consultants of Arizona, Inc.)

also tells us what type of machinery is required to make such a hole. It is not surprising that the more precise a hole it can make, the more expensive the equipment!

All dimensions require a tolerance (except for the basic, reference or stock dimensions described above). On a drawing, there are several places to look for tolerance specifications:

- associated with a dimension ($\pm$),
- in a feature control frame (which we describe below)
- in a drawing note, or
- in the block tolerance as the default if no other tolerance is applied.

It is possible to tolerance every dimension with a plus or minus tolerance, but the geometric dimensioning and tolerancing system provides more leeway for each part, which in turn leads to cost savings. The geometric tolerancing system takes into account not only variations in *size* of an object, but also permissible variations on the *position*, *form* and *orientation* of features. We now describe some of the tolerancing components of the GD&T system.

8.2.3.1 The 14 geometric tolerances

There are 14 characteristics specified in the ASME Y14.5M–1994 standard that can vary, and therefore have an associated

	TYPE OF TOLERANCE	CHARACTERISTIC	SYMBOL	SEE:
FOR INDIVIDUAL FEATURES	FORM	STRAIGHTNESS	——	6.4.1
		FLATNESS	▱	6.4.2
		CIRCULARITY (ROUNDNESS)	○	6.4.3
		CYLINDRICITY	⌀	6.4.4
FOR INDIVIDUAL OR RELATED FEATURES	PROFILE	PROFILE OF A LINE	⌒	6.5.2 (b)
		PROFILE OF A SURFACE	⌓	6.5.2 (a)
FOR RELATED FEATURES	ORIENTATION	ANGULARITY	∠	6.6.2
		PERPENDICULARITY	⊥	6.6.4
		PARALLESIM	//	6.6.3
	LOCATION	POSITION	⊕	5.2
		CONDENTRICITY	◎	5.11.3
		SYMMETRY	≡	5.13
	RUNOUT	CIRCULAR RUNOUT	↗ •	6.7.1.21
		TOTAL RUNOUT	↗↗ •	6.7.1.2.2
• ARROWHEADS MAY BE FILLED OR NOT FILLED				3.3.1

FIGURE 8.13 The 14 geometric tolerances and their symbols. (Reprinted from ASME Y14.3–1975 and ASME Y14.5–1994 (R2004), by permission of the American Society of Mechanical Engineers. All rights reserved.)

tolerance (Figure 8.13). For example, we can specify how much a surface can vary in flatness or how much variation is permissible in the location of a hole. These 14 characteristics are categorized into five groups: form, profile, orientation, location and runout. These groups are somewhat hierarchical. For example, a position tolerance is a refinement of an orientation tolerance, which is a refinement of a form tolerance, which is a refinement of the size tolerance on a feature. Thus, if a rectangular piece is .500 ± .004 inches in height, the minimum height is .496 and the maximum is .504, simply based upon the size dimensions. If each end of the part was made at one of these extremes, and the part would be within the size tolerance, the maximum out-of-flatness that the top surface could be is .008. Therefore, if a flatness tolerance is to be applied to this part, it must be *less than* .008 inches for it to make sense.

Form tolerances apply to individual features, for example, a surface in the case of straightness or flatness. All other tolerances apply to related features. For example, orientation and location tolerances specify permissible variation of a given feature with respect to a reference frame. Therefore, these tolerances require specification of a reference frame in order for them to be meaningful. The reference frames are defined by datums, which we will soon discuss below.

A full discussion of all of the 14 geometric tolerances is beyond our scope (see the notes in Section 8.4 for further reading.). Therefore, we will focus specifically on position tolerances to show how geometric tolerances are applied.

8.2.3.2 Feature control frames

Feature control frames are devices used to specify the particular geometric tolerance on the technical drawing. We have seen them in the drawings presented earlier. The feature control frame is either: *attached to a surface* via a leader line (e.g., the flatness tolerance associated with the screwdriver blade in the top view in Figure 8.11); *placed off an extension line from a surface* (e.g., the flatness and perpendicularity tolerances associated with the tip of the blade in the front view in Figure 8.11); or *associated with the size dimension of a particular feature* (e.g., the position tolerances on the hole in the top view of Figure 8.11).

The feature control frame is broken down into the three components depicted in Figure 8.14. We will define the parts from left to right. The first box (1) is for the geometric characteristic symbol, which tells us what tolerance is being specified. In this case, it is position.

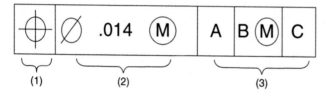

FIGURE 8.14 A feature control frame for an object that specifies the position of that object to a cylindrical tolerance zone of 0.014 in with respect to a reference frame determined by datums A, B, and C.

The second box (2) contains the actual permitted variation, or tolerance with some optional modifiers. In this particular case, the tolerance is .014 in. The diameter symbol in front of the number indicates that the tolerance zone shape is cylindrical. Figure 8.15

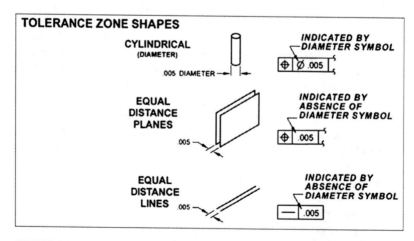

FIGURE 8.15 The tolerance zone shape depends on the type of tolerance being specified and the presence or absence of a diameter symbol before the tolerance in the feature control frame. (Reprinted by permission of Technical Documentation Consultants of Arizona, Inc.)

indicates the difference between the presence and absence of a diameter symbol in defining a tolerance zone shape. A position tolerance with a diameter symbol means that the position of the item being controlled must fit inside a cylindrical tolerance zone of a diameter specified by the tolerance. Lack of a diameter symbol indicates that the position must fall between two parallel planes; the distance between those planes is defined by the tolerance specified. The tolerance material condition modifier appears after the tolerance itself and specifies the conditions under which this tolerance applies. Material condition modifiers will be described fully below.

The last set of boxes (3) contain the datum references. These define the frame of reference from which the tolerance is measured. The datum references appear in a specific order that indicates their relative importance. Note that datum references may also have material condition modifiers (described in the next section).

The feature control frame described in Figure 8.14 can be read as follows, assuming it is associated with the size dimension for a hole: The hole has a permissible variation in *position* such that the center of the hole must fit within a *cylindrical tolerance zone* that is *.014 inches* in diameter when the hole is at *maximum material condition*, with respect to *datums A, then B, then C.*

8.2.3.3 *Material condition modifiers* Three *material condition modifiers* specify the state of the feature when the tolerance is applied. These are: maximum material condition, least material condition; and regardless of feature size. It is important to know under what conditions the part is toleranced, because use of these modifiers can lead to substantial cost savings.

Maximum material condition (MMC) is the condition in which a feature of size contains the maximum amount of material (weighs the most) within its size tolerance. MMC is indicated by the letter M enclosed in a circle, as in Figure 8.14. For a hole, this means the minimum diameter specified in the size tolerance. For a cylindrical shaft, this means the maximum diameter specified by the size tolerance. For example, a hole specified as .500 ± .005 inches in diameter would have an MMC size of .495 inches in diameter.

Least material condition (LMC), indicated by the letter L enclosed in a circle, is the condition in which a feature of size contains the least amount of material (weighs the least) within its size tolerance. The LMC size of the hole is the largest hole within the size tolerance, the LMC size of a shaft is the smallest shaft within the size tolerance. The same hole described above would have an LMC size of .505 inches.

Regardless of feature size (RFS) means just that: The tolerance is applied regardless of the size of the produced part. It is indicated by the absence of either of the MMC or LMC modifiers.

Why would we want to use these modifiers? These modifiers are extremely useful in that they can reduce the cost of manufacturing a part substantially. They take into account the fact that if a part is produced at the extremes of its permitted size variation, there is potential for more "wiggle room" in the placement of that part. For example, if a hole is produced at its largest possible size, its position can vary more than it would if it was produced at its smallest possible size — and still match up with a mating part. The material condition modifiers enable us to have a *maximum interchangeability of parts.*

This is important if we are trying to manufacture thousands of the same pieces, and we expect them all to fit together without specifying extremely tight tolerances.

If a maximum material condition modifier is placed in a feature control frame associated with the tolerance, this means that the specified tolerance applies *only at the MMC size of the feature*. Figure 8.16 shows a part with two holes controlled by a position tolerance. Since this tolerance is specified at MMC, it means that when the hole is produced at its MMC size (smallest hole, .514 in this example), the center of the hole must fit within a cylindrical tolerance zone of diameter .014. If however, the hole is produced larger than the MMC size, additional bonus tolerance is added to the allowable position variation. The bonus tolerance added is the difference between the MMC size of the hole and the actual size of the hole. This bonus tolerance is to account for the fact that a larger hole can vary more and still line up with a mating part. Essentially, it takes into account the additive effects of the variation in size and variation in position.

We illustrate the use of a least material condition modifier in Figure 8.17. The same example is shown, this time with the tolerance specified at LMC instead of MMC. In this case, when the hole is produced at its largest size, .520, the tolerance is specified as .014. As the hole gets smaller, bonus tolerance is added to allow the hole to vary more in position. The LMC modifier is used less than the MMC modifier but is useful when it is desirable that the position of a larger hole needs to be more tightly controlled, such as in the case where it is placed near the edge of a part.

RFS does not take advantage of size variations in the feature and specifies the same tolerance for all cases. This condition is assumed if no material condition modifier is used on the drawing, so we should be careful not to omit these symbols! RFS should be used only if the requirements are very strict, since manufacturing parts to RFS is much more expensive.

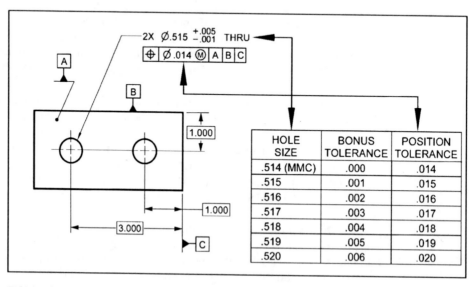

HOLE SIZE	BONUS TOLERANCE	POSITION TOLERANCE
.514 (MMC)	.000	.014
.515	.001	.015
.516	.002	.016
.517	.003	.017
.518	.004	.018
.519	.005	.019
.520	.006	.020

FIGURE 8.16 Maximum material condition modifier allows additional "bonus tolerance" if part is produced at a size other than MMC. In this example, as the produced hole gets larger, the allowable variation in its position also gets larger. (Reprinted by permission of Technical Documentation Consultants of Arizona, Inc.)

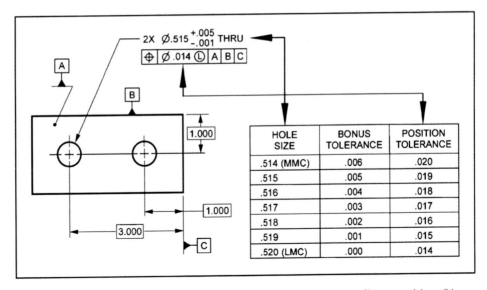

HOLE SIZE	BONUS TOLERANCE	POSITION TOLERANCE
.514 (MMC)	.006	.020
.515	.005	.019
.516	.004	.018
.517	.003	.017
.518	.002	.016
.519	.001	.015
.520 (LMC)	.000	.014

FIGURE 8.17   Least material condition modifier applies the tolerance in the case of least material condition and allows bonus tolerance for parts produced (in the case of a hole) smaller than the LMC size. This modifier is used much less frequently than the MMC modifier and can help to control hole position if the hole is located close to the edge of a part. (Reprinted by permission of Technical Documentation Consultants of Arizona, Inc.)

One final note on material condition modifiers. These conditions may only be applied to *features of size*. A feature of size can be a cylinder, a slot, or a hole, for example. Material condition modifiers may not be applied to surfaces, as there is no size associated with a surface. It wouldn't make sense, therefore, to have a material condition modifier associated with a tolerance in a flatness feature control frame applied to a surface.

8.2.3.4 *Datums*   In the GD&T system datums form the reference frame from which to locate tolerance zones specified in the feature control frames. A few definitions are in order before we proceed. A *datum symbol* is used to define the datum on the drawing. A datum symbol looks like this:

Any letter, except I and Q, may be used as a datum symbol. We must be careful about where we place the datum symbol because we want to be sure that the correct feature is specified as the datum. To specify a surface as a datum, the datum symbol may be placed off of an extension line or may be attached directly to the surface itself (Figure 8.18). To specify a feature of size as a datum, the datum symbol may be placed in line with the dimension line for the feature, or it may be attached directly to a cylindrical feature in the view where it appears as a cylinder. Datum symbols may also be attached to the feature control frame associated with the feature of size (see Figure 8.19).

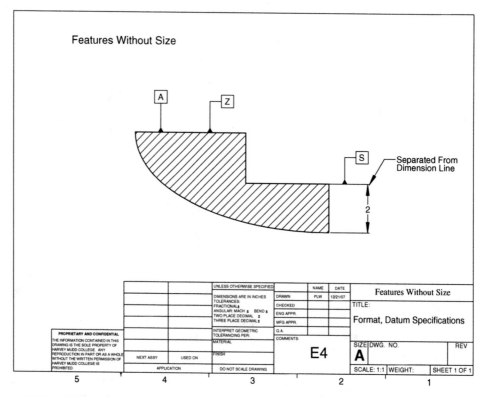

FIGURE 8.18 Specifying surfaces as datum features. The datum symbol may be placed directly on the surface (datums A, Z) or associated with a leader line pointed at the surface, or attached to the extension line from a surface (datum S). It must be separated from the dimension lines as shown. (Drawing courtesy of Joseph A. King.)

A *datum feature* is what the datum symbol is applied to, the actual feature on the part. A *datum simulator* is the manufacturing and inspection tooling used to simulate the datum during production. Simulators can be a precise surface or precise tooling in which to place the part. Locations of holes or other features are then determined from the datum simulator instead of the irregular surface or edge of the part itself. The datum simulator for a surface is a surface that the part may be placed upon, and is typically made from granite due to its smooth surface free of irregularities. The simulator for a feature of size is usually a chuck or a vise that clamps around an external feature.

So how do we choose the datums for a particular part? Considerations should include the function of the part, the manufacturing processes to be employed, inspection processes that may be used, and the part's relationship to other parts. For a rectangular object, three datum references must be chosen to refer to three perpendicular planes (see Figure 8.20). The *primary datum* (A) is listed first in the feature control frame and must make three points of contact on that surface. If our rectangular part is going to fit flush with another part, the largest contacting surface should be chosen as the primary datum. The primary datum creates a flat surface. The *secondary datum* (B) is usually the longest

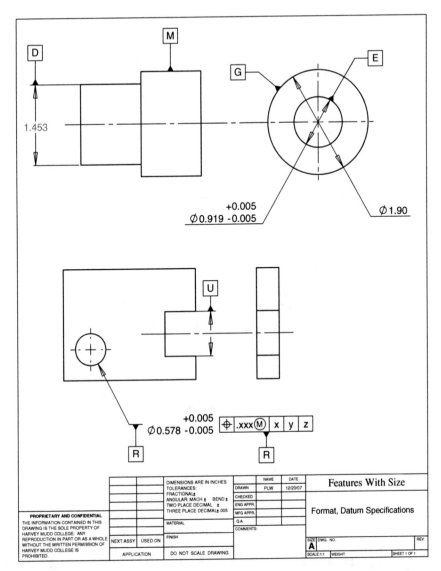

FIGURE 8.19 Specifying features of size (such as holes or shafts) as datum features. The datum symbol may be placed in line with the size dimension of the feature (datums U, D), associated with the feature control frame (datum R), or attached directly to a cylindrical feature in the view where it appears as a cylinder (datums E, G). (Drawing courtesy of Joseph A. King.)

side, or a side in contact with a mating part and requires two points of contact. This datum creates alignment and stability. The *tertiary datum* (C) is, then, the other side of the part. This datum requires one point of contact and prevents the part from sliding on datum B. In order to measure the accuracy of a part for testing, or to machine a hole located with respect to these datums, the part must first be set down on datum A, slid over to make

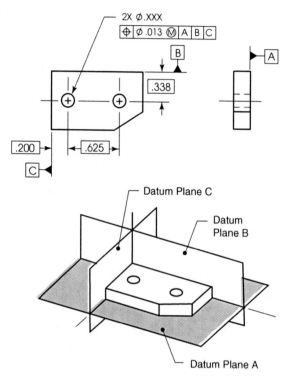

FIGURE 8.20   Specifying datums for a rectangular feature. The function of the part is important for specifying datums. The primary datum is usually chosen as the largest contacting surface. (Reprinted from ASME Y14.3–1975 and ASME Y14.5–1994 (R2004), by permission of the American Society of Mechanical Engineers. All rights reserved.)

contact with datum B, and then slid until it makes contact with datum C, while maintaining contact with datums A and B.

For a cylindrical object, two datum references are required (see Figure 8.21). One reference is the surface, the other is the axis determined by a particular feature of size. In Figure 8.21, the primary datum D is the bottom surface; it establishes a flat surface with three points of contact. The secondary datum E is established by the axis of the cylindrical part. This axis establishes two planes that bisect the axis. To measure or locate from this datum, the part must be contacted by a chuck at three points. To make this particular part, the cylinder would be placed on a precise surface and grabbed by a chuck to establish the axis, and then the holes would be located from there.

Many parts have large irregular surfaces that are not flat surfaces and not cylindrical parts. For these parts, it is impractical to define datums as we have described above. In these cases, it is permissible to identify datums using points, lines or areas instead of a whole surface. In these cases, it is permissible to identify datums using points, lines or areas instead of a whole surface. These points are called *datum targets* and specify where the workpiece contacts the tooling during manufacturing and inspection. A datum target is indicated by an "X" on the drawings, and the datum reference symbols are defined in circles. Since a whole surface is not in contact, the points of contact are

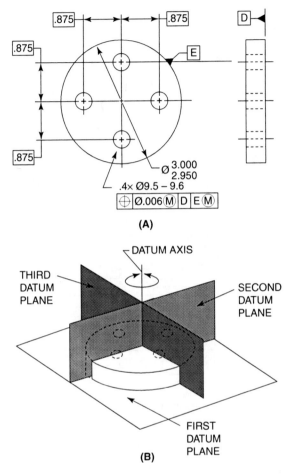

(A)

(B)

FIGURE 8.21   Specifying datums for a cylindrical feature. The primary datum is usually chosen as the flat surface to stabilize the part. The secondary datum is the axis described by the cylindrical feature. (Reprinted from ASME Y14.3–1975 and ASME Y14.5–1994 (R2004), by permission of the American Society of Mechanical Engineers. All rights reserved.)

typically numbered "A1," "A2," etc. Figure 8.22 shows a hammer handle from Harvey Mudd's E8 course with Xs marking the datum targets on this irregularly shaped handle surface. The profile of the surface is then permitted to vary with respect to those points.

One last note on datums and datum references. It is important to understand that not all types of tolerance specifications require a datum reference. Figure 8.14 lists all of the geometric tolerances. Note that the first set of tolerances are form tolerances and apply to *individual features*. The column on the far left in the figure distinguishes the tolerances that apply to individual features versus ones that apply to *related features*. For example, if a flatness tolerance is applied to a surface, that surface is specified to be flat but it is not

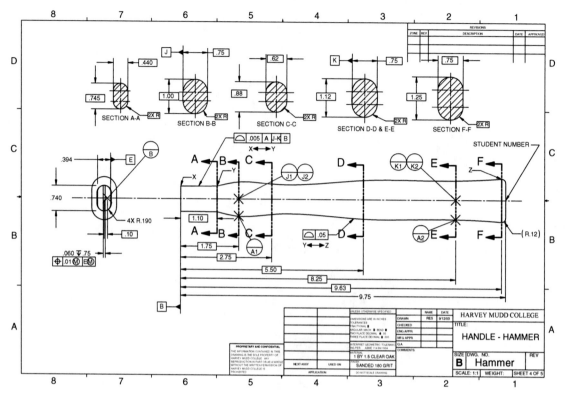

FIGURE 8.22 A drawing of the hammer handle. Note the use of datum targets (marked "X" on the drawing) instead of a datum feature.

with respect to any frame of reference. It would be inappropriate to specify a datum reference in this case. In contrast, a perpendicularity tolerance specifies that some feature be perpendicular to something; that something must be defined by one or more datum references.

8.2.3.5 Position tolerance

We have used position tolerance examples throughout the foregoing discussion, and we will now *put the pieces together* using the illustrative example shown in Figure 8.23. The part depicted in this drawing has a specified position tolerance on the hole. The specified tolerance defines a cylindrical tolerance zone (note the diameter symbol) .100 inches in diameter that extends through the part. The hole axis may be tilted, but it must fit within that tolerance zone. Since MMC is specified, this tolerance is required to be met at MMC only; that is, at the smallest hole size. As the hole gets larger, bonus tolerance is added, making the cylindrical tolerance zone for the hole axis larger as the hole gets larger. The theoretical center of the hole is located at specified distances from the datums; these specified distances are called out using basic dimensions (boxed). To make this part, the stock piece would be placed on a datum simulator surface (datum A), pushed up against datum simulator surface B, and slid along to make contact

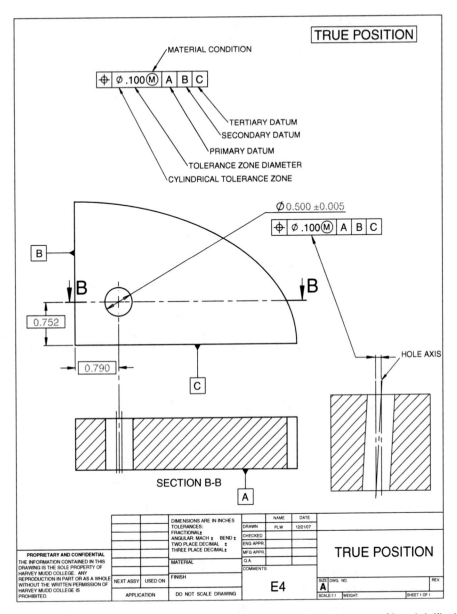

FIGURE 8.23   Putting it all together: Tolerancing the true position. (Drawing courtesy of Joseph A. King.)

with datum simulator surface C. The theoretical center of the hole would then be located from the datum simulators surfaces B and C using the basic dimensions on the drawing. Any drawing that has tolerances specified by a feature control frame will have basic dimensions defining distances between the datums and the theoretical position of the tolerance zone.

8.2.3.6 Fasteners How do we know how to specify position tolerance zones such that we can fasten two parts together? How do we ensure that fasteners will always fit? There are three types of fastener conditions:

- *Floating fasteners*: The fasteners pass through holes on two or more parts and are fastened with a nut on the other side. The fasteners do not need to come into contact with the part.

- *Fixed fasteners*: One of the two (or more) parts involved is tapped or press fit (fixed) and the other has a clearance hole. The fixed part fixes the location of the fastener.

- A *double fixed fastener*: Here both holes are fixed. This gives zero position tolerance at RFS and should be avoided due to expense.

How do we calculate the position tolerance of any given hole in two parts that are to be fastened together? For a floating fastener condition, we first determine the MMC size of the hole (smallest hole, H) and the MMC size of the fastener (largest fastener, F). The difference between these two numbers gives the amount of clearance available in the worst case scenario, when both fastener and hole are at MMC. The amount of tolerance in this case is simply this difference, that is, the tolerance $T = H - F$. For a floating fastener condition, this tolerance is applied to the position of the holes on both parts. For a fixed fastener condition, the tolerance is calculated in the same way, but now the tolerance must be distributed over the two parts. The rule of thumb is to give 60–70% of the allowable tolerance to the fixed/threaded part.

8.2.4 How do I know my part meets the specifications in my drawing?

All manufactured parts need to be evaluated to make sure that they are within the specifications. A *coordinate measurement machine* (CMM) is a device that can be programmed to examine a specific part for its adherence to the geometric tolerances specified on the drawings. Often companies will invest in one of these systems if they are manufacturing a large number of similar parts and need to know if each part meets the requirements. Figure 8.24 shows photographs of the hammer, described in the technical drawing in Figure 8.22, mounted on the CMM system at Harvey Mudd College. Note that the points at which the tooling makes contact with the hammer correspond to the datum target points we saw in the drawing. This system is used to grade student-machined tools at HMC, but is more widely used in industry for quality control of manufactured parts.

There is another, much cheaper, way of evaluating parts that fit together called *functional gaging*. As the name hints, this method evaluates the *function* of a given part, that is, will a part fit with its intended mating part? This approach further illustrates the power of the material condition modifiers. In order to understand functional gaging, we must first define another term: The *virtual condition* of a given feature is the combined effect of the size tolerance and the geometric tolerance on the part. If we imagine an external feature, such as a cylindrical shaft, the virtual condition is the MMC size of the shaft plus the geometric tolerance. It is the largest possible shaft with the largest possible variation in position, making it the worst-case scenario for the external part fitting into a mating part. For an internal feature, such as a hole, the virtual condition is also the

FIGURE 8.24 The hammer described by the drawing in Figure 8.22 on the coordinate measurement machine. Note the ruby tip used to measure the part and the precise granite surface of the machine. Note also that the tooling used to hold the hammer while it is being tested makes contact with the part at the datum target locations specified in the drawing.

worst-case scenario, the MMC size of the hole (smallest hole) minus the geometric tolerance. The virtual conditions of the two parts must match in order to ensure that two parts, specified on different drawings and manufactured to specifications, will always fit together. If the virtual condition of a hole in one part is matched to the virtual condition of a shaft in another part meant to fit together, this will ensure maximum interchangeability of parts. The result is powerful. It means that all parts that meet the drawing specifications will be interchangeable, that is, the two parts need not be made specifically to fit together. This clearly offers large cost savings in the manufacturing of parts.

Now, back to functional gaging. Virtual condition matching allows us to use this much cheaper way to evaluate manufactured parts. In our hole/shaft example above, we could simply manufacture one part with the shaft made at virtual condition and then use that part to evaluate the potentially hundreds of matching parts. The manufactured part with the hole to be tested would simply be placed over the shaft at virtual condition: If the hole goes over the shaft, the part is good; if not, the part is thrown out.

8.3 NOTES

Section 8.1: Much of the discussion of drawing is drawn from (Ullman, Wood, and Craig 1990) and (Dym 1994). The listing of kinds of design drawings is adapted from (Ullman 1997). The Chinese and German proverbs are from (Woodson 1966). The drawings shown in Figures 8.2–8.4 are adapted from (Boyer et al. 1991).

Section 8.2: Our brief overview of the basics of dimensioning and tolerancing has relied on the information found in (ASME 1994), (TDCA 1996), (Goetsch 2000) and (Wilson 2005).

8.4 EXERCISES

8.1 *Thought Exercise*: Do this exercise with a partner. **Do not read the other's instructions.**
To be read by Partner #1: You will be completing a simple paper design exercise: Design a handheld paper hole puncher. Design and document your item on paper in whatever manner

you feel is required such that someone could interpret and manufacture it. *To be read by Partner #2:* Take the drawing from your partner. Try to answer the following questions: What is being represented? What steps are needed to manufacture the object? What materials would you use? Could you create this object in a repeatable manner from this drawing? How do parts fit together? *Both Partners:* From this exercise, come up with a list of necessary components for a technical design drawing.

8.2 Draw the correct third-angle projection orthographic views for a block letter "F."

8.3 If a dimension on a drawing has no associated tolerance, where should the machinist look to determine the permissible variation?

8.4 Is a datum reference needed when specifying a flatness tolerance? Why (or why not)?

8.5 Sketch a functional gage for the part shown in Figure 8.16 to test the position of the holes.

8.6 Make a technical drawing of a rectangular part 3 in long, 2 in wide, with a thickness of 0.5 in. The part has one hole, located 0.75 in from the part's left side of the part and 0.75 in from its bottom. The hole has a diameter 0.25 in. All size dimensions can vary $\pm$ 0.01 in and the location of the hole can vary 0.005 in. Design for maximum interchangeability of parts.

COMMUNICATING THE DESIGN OUTCOME (III): ORAL AND WRITTEN REPORTS

How do we let our client know about our solutions?

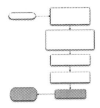

REPORTING IS an essential part of a design project: The project is not complete if its results have not been communicated to the client and to other stakeholders designated by the client. Final design results can be communicated in several ways, including oral presentations, final reports (that may include design drawings and/or fabrication specifications), and prototypes and models such as those discussed in Chapter 7. In this chapter we first consider some common guidelines for all reporting modes, and then we look in turn at final technical reports and at oral presentations. Thus, we complete our discussion of the final design task (15) of Figure 2.3.

Independent of the details, however, note that the primary purpose of such communication is to inform the client *about the design*, including explanations of how and why this design was chosen over competing design alternatives. It is most important to convey the *results* of the design process. The client is likely uninterested in the history of the project or in the design team's internal workings. Thus, final reports and presentations are *not* chronologies of a team's work. Rather, they should be lucid descriptions of design *outcomes*.

9.1 GENERAL GUIDELINES FOR TECHNICAL COMMUNICATION

There are some basic elements of effective communication that apply to writing reports, giving oral presentations, and even providing informal updates to your client. Thomas Pearsall summarized these common concepts as the seven principles of technical writing (see Figure 9.1), but they clearly apply more generally. And while Pearsall devoted more than one-half of his book to these principles, we will summarize them here as a prelude to the rest of this chapter.

Know your purpose. This is the writing analog of understanding objectives and functions for a designed artifact. Just as we want to understand what the designed object must be and must do, we need to understand the goals of a report or presentation. In many cases design documentation seeks to inform the client about the features and design elements of a selected design. In other cases the design team may be trying to persuade a client that a design is the best alternative. In still other cases a designer may wish to report how a design operates to users, whether beginners or highly experienced ones. If you don't know what purpose you are trying to serve with your writing or presentation, you may not produce anything or serve any purpose.

Know your audience. We have all sat through lectures where we didn't know what was going on or where the material was so simple that we already knew it. We can often take some action once we realize that the material is not set at a level that we find appropriate. Similarly, when documenting a design, it is essential that a design team structure its materials to its targeted audience. Thus, the team should ask questions like, "What is the technical level of the target audience?" and "What is their interest in the design being presented?" Taking time to understand the target audience will help ensure that its members appreciate your documentation. Sometimes you may prepare multiple documents and briefings on the same project for different audiences. For example, it is quite common for designers to close projects out with both a technical briefing and a management briefing. It is also common for designers to confine calculations or concepts that are of limited interest to a report's primary audience to specific sections of their reports, usually appendices.

---

1. *Know your purpose.*

2. *Know your audience.*

3. *Choose and organize the content around your purpose and your audience.*

4. *Write precisely and clearly.*

5. *Design your pages well.*

6. *Think visually.*

7. *Write ethically!*

FIGURE 9.1 Pearsall's seven principles of effective technical writing, which, we argue, are the seven principles of effective reporting and presentation for all modes of communication (after Pearsall 2001).

Choose and organize the content around your purpose and your audience. Once we are sure of the purpose of the report or presentation and its target audience, it only makes sense to try to select and organize its content so that it will reach its intended target. The key element is to structure the presentation to best reach the audience. In some cases, for example, it is useful to present the entire process by which the design team selected an alternative. Other audiences may only be interested in the outcome.

There are many different ways to organize information, including going from general overviews or concepts to specific details (analogous to deduction in logic), going from specific details to general concepts (analogous to induction or inference), sequencing design events chronologically (which we do not recommend), and describing devices or systems.

Once an organizational pattern is chosen, no matter which form is used, the design team should translate it into a written outline. As we will discuss below, this allows the team to develop a unified, coherent document or presentation and to avoid needless repetition.

Write precisely and clearly. This particular guideline sounds like "use common sense," that is, do something that everyone wants to do but few achieve. There are, however, some specific elements that seem to occur in all good writing and presentations. These include effective use of: short paragraphs (and other structural elements) that have a single common thesis or topic; short, direct sentences that contain a subject and a verb; and active voice and action verbs that allow a reader to understand directly what is being said or done. Opinions or viewpoints should be clearly identified as such. These elements of style should be learned so that they can be correctly applied. Young designers may have practiced these skills more in humanities and social science classes than in technical courses. This is acceptable, and even welcome, so long as the designer remembers that the goals of both technical and nontechnical communications remain the same.

Design your pages well. Whether writing a technical report or organizing supporting materials for a verbal briefing or presentation, effective designers utilize the characteristics of their media wisely. In technical reports, for example, writers judiciously use headings and subheadings, often identified by different fonts and underlining, to support and extend the organizational structure of the report. Dividing a long section into several subsections helps readers to understand where the long section is going, and it sustains their interest over the journey. Selecting fonts to highlight key elements or to indicate different types of information (such as new, important terms) guides the reader's eye to key elements on the page. White space on a page helps keep readers alert and avoids a forbidding look in documents.

Similarly, careful planning of presentation support materials such as slides and transparencies can enhance and reinforce important concepts or elements of design choices. Using fonts that are large enough for the entire audience to see is an obvious, but often overlooked, aspect of presentations. Just as white space on a page invites readers to focus on the text without being distracted, simple and direct slides encourage readers to listen to the speaker without being distracted visually. Thus, text on a slide should present succinct concepts that the presenter can amplify and describe in more detail. A slide does not have to show every relevant thought. It is a mistake to fill slides with so many words (or other content) that audiences have to choose between reading the slide and listening to the speaker, because then the presenter's message will almost certainly be diluted or lost.

Think visually. By their very nature, design projects invite visual thinking. Designs often start as sketches, analyses often begin with free-body or circuit diagrams, and plans for realizing a design involve graphics such as objectives trees and work breakdown structures. Just as designers often find that visual approaches are helpful to them, audiences are helped by judicious use of visual representation of information. These can range from the design tools discussed throughout this book, to detailed drawings or assembly drawings, to flow charts and cartoons. Even tables present an opportunity for a design team to concentrate attention on critical facts or data. In fact, given the enormous capabilities of word processing and presentation graphics software, there is no excuse for a team not to use visual aids in its reports and presentations. On the other hand, a team should not be seduced by their graphics capabilities by, for example, clouding their slides with artistic backgrounds that make the words illegible. The key to success here, as it is with words, is to know your purpose and your audience, and to use your medium appropriately.

Write ethically! Designers often invest themselves in the design choices they make, in time, effort, and even values. It is, therefore, not surprising that there are temptations to present designs or other technical results in ways that not only show what is favorable, but that suppress unfavorable data or issues. Ethical designers resist this temptation and present facts fully and accurately. This means that that *all* results or test outcomes, even those that are not favorable, are presented and discussed. Ethical presentations also describe honestly and directly any limitations of a design. Further, it is also important to give full credit to others, such as authors or previous researchers, where it is due. (Remember that this discussion of the seven principles began with an acknowledgment to their originator, Thomas Pearsall, and that each chapter of the book ends with references and citations.) We will talk more about engineering ethics in Chapter 12, where will also describe a landmark case involving an engineer widely acclaimed for both his work and for his ethics.

Now we turn our attention to specific forms of documentation.

9.2 ORAL PRESENTATIONS: TELLING A CROWD WHAT'S BEEN DONE

Most design projects call for a number of meetings with and presentations to clients, users, and technical reviewers. Such presentations may be made before the award of a contract to do the design work, perhaps focusing on the team's ability to understand and do the job in the hope of winning the contract in a competitive procurement. During the project, the team may be called upon to present their understanding of the project (e.g., the client's needs, the artifact's functions, etc.), the alternatives under consideration and the team's plan for selecting one, or simply their progress toward completing the project. After a design alternative has been selected by the team, the team is often asked to undertake a design review before a technical audience to assess the design, identify possible problems, and suggest alternate solutions or approaches. At the end of a project, design teams usually report on the overall project to the client and to other stakeholders and interested parties.

Because of the variety of presentations and briefings that a team may be called upon to make, it is impossible to examine each of them in detail. However, there are elements common and key to most of them. Foremost among these are the needs to: identify the audience, outline the presentation, develop appropriate supporting materials, and practice the presentation.

9.2.1 Knowing the audience: Who's listening?

Design briefings and presentations are given to many types of audiences. For example, some projects call for the work to be reviewed periodically by technical experts. Others are concerned with a design's implications for management. Some may be concerned with how a design will be manufactured. Consider the new beverage container whose design we begun in Chapter 3. Our design work might have to be presented to logistics managers who are concerned with how the containers will be shipped to warehouses around the country. The marketing department, concerned with establishing brand identity with the design, might want to hear about our design alternatives. Similarly, the manufacturing managers will want to be briefed about any special production needs. Thus, as noted in our review of Pearsall's seven principles, a team planning a briefing should consider factors such as varying levels of interest, understanding, and technical skill, as well as the amounts of available time. We can assume that most attendees at a meeting are interested in at least some aspect of a project, but it is generally true that most are only interested in particular dimensions of that project. A team usually can identify such interests and other dimensions simply by asking the organizer of the meeting.

Once the audience has been identified, a team can tailor its presentation to that audience. As with other deliverables, the presentation must be properly organized and structured. The first step is to articulate a rough outline; the second is to formulate a detailed outline; and the third is to prepare the proper supporting materials, such as visual aids or physical models.

9.2.2 The presentation outline

Just as with the final report discussed in Section 9.3, a presentation must have a clear structure. We achieve this structure by developing a rough outline. This presentation structure and organization, which should be logical and understandable, then guides the preparation of supporting dialogue and discussion. And because a design presentation is neither a movie nor a novel, it should not have a "surprise ending." A sample presentation outline would include the following elements:

- *A title slide* that identifies the client(s), the project, and the design team or organization responsible for the work being presented.

- *An overview of the presentation* that shows the audience the direction that the presentation will take.

- *A problem statement,* including the initial statement given by the client and an indication of how that problem statement changed as the team came to understand the project.

- *Background material on the problem*, including relevant prior work and other materials developed through team research.
- *The key objectives of the client and users* as reflected in the top level or two of the objectives tree.
- *Functions that the design must perform*, focusing on basic functions and means for achieving those functions, but possibly also including issues of unwanted secondary functions.
- *Design alternatives*, particularly those that were still considered at the evaluation stage.
- *Highlights of the evaluation procedure and outcomes*, including key metrics or objectives that bear heavily on the outcome.
- *The selected design*, explaining why this design was chosen.
- *Features of the design*, highlighting aspects that make it superior to other alternatives and any novel or unique features.
- *Proof-of-concept testing*, especially for an audience of technical professionals for whom this is likely to be of great interest.
- *A demonstration of the prototype*, assuming that a prototype was developed and that it can be shown. Videotapes or still photos may also be appropriate here.
- *The conclusion(s)*, including the identification of any future work that remains to be done.

There may not always be enough time to include all of these elements in a talk or presentation, so a team may need to limit or exclude some of them. This decision will also depend at least in part on the nature of the audience.

Once the rough outline has been articulated, a detailed outline of the presentation should also be developed. This is important both to ensure that the team understands the point that is being made at all times in the presentation and to develop corresponding bullets or similar entries in its slides or transparencies. Bullets generally correspond to entries in the detailed outline.

Preparing a detailed outline for the presentation may seem like a great deal of work, just as developing a topic sentence outline (TSO; see Section 9.3) for a report appears at first to be cumbersome. And, ironically, team members with public speaking experience may be most resistant to such tasks, most likely because they have already internalized a similar method of preparation. However, since presentations represent the entire team, every member of a team should review the structure and details of their presentations, as well as the detailed outline required by such reviews.

9.2.3 Presentations are visual events

Just as the team needs to know the audience, it should also try to know the setting in which it will be presenting. Some rooms will support certain types of visual aids, while others will not. At the earliest stages of the presentation planning, the design team should find out what devices (e.g., 35 mm slide projectors, overhead projectors, computer connections) are available and the general setting of the room in which it will be presenting. This includes its size and capacity, lighting, seating, and other factors. Even if a particular device or setup is said to be available, it is always wise to bring along a backup, such as transparencies or foils to back up a slide presentation.

There are other tips and pointers to keep in mind about visual aids, including:

- Avoid using too many slides or graphics. A reasonable estimate of the rate of slides at which slides can be covered is 1–2 slides per minute. If too many slides are planned, the presenter(s) will find themselves rushing through them in the hope of finishing. This makes for a far worse talk than a smaller selection wisely used.

- Be sure to introduce yourself and your teammates on the title slide. This is also an appropriate time for a brief overall description of the project and acknowledgment of the client. Inexperienced speakers often have the tendency to flash the title slide and move on, instead of using it as an opportunity to introduce the project and people involved.

- Beware of "clutter." Slides should be used to highlight key points; they are not a direct substitute for the reasoning of the final report. The speaker should be able to expand upon the points in the slides.

- Make points clearly, directly, and simply. Slides that are too flashy or clever tend to detract from a presentation.

- Use color skillfully. Current computer-based packages support many colors and fonts, but their defaults are often quite appropriate. Also, avoid weird or clashing colors in professional presentations, and be sure to keep in mind that certain color combinations are difficult to read for audience members who are color blind.

- Do not reproduce completed design tools (e.g., objectives tree, large morph charts) to describe the outcomes of the design process. Instead, highlight selected points of the outcomes. This is a situation where it makes more sense to refer the audience to the final report for more detailed information.

It is worth remembering that audiences tend to read visual aids as the speaker is talking. Therefore, the speaker does not need to read or quote those slides. The visual aids can be simpler (and more elegant) in their content because the visual aids reinforce the speaker, rather than the other way around.

Effective visuals don't replace effective speakers; they enhance them.

Finally, if design drawings are being reproduced and shown, the size and distance of the audience must be considered carefully. Many line drawings are difficult to display and to see and interpret in large presentations.

9.2.4   Practice makes perfect, maybe...

Presenters and speechmakers are usually effective because they have extensive experience. They have given many speeches and made many presentations, as a result of which they have identified styles and approaches that work well for them. Design teams cannot conjure up or create such real-world experience, but they can practice a presentation often enough to gain some of the confidence that experience breeds. To be effective, speakers typically need to practice their parts in a presentation alone, then in front of others, including before a "rump" audience including at least some people who are not familiar with the topic.

Another important element of effective presentation is that speakers use words and phrases that are natural to them. Each of us normally has an everyday manner of speech with which we are comfortable. While developing a speaking style, however, we have to keep in mind that ultimately we want to speak *to* an audience in *their* language, and that we want to maintain a professional tone. Thus, when practicing alone, it is useful for a presenter to try saying the key points in several different ways as a means of identifying and adopting new speech patterns. Then, as we find some new styles that work, we should repeat them often enough to feel some ownership.

Practice sessions, whether solitary or with others, should be timed and done under conditions that come as close as possible to the actual environment. Inexperienced speakers typically have unrealistic views of how long their talk will last, and they also have trouble setting the right pace, going either too fast or too slow. Thus, timing the presentation — even setting a clock in front of the presenter — can be very helpful. If slides (or transparencies or a computer) are to be used in the actual presentation, then slides (or transparencies or a computer) should be used in practice.

Speech coaches and athletic coaches say that you play how you practice!

The team should decide in advance how to handle questions that may arise. This should be discussed with the client or the sponsor of the presentation before the team has finished practicing. There are several options available for handling questions that are asked during a talk, including deferring them to the end of the talk, answering them as they arise, or limiting questions during the presentation to clarifications of facts while deferring others until later. The nature of the presentation and the audience will determine which of these is most appropriate, but the audience should be told of that choice at the start of the presentation. When responding to questions, it is often useful for a speaker to repeat the question, particularly when there is a large audience present or if the question is unclear. The presenter or the team leader should refer questions to the appropriate team members for answers. If a question is unclear, the team should seek to clarify it before trying to answer it. And as with the presentation itself, the team should practice handling questions that it thinks might arise.

There are several ways of preparing for questions while practicing talks, including:

- generate a list of questions that might arise and prepare for them;
- prepare in advance supporting materials for points that are likely to arise (e.g., backup slides that may include computer results, statistical charts and other data that may be used to answer anticipated questions); and
- be prepared to say "I don't know" or "We didn't consider that." This is a very important point. To be caught *pretending* to know undermines the presenter's (and the team's) credibility and invites severe embarrassment.

A final note about selecting speakers is in order. Depending on the nature of the presentation and the project, the team may want to have all members speak (for example, for a course requirement), it may want to encourage less experienced members to speak in order to gain experience and confidence, or it may want to tap its most skilled and confident members. As with so many of the presentation decisions, choosing a "batting order" will depend on the circumstances surrounding the presentation. This means that, as with all of the other matters we've touched upon, the team should carefully consider and consciously decide its speaking order.

9.2.5 Design reviews

A design review is a unique type of presentation, quite different from all the others that a design team is likely to do. It is also particularly challenging and useful to the team. As such, a few points about design reviews are worth noting.

A design review is typically a long meeting at which the team presents its design choices in detail to an audience of technical professionals who are there to assess the design, raise questions, and offer suggestions. The review is intended to be a full and frank exploration of the design, and it should expose the implications of solving the design problem at hand or even of creating new ones. A typical design review will consist of a briefing by the team on the nature of the problem being addressed, which is followed with an extensive presentation of the proposed solution. In the cases of artifacts, the team will often present an organized set of drawings or sketches that allows its audience to understand and question the team's design choices. In some cases, these materials may be provided to the attendees in advance.

A design review is often the best opportunity that the team will have to get the undivided attention of professionals about their design project. It is often also scary and worrisome for the design team because its members may be asked to defend their design and answer pointed questions. A design review thus offers both a challenge and an opportunity to the team, giving it a chance to display its technical knowledge and its skills in constructive conflict. Questions and technical issues should be fully explored in a positive, frank environment. To benefit from the design review, the team should try to withhold the natural defensiveness that comes from having its work questioned and challenged. In many cases the team can answer the questions raised, but sometimes it cannot. Depending on the nature of the meeting, the team may call upon the expertise of all of the participants to suggest new ways to frame the problem or even the design itself.

Not surprisingly, such reviews can last several hours, or even a day or two. One important decision for the team is to determine, during the review, when a matter has been adequately covered and move on. This is a real challenge, since there is a natural temptation to move on quickly if the discussion suggests that a design must be changed in ways the team doesn't like. There may be a similar temptation if the team feels that the review participants have not really "heard" the team's point of view. It is important to resist both urges: Time management should not become a cover for hiding from criticism or belaboring points.

A final point about design reviews is the need to remember that conflict in the realm of ideas is generally constructive, while personality-oriented criticism is destructive. Given the heat and light that sometimes arise at design reviews, team leaders and team members (as well as the members of the audience) must continually maintain the review's focus on the design, and not on the designers.

9.3 THE PROJECT REPORT: WRITING FOR THE CLIENT, NOT FOR HISTORY

The usual purpose of a final or project report is to communicate with the client in terms that ensure the client's thoughtful acceptance of a team's design choices. The client's interests demand a clear presentation of the design problem, including analyses of the needs to be met, the alternatives considered, the bases on which decisions were made, and, of course, the decisions that were taken. The results should be summarized in clear, understandable language. Highly detailed or technical materials are often placed in appendices at the end of the report, in order to support clarity. In fact, it is not unusual (and in large public-works works projects it is the norm) for all of the technical and other supporting materials to be moved to separate volumes. This is especially important when the client and the principal stakeholders are not engineers or technical managers, but perhaps members of the general public.

The process of writing a final report, like so much of design, is best managed and controlled with a structured approach. The design process and report writing are strikingly similar, especially in their early, conceptual stages. It is very important to delineate objectives clearly, both for the designed object and for the project report. It is very important to understand the "market," that is, to understand both the user needs for the design and the intended audience of the final report. It is very important to be reflective and analytical and to recognize that analysis isn't limited to applying known formulas. We have described several tools that can support our thinking during the design process. Similarly, writing is also an analytical thinking process.

As with the design process, structure is not intended to displace initiative or creativity. Rather, we find that structure can help us learn how to construct an organized report of the design results. In this case, one structured process that a design team might follow would include the following steps:

- determine the purpose and audience of the technical report;
- construct a rough outline of the overall structure of the report;
- review that outline within the team and with the team's managers or, in case of an academic project, with the faculty advisor;
- construct a topic sentence outline (TSO) and review it within the team;
- distribute individual writing assignments and assemble, write, and edit an initial draft;
- solicit reviews of the initial draft from managers and advisors;
- revise and rewrite the initial draft in response to its reviews; and
- prepare the final version of the report and present it to the client.

We now discuss these steps in greater detail.

9.3.1 The purpose of and audience for the final report

We have already discussed determining the purpose and audience of the report in general terms. Several points should be noted in the case of a final report. The first is that the report is likely to be read by a much wider audience than simply by the client's liaison with whom the team has been interacting. In this respect, the team needs to determine whether or not the liaison's interests and levels of technical knowledge are representative of the audience for the final report. The liaison may, however, be able to guide the team to a better understanding of the expected reader(s), and may highlight issues that may be of particular concern.

Another important element here is for the team to understand what the report's recipient hopes to do with the information in the final report. If, for example, the intent of the project was to create a large number of conceptual design alternatives, the audience is likely to want to see a full presentation of the design space that was explored. If, on the other hand, the client simply wanted a solution to a particular problem, s/he is much more likely to want to see how well the selected alternative meets the specified need.

A project report often has several different audiences, in which case the team will have to organize information to satisfy each of these target groups. This may include using technical supplements or appendices, or it may call for a structure that begins with general language and concepts, and then explores these concepts in technical subsections. The team, however, should write clearly and well for each audience, regardless of the organizational principle selected.

9.3.2   The rough outline: Structuring the final report

Only a fool would start building a house or an office building without first analyzing the structure being built and organizing the construction process. Yet many people sit down to prepare a technical report and immediately begin writing, without trying to lay out in advance all of the ideas and issues that need to addressed, and without considering how these ideas and issues relate to one another. One result of such unplanned report writing is that the report turns into a project history, or worse, it sounds like a "What I Did Last Summer" essay: First we talked to the client, then we went to the library, then we did research, then we did tests, etc., etc., etc. While technical reports may not be as complex as high-rise office buildings or airplanes, they are still complicated enough that they cannot simply be written as simple chronologies. Reports must be planned!

The first step in writing a good project or final report is building a good rough outline that outlines the report's overall structure. That is, we identify the major sections into which the report is divided. Typically, some of these sections are:

- Abstract
- Executive summary
- Introduction and overview
- Problem statement and problem definition or framing, including relevant prior work or research
- Design alternatives considered
- Evaluation of design alternatives and basis for design selection
- Results of the alternatives analysis and design selection
- Supporting materials, often set out in appendices, including:
  - Drawings and details
  - Fabrication specifications
  - Supporting calculations or modeling results
  - Other materials that the client may require

This outline looks like a table of contents, as it should, because a final report of an engineering or design project must be organized so that a reader can go to any particular section and see it as clear and coherent stand-alone document. It is not that we think things should be taken out of context. Rather, it is that we expect each major section of a report to make sense all by itself; that is, it should tell a complete story about some aspect of the design project and its results.

Having identified a rough outline as the starting point for a final report, *when* should that outline be prepared? Indeed, when should the final report be written? It evident that we can't write a *final* report until we have completed our work and identified and articulated a final design. On the other hand, as with the design process, it is very helpful to have an idea of where we're going with a final report so that we can organize and assemble it along the way. It can be very helpful to develop a general structure for the final report early in the project. We can then track and appropriately file or label key documents from the project (e.g., research memoranda, drawings, objectives trees) according to whether their contents would appear in the final report. Thinking about the report early on also emphasizes thinking about a project's *deliverables*, that is, those items that the team is contracted to deliver to the client during the project. Organizing the final report early on may make the final stages or endgame of the project much less stressful, simply because there will be fewer last-minute things to identify, create, edit, and so on, so they can be inserted into the final report.

9.3.3 The topic sentence outline: Every entry represents a paragraph

A cardinal rule of writing states that *every single paragraph* of a piece should have a topic sentence that indicates that paragraph's intent or thesis. Once the rough outline of a report has been established, it is usually quite useful to build a corresponding, detailed *topic sentence outline* (TSO) that identifies the themes or topics that, collectively, tell the story told within each section of the report. Thus, if a topic is identified by an entry in the TSO, we can assume that there is a paragraph in which that topic is covered.

The TSO enables us to follow the logic of the argument or story and assess the completeness of each section being drafted, as well as of the report as a whole. Suppose there is only one entry in a TSO for something that we consider important, say, the evaluation of alternatives. One implication of this is that the final report will have only one paragraph devoted to this topic. Since the evaluation of alternatives is a central issue in design, it is quite likely that there should be entries on a number of aspects, including the evaluation metrics and methods, the results of the evaluation, key insights learned from the evaluation, the interpretation of numerical results—especially for closely rated alternatives and the outcome of the process. Thus, a quick examination of this TSO shows us that a proposed report is not going to address all of the issues that it should.

Each entry in a topic sentence outline corresponds to a single paragraph.

For the same reason, of course, TSOs help identify appropriate cross-references that should be made between subsections and sections as different aspects of the same idea or issue are addressed in different contexts. The format of a TSO also makes it easier to eliminate needless duplication because it is much easier to spot repeated topics or ideas. In Section 9.4 we show examples of a TSO that demonstrates some of these points.

Writing in this way is hard, but TSOs do provide a number of advantages to a design team. One is that a TSO forces the team to agree on the topics to be covered in each section. It quickly becomes clear if a section is too short for the material, or if one of the co-authors (or team members) is "poaching" on another section that was agreed upon in the rough outline. Another advantage of a good TSO is that it becomes easier for team members to take over for one another if something comes up to prevent a "designated writer" from actually writing. For example, a team member may suddenly find that the prototype is not working as planned and she needs to do some more work on it. TSOs also make life easier for the team's report editor (see the next section) to begin to develop and use a single voice.

Notwithstanding the definition of the abbreviation TSO, the entries in a TSO do not really have to be grammatically complete sentences. However, they should be complete enough that their content is clear and unambiguous.

9.3.4 The first draft: Turning several voices into one

One advantage of agreed-upon rough and topic sentence outlines is that their structure allows teams members to write in parallel or simultaneously. However, this advantage comes at a price, most notably that of corralling the efforts of several writers into a single, clear, coherent document. Simply put, the more writers, the greater the need for a single, authoritative editor. Thus, one member of the team should enjoy the rights, privileges and *responsibilities* pertaining to being the editor. Further, the team should designate an editor as soon as the planning of the report begins, hopefully at or near the onset of the project.

The larger the writing team, the greater the need for a single editor.

The editor's role is to ensure that the report flows continuously, is consistent and accurate, and speaks in a single voice. *Continuity* means that topics and sections follow a logical sequence that reflects the structure of the ideas in the rough outline and the TSO. *Consistency* means that the report uses common terminology, abbreviations and acronyms, notation, units, similar reasoning styles, and so on, throughout the report and all of its appendices. It also means, for example, that the team's objectives tree, pairwise comparison chart, and evaluation matrix all have same elements; if not, discrepancies should be noted explicitly and explained.

Accuracy requires that calculations, experiments, measurements, or other technical work are done and reported to appropriate professional standards and current best practices. Such standards and practices are often specified in contracts between a design team and its client(s). They typically provide that stated results and conclusions must be supported by the team's prior work. Accuracy, as well as intellectual honesty, also requires that technical reports do not make unsupported claims. There is often a temptation in a project's final moments to add to a final report something that wasn't really done well or completely. This is a temptation that should be avoided.

The *voice* or style of a report reflects the way in which a report "speaks" to the reader, in ways very similar to how people literally speak to each other. It is essential that a technical report *speak with a single voice* — and ensuring that single voice is one of the editor's most important duties. This mandate has several facets, the first of which is that the report has to read (or "sound") as if it was written by one person, even when its sections were written by members of a very large team. A president of the United States sounds like the same, familiar person, even while using several speechwriters. So, similarly, a technical report must read in a single voice. Further, that voice should normally be more formal and impersonal than the voice of this book. Technical reports are not personal documents, so they should not sound familiar or idiosyncratic. Also, that single voice can either be active or passive, as modern practice renders both acceptable. It is important only that the voice of the report be the same from the opening Abstract through the closing Conclusions and to the last Appendix.

Good technical reports speak in a single voice.

Clearly, there are serious issues for the team dynamics of the writing process. Team members have to be comfortable surrendering control of pieces they have written, and they have to be willing to let the editor do her job. We will discuss aspects of the team dynamics of report writing in Section 9.5 below.

9.3.5   The final, final report: Ready for prime time

A good review process ensures that a draft final report gets thoughtful reconsideration and meaningful revision. Draft reports benefit from careful readings and reviews by team members, managers, client representatives or liaisons, faculty advisors, as well as people who have no connection with the project. This means that as we are trying to wrap up our project report, we need to incorporate reviewers' suggestions into a final, high-quality document. There are a few more points to keep in mind.

A final report should look professionally done and *polished*. This does not mean that it needs glossy covers, fancy type and graphics, and an expensive binding. Instead, it means that the report is clearly organized, easy to read and understand, and that its graphics or figures are also clear and easily interpreted. The report should also be of reproducible quality because it is quite likely to be photocopied and distributed within the client's organization, as well as to other individuals, groups, or agencies.

We should also keep in mind that a report may go to a very diverse audience, not simply to peers. Thus, while the editor needs to ensure that the report speaks with a single voice to an anticipated audience, she should try as much as possible to also ensure that the report can be read and understood by readers that may have different skill levels or backgrounds than either the design team or the client. An *executive summary* is one way to address readers who may not have the time or interest to read all of the details of the entire project.

Finally, the final report will be read and used by client(s), who will, one hopes, adopt the team's design. This means that the report, including appendices and supporting materials, are sufficiently detailed and complete to stand alone as the final documentation of the work done.

9.4   FINAL REPORT ELEMENTS FOR THE DANBURY ARM SUPPORT

As required in most design projects, the student teams responsible for the arm support design for the Danbury School reported their results in the form of final reports and oral presentations. In this section we look briefly at some of the intermediate work products associated with their reports to gain further insight into some of the dos and don'ts discussed in Section 9.3.

9.4.1 Rough outlines of two project reports

The two teams we have followed each prepared a rough outline as a first step in laying out the report structure. Table 9.1 shows the rough outline of one of the teams; Table 9.2 shows that for the other team.

These two outlines display both similarities and differences. The first team, for example, dedicated several sections to justifying their final design, while the second team organized around process. Both teams relegated sketches and drawings to appendices, although the second team put building instructions in the report body. This reflects the freedom that teams have to decide on an appropriate structure to convey their design results. This freedom, however, does not excuse them from having a logical ordering that allows the reader to understand the nature of the problem or the benefits of their solution.

As a second point on the structure of these final reports, note just how much of each could have been written during the course of the project. Each team used the formal

TABLE 9.1 Rough outline for the report of one of the Danbury arm support teams.

I. Introduction

II. Description of problem definition
    a. Problem statement
    b. Design objectives and constraints

III. Generation of design alternatives
    a. Morphological chart
    b. Description of design alternatives
    c. Description of subcomponents

IV. Design selection process
    a. Metrics description
    b. Metrics application

V. Final design
    a. Detailed description
    b. Prototype details

VI. Testing the design

VII. Conclusions
    a. Strengths and weaknesses of final design
    b. Suggestions for more advanced prototype
    c. Recommendations to the client

VIII. References

Appendix: Work breakdown structure

Appendix: Pairwise comparison chart

The rough outline should show the overall structure of the report in a way that allows team members to divide up work with little or no unintended duplication. The structure should also proceed in a clear and logical manner. Does it for this report?

TABLE 9.2   The rough outline for the report of the second of the Danbury arm support teams

Introduction

   I. Problem statement

  II. Background information on cerebral palsy, motivation for project

 III. Design plan

    a. Work breakdown structure

    b. Definition of objectives and constraints, including objectives tree

    c. Definition of functions and means, morphological chart

 IV. Design research

    a. Summary of devices currently available

    b. Evaluation of these devices for suitability in this project

  V. Description and evaluation of design alternatives

    a. Details and drawings of each alternative

    b. Metrics for choosing between designs

 VI. Final design

    a. Detailed description of chosen alternative

    b. Description of prototype and how it works

 VII. Testing the design

    a. Description of three test sessions at Danbury

    b. Conclusions and refinements of design based on testing

 VIII. Design evaluation

    a. Consideration of constraints

    b. How well design meets objectives

    c. Functional analysis

    d. Details on proposed design changes based on testing and evaluation

 IX. Works cited

Appendix: Work breakdown structure

Appendix: Research on dashpots

As with the outline presented in Table 9.1, this outline also show the overall structure, and it is clear that the report focuses on reporting detailed testing and evaluation of the chosen design

design tools discussed in previous chapters to document its decision process. Thus, the teams could — and should — have tracked and organized their outcomes in order to facilitate the writing of their final reports.

Finally, neither outline will adequately translate directly into a report. There are issues that could be considered in more than one section, and others could not be covered at all. Unless a team continues with a TSO or some other detailed plan, its first draft of the final report will need an unnecessarily high degree of editing.

9.4.2   One TSO for the Danbury arm support

Table 9.3 shows an excerpt from the topic sentence outline prepared by one of the student design teams. Notice that while each entry is not in itself a complete sentence, the specific point of that entry is easy to see. At this level of detail, it is relatively easy to identify redundant or inadequately covered points.

TABLE 9.3 An excerpt from the TSO for one section of the final report outline (shown in Table 9.2) developed by the first of the Danbury arm support project teams

III. Design plan
   A. After clarifying the problem statement, the team began the process of designing the device
      a. Paragraph describing the overall approach to the design
         i. Work breakdown structure
         ii. Objectives and constraints
         iii. Defining functions and means
         iv. Creating and evaluating design alternative
   B. The work breakdown structure consists of the tasks for the design process and their deadlines
   C. In order to implement the design, the team needed to define objectives and constraints
      a. Paragraph defining objectives
         i. Objectives are things one wants the design to achieve
         ii. Objectives have a hierarchy
         iii. List of ranked objectives
      b. Paragraph on liaison reaction to ranked objectives list
         i. Liaisons added an objective and ranked it
      c. Paragraph on organization of objectives
         i. Objectives sorted into three categories: user-friendly, primary device functions, and features
         ii. Objectives divided into sub-objectives
         iii. Listing of sub-objectives
         iv. Objectives tree
      d. Paragraph on evaluating objectives using metrics
      e. Paragraph defining constraints
         i. Constraints are limits on the design
         ii. List of constraints and description
      f. Paragraph on liaison input and reaction to constraints
         i. Initial constraints
         ii. Constraints added after reaction from liaisons

The TSO enables the team to see what will be covered not only within each section, but also within each paragraph of the report. It also permits team members to take issue with or make suggestions about a section before writing and "wordsmithing" efforts are invested. For example, the team's definitions of metrics are not clear and could be challenged by a nit-picking reader (such as a professor or a technical manager). It is also unclear that all the ideas conveyed in some of the paragraphs couldn't be better covered by separating them into two paragraphs. Further, it is not clear why attention was specifically called to the liaison's definitions of constraints. Don't the objectives and the revised problem statement also emerge out of the team's questioning of and discussions with the client? Is it perhaps because elsewhere in that same report the team implies that it alone developed the list of objectives? So perhaps this TSO indicates a possible confusion about how the design process was executed or, at least, how its execution was reported.

Notice also that the team has adopted a historical approach to the process, which is very much at odds with recommended style. For example, "After clarifying the problem statement, the team began the process of designing the device" is a red flag that the team is documenting the passing of time and events, not the design process.

Fortunately, because the team has invested effort in a TSO, it is relatively easy to make changes.

9.4.3 The final outcome: The Danbury arm support

We have followed two teams working on the Danbury arm support project, and they did finally finish their work. They produced final reports, formal oral presentations, and prototypes. We have discussed the development of the two teams' reports above, and they were as different as were their designs. For instance, the reports were, respectively, 18 and 61 pages long! We have also seen some drawings and photos of their prototypes (see Figures 5.9–5.11), and we show two more in Figures 9.2 and 9.3.

It is also worth noting that, as we described previously on a running basis, the two teams generally followed the design process outlined in Figure 2.3. On the other hand, many of the teams' work products differed significantly, and this reflects in part the fact that design is an *ill structured* process. As we have pointed out frequently, there are significant roles for mathematics and physics to play, but the design process is a basically thoughtful, yet decidedly non-algorithmic process.

Finally, while the two designs have obvious similarities, they also have clear differences. For example, their mounting structures differ, as do their use of damping devices. This is not a surprise to designers or to engineering faculty who teach design, and it should not surprise you. As we have also noted previously, design is an *open-ended* activity, that is, there is no single — or even guaranteed — solution to a design problem. Dealing with the uncertainty implied by the absence of a guaranteed, unique outcome is the reason that design is both challenging and exciting. The only thing that is guaranteed is

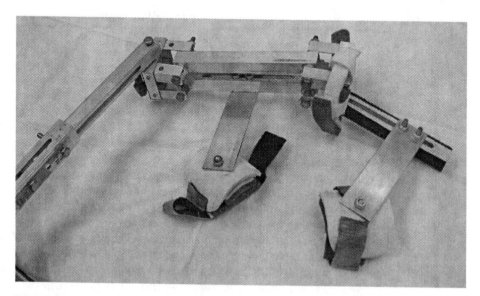

FIGURE 9.2 This photograph shows one prototype for the arm support design done for the Danbury School. Note its resemblance to the sketches in Figure 5.9.

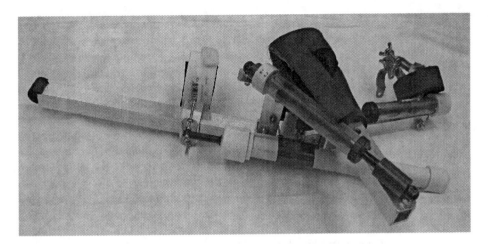

FIGURE 9.3 This photograph shows a second prototype for the arm support design done for the Danbury School. Note its resemblance to the drawings in Figure 5.10, and compare its clarity with the student team's photos displayed in Figure 5.11.

the satisfaction and excitement experienced by designers, clients and users when a good design is achieved.

9.5 MANAGING THE PROJECT ENDGAME

In this section we present some final remarks on managing the documentation activities and on closing out the project. Of particular importance to teams that want to use their experience on one project as a basis for improving their performance on the next project is a discussion of the post-project audit.

9.5.1 Team writing is a dynamic event

Most of us have considerable experience in writing papers by ourselves, all alone. This may include school term papers, technical memoranda, lab write-ups, and creative writing or journalistic experiences. Documenting a design in a team setting is, however, a fundamentally different activity than writing a paper alone. These differences turn on our dependence on co-authors, on the technical demands, and on the need to ensure a uniform style.

When writing as a team, we can only be sure what others are writing if all of our writing assignments and their associated content are explicit. Thus, even if any one team member's personal writing style does not include detailed outlines and multiple drafts, *all members* should do outlines and rough drafts because they are essential to the team's success. This makes issues of responsibility and cooperation important in a way that is unique for most writers. In Chapter 10 we will introduce the linear responsibility chart (LRC) as a means to ensure that work is allocated fairly and productively. The LRC should be revised and updated as part of the documentation phase.

Even with a fair allocation of work, the ultimate quality of the final report, an oral presentation, and other forms of documentation will reflect on the team as a whole and on each of its members. It is important, therefore, to allow sufficient time for each team member to read report drafts carefully. Equally important, the team must create an atmosphere in which the comments and suggestions of others are treated with respect and consideration. No one on the team should be exempt from reading the final report drafts; no one's views should be "dissed." Given the pressure under which final deliverables are often prepared, the atmosphere is, in some sense, a test of the group's culture and attitudes. The interpersonal dynamics should be monitored closely and managed carefully.

As we noted in Section 9.3, the team must also agree on a single voice. This is often quite difficult for teams, especially when several days' work is rewritten after so much work has been done. It is hard for any writer, and still harder for those who consider themselves skilled writers, to sublimate their styles (and egos) to satisfy a team and its designated editor. Once again, each member of a team has to keep the team's overall goals in mind.

Oral presentations also demand that a team divide up its work fairly. Each team member must recognize that other members may be presenting the team's work. In many cases, the presenter of a particular piece of the project may have had little to do with the element of the work being presented, or may even have opposed that approach. Once again, then, a central issue here is the need for mutual respect and appropriate action by the members of the team.

9.5.2 Project post-audits: Next time we will . . .

In actual practice, most projects end not with the delivery to the client, but with a *project post-audit:* An organized review of the project, including the technical work, the management practices, the work load and assignments, and the final outcomes. This is an excellent practice to develop, even for student projects or activities in which the team is disbanding completely. There is an old Kentucky saying to the effect that the second kick of the horse has no real educational value. The post-project audit is an opportunity to understand the horse better and learn where to stand next time.

The key issue in post-project audit is focus on doing an even better job next time around. As a practical matter, the post-project audit may be as simple as a meeting that takes one or two hours, or it may be part of a larger formal process that is directed by the design team's parent organization. Regardless of the scope or formal mechanism, if any, the basic post-audit process is simple:

- review the project goals;
- review the project processes, especially in terms of ordering of events;
- review the project plans, budgets, and use of resources; and
- review the outcomes.

Reviewing the project goals is particularly important for design projects, since design is a goal-oriented activity. If the project was supposed to solve problem A, then even an idea that results in earning a patent for solving problem B may not always be viewed as a success. We can only evaluate a project in terms of what it set out to do. To this end, many of the problem definition tools and techniques should be reviewed as part of the post-audit.

Closely tied to reviewing the results of using the design and management tools is the useful idea of having the team consider the effectiveness of the tools themselves. Just as a toolbox may contain many items that are only useful some of the time, many of the formal methods and techniques that are presented in this book and elsewhere will be more effective in some situations than in others. No catalog of successes or failures by the authors will have the same purchase with the team as their own experience. Reflecting on what worked and what didn't, coming to grips with why a tool did or did not work, are both important elements of the post-project audit.

Analogously, reviewing the manner in which a team managed and controlled its work activities is also important to avoiding that "second kick of the horse." Most people learn how to organize activities, determine their sequence, assign the work, and monitor progress only through experience and practice. Such experiences are much more valuable if they are reviewed and reconsidered after the fact. As with the design tools, the management tools are not equally useful in every setting (although some, such as the work breakdown structure, appear useful in almost every situation). In commercial settings, reviewing both budgets and work assignments is critically important for planning future projects.

The last post-project audit step is a review of the outcome of the project, in terms of the goals and the processes used. While it is certainly useful to know whether or not the goals were achieved, it is important for the team members to ascertain whether this is a consequence of excess resources, good planning and execution, or simply good luck. In the long run, only teams that learn good planning and execution are likely to have repeated successes.

Our final note is that the post-project audit is not, in and of itself, a tool for assigning blame or for pointing fingers. Many project and institutional settings have formal mechanisms for peer review and supervisory evaluation of team members, and they can be valuable means for highlighting individual strengths, weaknesses, and contributions. They can also provide team members with important insights that they can use to improve their work in design teams. However, individual performance reviews are not central to, or even a desirable aspect of the post-project audit. The audit is intended to show what the team and organization did right to make the project successful, or what must be done differently if the project was not successful.

9.6 NOTES

Section 9.1: As noted in the text, the seven principles of technical writing are drawn from (Pearsall 2001). In addition to Pearsall, there are a number of excellent books to support technical writing, including (Pfieffer 2001), (Stevenson and Whitmore 2002), and the classic (Turabian 1996). There is no better reference to effective use of graphics than (Tufte 2001), a classic that belongs in the library of every engineer.

Section 9.5: The final results for the Danbury arm support design project from (Attarian et al. 2007) and (Best et al. 2007).

CHAPTER 10

LEADING AND MANAGING THE DESIGN PROCESS

Who's in charge here? And you want it when?

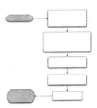

IT SHOULD be clear from what we have described thus far that design is an activity that can consume significant time and resources. In this chapter we will explore some techniques that a design team can use to manage its time and its other resources. We also introduce ways to manage and control a design project, emphasizing tools that are well suited to the particular situation of small design teams. Thus, we present a discussion of the management tasks 1–13 that were identified in Figure 2.4.

10.1 GETTING STARTED: ORGANIZING THE DESIGN PROCESS

Just as there are many models describing the design process, there are also many depictions of project management. We show (again) a brief road map of the project management path for design projects in Figure 10.1, in direct analogy with Figure 2.3 for the design process. This figure highlights the fact that project management follows a path that encompasses:

- *project definition*, developing an initial understanding of the design problem and its associated project;
- *project framework*, developing and applying a plan to do the design project;
- *project scheduling*, organizing that plan in light of time and other resource constraints; and as the project unfolds,
- *project tracking*, evaluation, and control, keeping track of time, work, and cost.

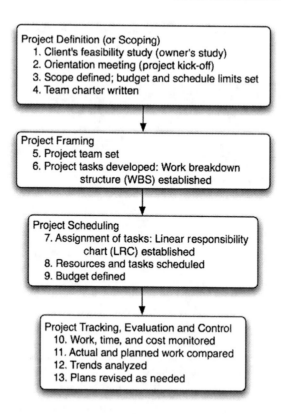

Project Definition (or Scoping)
   1. Client's feasibility study (owner's study)
   2. Orientation meeting (project kick-off)
   3. Scope defined; budget and schedule limits set
   4. Team charter written

Project Framing
   5. Project team set
   6. Project tasks developed: Work breakdown
       structure (WBS) established

Project Scheduling
   7. Assignment of tasks: Linear responsibility
       chart (LRC) established
   8. Resources and tasks scheduled
   9. Budget defined

Project Tracking, Evaluation and Control
   10. Work, time, and cost monitored
   11. Actual and planned work compared
   12. Trends analyzed
   13. Plans revised as needed

FIGURE 10.1 This is the same orderly process presented in Figure 2.4. Here it is worth noting management task 4 in which the team writes a team charter, a relatively recent innovation in the management of team projects. Team charters are discussed in Section 10.4. Adapted from (Orberlander 1993).

In Sections 10.3–10.9, we detail a number of tools to help us proceed along this path. At this time, however, it is worthwhile to note how this model affects the team at the very beginning of a project. Many of the activities associated with project definition, such as the feasibility study, the orientation meeting, and setting the overall schedule and budget, may be beyond the control of the design team. Nevertheless, the team will usually devote its initial meetings and activities to trying to better understand these issues. One activity that cannot be deferred is the organization and development of the project team. We turn to that now.

10.1.1 Organizing design teams

Design is an activity that is increasingly done by teams, rather than by individuals acting alone. For example, new products are often developed by teams that include designers, manufacturing engineers, and marketing experts. These teams are assembled to gather the diverse skills, experiences, and viewpoints needed to successfully design, manufacture, and sell new products. This dependence on teams is not surprising if we reflect on the stages, methods, and means for design that we have been discussing. Many of the activities and methods are devoted to applying different talents and skills to achieving a common understanding of a problem. Consider, for example, the difference between laboratory testing and computer-based analysis of a structure. Both require common knowledge of structural mechanics, yet years of investment are required to master the specific testing and laboratory skills or the analysis and computer skills. Thus, there may be considerable value in constructing teams whose members have all the needed skills and can work together successfully. In this section we briefly introduce some aspects of team formation and performance, and we then relate them to one of the means for generating ideas discussed above, namely brainstorming.

10.1.1.1 *Stages of group formation*
Groups and teams are such an important element of human enterprise that we should not be surprised to learn that they have been extensively studied and modeled. One of the most useful models of group formation suggests that almost all groups undergo five stages of development that have been rather memorably named as:

- *forming,*
- *storming,*
- *norming,*
- *performing,* and
- *adjourning.*

We will use this five-stage model to describe some of the elements of group dynamics that are often encountered in engineering design projects.

Forming: Most of us experience a number of feelings simultaneously when we are initially assigned to a team or group. These feelings range from excitement and anticipation to anxiety and concern. We may worry about our ability — or that of our teammates — to perform the tasks asked of us. We may be concerned about who will show the leadership needed to accomplish the job. We may be so eager to get started that we rush into assignments and activities before we are really ready to begin. Each of these feelings and concerns are elements of the *forming* stage of group development, which has been characterized by a number of aspects and behaviors, including:

- becoming oriented to the (design) task at hand,
- becoming acquainted with the other members of the team,
- testing group behaviors in an attempt to determine if there are common viewpoints and values,
- becoming dependent upon whoever is believed to be ''in charge'' of the project or task, and
- attempting to define some initial ground rules, usually by reference to explicitly stated or externally imposed rules.

In this stage, the team members may often do or say things that reflect their uncertainties and anxieties. It is important to recognize this because judgments made in the forming stage may not prove to be valid over the life of a project.

Storming: After the initial or forming stage, most groups come to understand that they will have to take an active role in defining the project and the tasks needed to complete it. At this point the team may resist or even resent the assignment, and it may challenge established roles and norms. This period of group development is known as the *storming* phase and is often marked by intense conflict as team members decide for themselves where the leadership and power of the team will lie, and what roles they must individually play. At the same time, the team will usually be redefining the project and tasks, and discussing opinions about the directions the team should explore. Some characteristics of the storming phase are:

- resistance to task demands,
- interpersonal conflict,
- venting of disagreement, often without apparent resolution, and
- struggle for group leadership.

The storming phase is particularly important for the design team because there is often already a high level of uncertainty and ambiguity about client and user needs. Some team members may want to rush to solutions and will consider a more thoughtful exploration of the design space simply as stubbornness. At the same time, most design teams will not have as clear a leadership structure as, for example, a construction, manufacturing, or research project. For these reasons, it is important for effective teams to recognize when the team is spending too long in the storming phase and to encourage all team members to move to the next phases, norming and performing.

Norming: At some point, most groups do agree on ways of working together and on acceptable behaviors, or norms, for the group. This important period in the group's formation defines whether, for example, the group will insist that all members attend meetings, whether insulting or other disrespectful remarks will be tolerated, and whether or not team members will be held to high or low standards for acceptable work. It is particularly important that team members understand and agree to the outcome of this so-called *norming* phase because it may well determine both the tone and the quality of subsequent work. Some characteristics of the norming phase include:

- clarification of roles in the group,
- emergence of informal leadership,
- development of a consensus on group behaviors and norms, and
- emergence of a consensus on the group's activities and purpose.

Significantly, norming is often the stage at which members decide just how seriously they are going to take the project. As such, it is important for team members who want a successful outcome to recognize that simply ignoring unacceptable behavior or poor work products will not be productive. For many teams, the norms of behavior that are established during the norming stage become the basis for behavior for the remainder of the project.

Many organizations use a *team charter*, which we discuss below, to help document or formalize the norms of the team, and to also articulate the overall scope and time scale of the project.

Performing: After the team has passed through the forming, storming, and norming stages, it should reach the stage of actively working on its project. This is the *performing* phase — the stage that most teams hope to reach. Here team members focus their energies on the tasks themselves, conduct themselves in accordance with the established norms of the group, and generate useful solutions to the problems they face. The characteristics of the performing phase include:

- clearly understood roles and tasks,
- well defined norms that support the overall goals of the project,
- sufficient interest and energy to accomplish tasks, and
- development of solutions and results.

This is the stage of team development in which it becomes possible for the goals of the team to be fully realized.

Adjourning: The last phase that teams typically pass through is referred to as *adjourning*. This stage is reached when the group has accomplished its tasks and is preparing to disband. Depending on the extent to which the group has forged its own identity, this stage may be marked by members feeling regret that they will no longer be working together. Some team members may act out some of these concerns in ways that are not consistent with the group's prior norms. These feelings of regret typically emerge after teams (or any groups) have been working together for a very long time, much longer than an academic semester or two because such complete group identity usually develops after a long time.

One final point about these stages of group formation should be made. Teams will typically pass through each of them *at least once*. If the team undertakes significant changes in composition or structure, such as a change in membership or a change in team leadership, it is likely that the team will revisit the storming and norming phases again.

10.1.1.2 *Team dynamics and brainstorming* Much earlier in the book (in Section 2.3.3) we discussed means of gaining information, generating and evaluating ideas, and obtaining feedback. Some of these means were based on putting the team and other stakeholders into situations that would encourage a free flow of ideas. Brainstorming and synectics in particular are based on the idea that one person's ideas may serve to stimulate other team members to come up with better alternatives. We now briefly summarize brainstorming and relate it to our discussion of the stages of group formation. We will see that our warnings to defer solutions until after the problem is sufficiently understood are consistent not only with our models of the design process, but also with our understanding of how teams can best function.

Brainstorming is a classic technique for generating ideas and solutions to problems. Brainstorming consists of the members of a group offering individual ideas *without any concurrent evaluation of the ideas*. Typically, a team will form a circle or sit around a table and, after a brief review of the problem for which ideas are being sought, offer ideas one after another. One or more members of the team acts as "scribe," writing down each idea offered for later discussion and review. Each member of the group should offer an idea when his or her turn comes, even if the idea is poorly formed or even silly. At some point it becomes allowable for a team member to pass, but this must be done explicitly.

One of the anticipated outcomes of such brainstorming is an exhaustive list of potential solutions to the problem. Another hoped-for result is that one member's idea, even if impractical, may stimulate another member to leverage that first idea into a more useful one. It is extremely important that the participants *separate the generation of ideas from their evaluation*. Brainstorming is a technique for generating ideas which, by its nature, will lead to ideas that are subsequently rejected. At its core, brainstorming is an activity based on respect for the ideas of others, even to the point of being willing to suspend judgment on them temporarily. If a team focuses on evaluating ideas on the spot, it is likely to limit the willingness of team members to offer ideas. The team thus limits the extent to which creative changes or "piggybacked" ideas can emerge from earlier offerings that are now less forthcoming.

Separate the generation of ideas from their evaluation.

Our previous discussion of group formation stages is relevant in helping us to understand when a team can effectively engage in activities like brainstorming. Clearly in the forming and storming phases, trust and confidence are likely to be absent from the team's dynamics. Indeed, the team may still be trying to define what the real task of the team is. The team is not likely to be in agreement about how seriously it will attempt to meet the team's nominal goals. As such, the team is almost certainly going to be unable to undertake effective brainstorming just yet. On the other hand, at the norming stage, the team is likely to be developing a consensus about norms of behavior, which may make it possible for the team to set the sort of respect-based behavior that brainstorming requires. The team is most able to engage in brainstorming during the performing stage. This implies that models of design that allow for considerable early research and problem definition are most likely to be consistent with the underlying dynamics of how the team will perform.

10.1.2 Leading versus managing in design teams

A common misunderstanding among people is that leading and managing are the same activity, particularly in professional settings. This can result in all sorts of problems, including a failure to utilize fully all of the talents of a team, so we should be very explicit about the differences. In Chapter 1 we defined management as "the process of achieving organizational goals by engaging in the four major functions of planning, organizing, leading, and controlling." This definition clearly highlights the fact that leadership is only one of the functional processes of management. We also defined leading as "the ongoing activity of exerting influence and using power to motivate others to work toward reaching organizational goals." Successful managers certainly have or develop leadership skills, using influence and power to encourage their team to realize team goals. At the same time, however, many leaders are not concerned with planning, organizing or controlling the team's progress toward their goals; that is, not all leaders are managers.

This distinction can be very empowering for a good team. Many people's leadership skills derive from areas generally unrelated to what we think of as management. For example, a technically strong engineer may have her opinions and views taken seriously simply out of respect for her professional insights, even if she holds no formal management role. Similarly, there are individuals who have what might be called "moral authority," which comes from living and acting in exemplary ways. These people's perspectives are highly regarded, even when they are discussing matters outside of their normal areas of expertise. Effective teams nurture leadership in all its forms, recognize the ability of others to "lead from within," and allow themselves to be motivated by more than formal authority structures.

At the same time, effective teams also recognize the benefits of being well managed, and so they encourage their leaders and managers to plan, organize, and control the team's actions to help realize its goals. Put most simply, good teams need leaders and managers, and they may or may not be the same people.

10.1.3 Constructive conflict: Enjoying a good fight

Whenever people get together to accomplish tasks, conflict is an inevitable by-product. Much of this conflict is healthy, a necessary part of exchanging ideas, comparing alternatives, and resolving differences of opinion. Conflict can, however, be unpleasant and unhealthy to a group, and it can result in some team members feeling shut out or unwanted by the rest of the group. Thus, a solid understanding of the notions of constructive and destructive conflict is an essential starting point for team-based projects. Even in those cases where team members have been exposed to conflict management skills and tools, it is useful to review them at the start of every project.

The notion of constructive conflict had its origins in research on management conducted in the 1920s. It was observed that the essential element underlying all conflict is a set of *differences*: differences of opinion, differences of interests, differences of underlying desires, etc. Conflict is unavoidable in interpersonal settings, so it should be understood and used to increase the effectiveness of all of the people involved. To be useful, however, conflict must be constructive. *Constructive conflict* is usually based in the realm of ideas or values. On the other hand, *destructive conflict* is usually based on the personalities of the people involved. If we were to list situations where conflict is useful or healthy, we might find such items as "generating new ideas" or "exposing alternative viewpoints." A similar list of situations in which conflict reduces a team's effectiveness would probably include items such as "hurting feelings" or "reducing respect for others."

Constructive conflict is based on ideas and values.

The difference between *destructive, personality-based conflict* and *constructive, idea-based conflict* must be recognized by a team from the outset. While a team is establishing norms, and even before these have been formalized or agreed to in the "norming" phase, the team must establish some basic ground rules that prohibit destructive conflict, and it must enforce them by responding to violations of these ground rules. A team must not permit destructive conflict, including insults, personally denigrating remarks, and other such behaviors, from the outset or it will become part of the team's culture.

Once we note this difference between constructive and destructive conflict, it is useful to recognize various ways that persons can react to conflict. Five basic strategies for resolving conflicts have been identified:

- *avoidance*: ignoring the conflict and hoping it will go away
- *smoothing*: allowing the desires of the other party to win out in order to avoid the conflict
- *forcing*: imposing a solution on the other party
- *compromise*: attempting to meet the other party "halfway"
- *constructive engagement*: determining the underlying desire of all the parties and then seeking ways to realize them

The first three of these — avoidance, smoothing, and forcing — all turn on the notion of somehow making the conflict "go away." Avoidance rarely works, and serves to undercut the other party's respect for the person who is hiding from the conflict. Smoothing may be appropriate for matters where one or both of the parties in conflict really don't care about the issue at hand, but it will not work if the dispute is over serious, important matters. Once again, the respect of the person "giving in" may become lost over time. Forcing is only likely to be effective if the power relationships are clear, such as in a "boss-subordinate" situation, and even then, the effects on morale and future participation may be very negative. Compromise, which is a first choice for many people, is actually a very risky strategy for teams and groups. At its core, it assumes that the dispute is over the "amount" or "degree" of something, rather than on a true underlying principle or difference. While this may work in cases such as labor rates or times allocated for something, it is not likely to be effective in matters such as choosing between two competing design alternatives. (We cannot, for example, compromise between a tunnel and bridge by building a suspension tunnel.) Even in those cases where compromise is possible, we should expect that the conflict will be likely to recur after some period of time. Labor and management, for example, often compromise on wage rates — only to find themselves revisiting the very same ground as soon as the next contract opens up. That leaves us with constructive conflict as the only tool that holds the possibility of stable solutions to important conflicts.

Constructive conflict takes as its point of departure an honest telling of and listening to each party's underlying desire. Each side must reflect on what it truly wants from the conflict and honestly report that to the other parties. Each side must also listen carefully to what the other party really seeks. In many cases, the conflict is not based on the apparent problem, but rather arises because each party's underlying desires are different.

The originator of the idea of constructive conflict, Mary Parker Follett, told the following paradigmatic anecdote. She was working in a library at Harvard on a wintry day, with the windows closed. Another person came into the room and immediately opened one of the windows. This set the stage for a conflict and for identifying a way to resolve that conflict. Each of the five resolution alternatives outlined above was available, but most of them were unacceptable. Doing nothing or smoothing would have left Ms. Follett uncomfortably cold. Compromising by opening the window halfway did not appear to be a viable alternative. Instead, she chose to speak with the other person and express her desire to keep the window closed in order to avoid the chill and draft. The other party agreed that this was a good thing, but noted that the room was very stuffy, which in turn bothered his sinuses. Both agreed to look for a reasonable solution to their underlying desires. They were fortunate to find that an adjacent work area also had windows that could be opened, thus allowing for fresh air to enter indirectly without creating a draft. Obviously, this solution was possible only because the configuration of the library allowed it. Nevertheless, they would not have even looked for this outcome except for their willingness to discuss their underlying desires. There are many cases where this will not work, such as when two persons wish to each marry the same third person. There are, however, many cases where constructive engagement will work, both to increase the solution space available to the parties in conflict and to heighten the understanding and respect of the other party. Even when the team is forced to revert to one of the "win-lose" strategies, the team should always first consider constructive engagement for resolving important conflicts.

10.2 MANAGING DESIGN ACTIVITIES

Much of the environment in which design takes place is created by the designer, who decides, among other things, which activities are to be performed, who is going to perform them, and the order in which they are to be completed. Thus, creating and controlling the design environment is part of a process that we term *managing design*. In this chapter we will look at some of the tools that are available to the design team for planning, organizing, leading, and controlling design projects. Before we do so, however, it is useful to consider briefly some aspects of design that make design projects hard to manage. We will then see why the careful application of appropriate tools can help the designer.

A project, whether it is about design or aimed toward some other goal, can be characterized as "a one-time activity with a well-defined set of desired ends." Successful project management is usually judged in terms of scope, budget, and schedule. That is, the project must accomplish the goals (in our case a successful design), be completed within the resource limits available, and it must be done "on time." Consider for a moment our beverage container example. The design that is developed must meet the concerns of the beverage company, including developing an attractive, unbreakable container that is easily and cheaply made. The designers must meet an agreed-upon budget or the design firm may not be able to stay in business over the long run. The schedule might be dictated by marketing concerns, such as producing the new container in time to sell the new juice in the coming school year. In this case, the scope of a successful design effort will balance all three sets of concerns, including achieving the client's desire to introduce a new product within the budgetary constraints of the design firm, and meeting the timing dictated by the client's marketing plan. On the other hand, student design projects in a course environment have budget concerns measured in student hours, because students have other commitments, e.g., other courses and extracurricular activities. The schedule would likely reflect the timing of the course, for example, that the design be completed within a quarter or semester. The scope of this version of the project would address the set of client, user, and faculty concerns. In either context, design firm or design course, it is the *3Ss* — the triad of

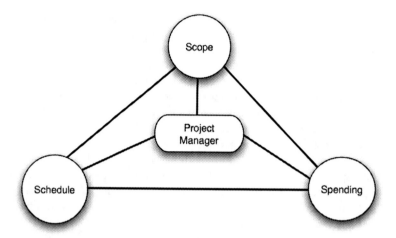

FIGURE 10.2 Project management can be thought of as balancing the three Ss: scope, scheduling, and spending.

Project management is concerned with scope, spending, and scheduling.

scope, spending, and scheduling — that provides the basis for the tools we use to manage projects. Figure 10.2 is a graphical representation of this 3S model. Notice that one of the key roles of the project managers is to maintain a balance between each S.

Having identified the 3Ss of projects, we might ask whether design projects are different than other types of projects, such as construction projects, and if they are, how? It turns out that there are several important differences, the first of which lies in the definition of a project's scope. For many projects, an experienced project manager knows exactly what constitutes a success. A stadium construction project has plans that must be followed, including architectural renderings, detailed blueprints, and volumes of detailed parts and fabrication specifications. There are also accepted practices in the construction industry, so that both the construction firm and the project manager understand the scope of this construction project. On the other hand, the designer often cannot know what constitutes a success until a design project is well underway because she may not have had a full range of discussions with the client and with users. Thus, she may not be able to clarify all of the project's objectives and reconcile all of the stakeholders' views. Similarly, while a construction project is expected to have only one outcome, a design project might yield many acceptable designs.

Scheduling is also different for design projects. In a construction project, the project manager can plan to do certain activities, knowing how long each will take, and can determine a logical ordering for them. For example, digging a hole for a foundation might take two weeks, but it clearly must be done before forms are built (three weeks) and concrete is poured into the forms (two days). Assembling and organizing such planning and scheduling data allows a project manager to determine how much time the project will take. In many cases, the manager of a design project asks how much time is available, rather than summing up how long the tasks take. This approach to scheduling represents the design team's intent to use all of the available time to generate and consider many viable design alternatives, while still trying to meet the constraint of finishing the project by a specified, agreed-upon time.

These differences lead us to wonder whether project management techniques are appropriate for use in engineering design projects. After all, if the scope is ambiguous and the timing is up to the client (and may even seem arbitrary), how relevant are tools developed for well understood projects? It turns out that the uncertainties and external impacts associated with design projects tend to make some of the management tools even more useful and necessary. We will see that project management tools can be useful for gaining consensus by the design team about what must be done, who will do it, and when things must be done — even if our expectations about the form of the final design are initially up in the air.

The team nature of design projects also lends support to the use of project management tools. There are important issues that must be addressed when the team moves into the performing stage, including the need to effectively communicate the activities, schedule, and progress to each team member and to other stakeholders. Work must be allocated fairly and appropriately. And we must ensure that tasks have been done properly and in a sequence that allows team members who depend on prior work to plan their actions. The management tools we introduce here are helpful in each of these circumstances.

10.3   AN OVERVIEW OF PROJECT MANAGEMENT TOOLS

Recall from Chapter 2 that we can model the process by which we eventually move from a client's problem to a detailed design of a solution. This process uses a number of formal design methods and several means of gathering and organizing information in order to generate alternatives and evaluate their effectiveness. While the methods and means discussed can be assigned to the steps in the design process, design is not a simple "cookbook" process. Similarly, managing the design process also requires more than rote application of project management tools. In this section we briefly describe the tools that we can use to plan a design project, organize our design activities, agree on our responsibilities for the project, and monitor our progress.

We noted in Chapter 1 that management consisted of four functions:

- *Planning* a project leads immediately back to the 3S model of project requirements. We must define the scope of the project, determine how much time we have to accomplish our goals (scheduling), and assess the level of resources (spending) we can apply to the project.

- *Organizing* a project consists largely of determining who is responsible for each task area or activity of the project, and which other human resources can be called upon to work on the tasks.

- *Leading* a project means motivating a team by showing that the tasks can be understood, the division of work is fair, and the level of work can produce satisfactory progress toward the team's goals.

- *Controlling* can only be done in the context created by the 3Ss and the plans that support those 3Ss. Tracking progress is meaningful only against a stated set of goals. Further, we can change our plans or take corrective action only if the team is confident that the plans they have developed are, in fact, going to be used.

A tool that is increasingly used to set forth the overall project goals, time, and budget constraints, and the key participants is the *team charter*. A team charter is, in a sense, the team's agreement with other stakeholders on what the project is about, including what constitutes project success, and what some of the limits on the project consist of.

The primary tool that is used to determine the scope of our activities is the *work breakdown structure* (WBS). The WBS is a hierarchical representation of all of the tasks that must be accomplished to complete a design project. Project managers use WBSs to determine which tasks must be done. They will generally break the work down (hence the name) into pieces sufficiently small that the resources and time needed for each task can be estimated with confidence.

We use the *linear responsibility chart* (LRC) to identify which team member has the primary responsibility for the successful completion of each task in the WBS, and to identify others who must participate in finishing that task. The LRC uses a matrix format to match each of the tasks requiring management responsibility with the members of the team, the client, users, and other stakeholders. This is particularly important for team-based activities, both to identify clearly who is responsible for each task and to indicate all individuals that should be involved (e.g., teammates, the client, or an external expert).

We can schedule activities in several ways:

- A *team calendar* shows all of the time that is available to the design team, with highlights that indicate deadlines and time frames within which work must be completed.

- A *Gantt chart* is a horizontal bar graph that maps various design activities against a time line.

- An *activity network* graphs the activities and events of the project, and shows the logical ordering in which they must be performed.

Team calendars are used by virtually all teams, particularly those doing design work. Gantt charts and activity networks, on the other hand, are rarely used in student projects in first courses in design, and so are not covered here. They are important tools for professional engineers, and are well discussed in most project management textbooks. They certainly should be learned as part of an engineering education.

A *budget* is a list of all of the items that will incur an economic cost, organized into some set of logically related categories (e.g., labor, materials, etc.). The budget is the key tool for managing spending activities in a project. Note that there is an important distinction between the budget for doing the design, or design activities, and the budget needed to produce or build the artifact being designed. Our concern is primarily with the budget needed to do a design.

There are other control methods that are used to manage projects, but many are simply not well suited to design projects. For example, *earned value analysis* relates costs and schedules to planned and completed work. While earned value analysis is useful for certain large-scale projects with effective reporting systems, it is overkill for the smaller, team-based projects that we discuss. On the other hand, the *percent-complete matrix* (PCM) that relates the extent of the work done to the total level of all work to be done is a more useful and appropriate tool. We will develop a version of the PCM that is appropriate for smaller team design activities.

In the following sections we discuss and provide examples of the tools mentioned above. Just as each of the design methods and means will not be appropriate for every design project, these management tools will not have to be used by every team on every project. However, since the tools are important to good, team-based design work, and since we can never predict with certainty the kinds of design activities we will undertake in the future, these management tools are worth having in our individual arsenals.

10.4 TEAM CHARTERS: WHAT EXACTLY HAVE WE GOTTEN OURSELVES INTO?

The team charter is, in many ways, the management analog to the revised problem statement that we discussed in Chapter 3. Recall that at the outset of the project there may be errors, biases and incomplete information in the initial problem statement, and that an effective designer works to restate the problem in a way that clearly states the client's needs while offering a large design space within which to meet those needs. In a similar vein, the project manager needs to understand:

- the goals of the project, including both minimally acceptable goals and ''stretch'' goals;
- how those goals align with larger organizational goals;
- the authority for the project;
- the project deliverables;
- the time frame for the project, including any limits on the schedule;
- the resources available for the project; and
- any unusual circumstances associated with the project.

This information, once gathered, can be written up and distributed to the client, managers within the larger organization, and any team members. In many organizations, drafting of the charter is begun before the team or even the project manager is selected, and is completed by the project manager. Often the signatures of the sponsor, senior managers of the design firm and the project manager complete the chartering process.

The written charter can be used for a number of purposes throughout the project's life, including describing the project to prospective team members, gaining commitment from team members, and setting conflicts about resources, timing or scope. Because of its written nature, it can also be used later in the project to avoid ''scope creep'' or ''mission growth,'' the tendency for project goals to be expanded as the problem becomes better understood, or as team interests emerge.

Team charters are a relatively new phenomenon in the engineering and design community, but the information in the charter is usually readily available to project managers early in the project. As noted above, the revised problem statement usually includes project goals, deliverables, and schedule constraints. Wise project managers know to negotiate resources early in the process of agreeing to take on a project, and the authority and special circumstances are likely to come out during early discussions as well.

Young design engineers may not be familiar with the relationship between project goals and organizational goals. Clarifying this relationship is important, both for the success of the project and for the professional development of the engineer. If the project goals and organizational goals are not aligned, it is quite possible that the project will be an apparent success but ultimately go nowhere. If the project and organizational goals are aligned, it is possible that a project that appears relatively unexciting may be part of a larger, more exciting program.

The length of a team charter can, in many cases, be as little as a page or two, another similarity with the revised project statement. The key is that the project manager and the team recognize the charter's value as a negotiating tool early and a formal record later.

As noted above, the team charter may begin with the members of the organization who select the project manager and the project team. This is because they will likely understand the organization's goals and the general needs of the sponsor or client. As the team develops the revised project statement and the work breakdown structure, the project manager is able to add information to the team charter which can then be used to guide the project.

10.5   WORK BREAKDOWN STRUCTURES: WHAT MUST BE DONE TO FINISH THE JOB

Most of us would be a bit overwhelmed if asked to describe exactly how to start and drive a car, even though this is a common task. We might start with something like "first you get in the front seat" — which pre-supposes that you know how to get into the car! If forced to describe this task to someone from a country with few cars, or where many people rely exclusively on public transportation, we might want to break down the task into a number of task groups, such as getting into the car, adjusting the seat and mirrors, starting the engine, driving the car, and stopping the car. We might even want to review the entire plan before starting the car so that a driving student would know how to stop before starting out. This decomposition of tasks or concepts is the central idea of the work breakdown structure (WBS). When we are confronted with a very large or difficult task, one of the best ways to figure out a plan of attack is to break it down into smaller, more manageable subtasks.

A more compelling example is that of a team that has been asked to design a spacecraft. The team will have to design across several specialties, including propulsion, communications, instrumentation, and structures. Here the design team leader will work very hard to ensure that the team's propulsion experts are actually assigned to propulsion tasks, so that the experts work on tasks relevant to their expertise. In order to do this properly, the team leader must determine just what those tasks are. The WBS is a listing of all the tasks needed to complete the project, organized in a way that helps the project leader and the design team understand how all of the tasks fit into the overall design project.

The WBS is the most important management tool for design projects — it organizes all of a project's tasks.

We show a work breakdown structure for the beverage container design example in Figure 10.3. At the top level it is organized in terms of seven basic task areas:

- Understand customer requirements
- Analyze function requirements

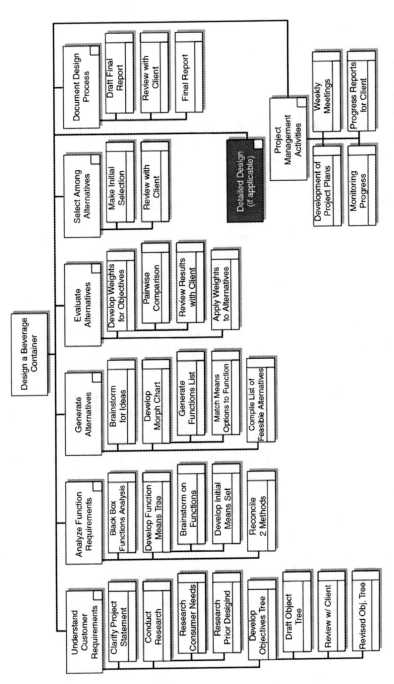

FIGURE 10.3 A work breakdown structure (WBS) for the beverage container design project. Because the design project is just beginning, the structure necessarily takes on a formal and somewhat generic framework. Note, however, that the designers are already aware of some details, such as the distinction between identifying consumer needs and prior designs.

- Generate alternatives
- Evaluate alternatives
- Select among alternatives
- Document the design process
- Manage the project
- Detailed design

We see also that each of these top-level tasks can be broken down in greater detail. Because of page size limitations, in this example we show great detail only for some of the tasks (e.g., understanding customer requirements). If we were actually part of a team carrying out this project, we would likely go into much greater depth in all of the areas. Also, we should note that this method of organizing the work is not the only way to structure a WBS. We will see several alternative organizing frameworks later in this chapter.

Several observations about the WBS depicted in Figure 10.3 are in order. First, the basic principle for a WBS is that each item that is taken to a lower level is *always* broken down into *two or more subtasks* at that lower level. If the task is not broken down (so the lower level is a lone entry), then either the lower level is incomplete or it is just a synonym for the upper level. Second, if we cannot determine how long an activity will take or who will do that activity, then the key WBS rule is that we should break it down further. In fact, experienced project managers will be more inclined to have shorter, less detailed WBSs than relatively inexperienced managers because their greater experience makes them more likely to be able to aggregate subtasks into identifiable, measurable tasks.

Our third observation is that a WBS should be *complete* in the sense that any task or activity that consumes resources or takes time should be included either in the WBS explicitly or as a known component of another task. That is why the tasks of documentation and management are shown in Figure 10.3. Activities such as writing reports, attending meetings, and presenting results are essential to the completion of the project, and failure to plan for them as work will certainly result in problems later. Estimating who and what is required, and for how long, is a valuable discipline for any project, whether in a design course or in the "real world," both for developing the design and for ensuring that there is sufficient time to document and present results.

Our final observation on the WBS is that any part of its hierarchy of tasks should *add up*, that is, the time needed to complete an activity at a top level should be the sum of the times for tasks listed at the level below. Thus, breaking work down to the next level below must be done thoroughly and completely.

The last two observations about WBSs provide us with two criteria for evaluating their utility:

- *Completeness* means that the WBS must account for all of the activities that consume resources or take time.
- *Adequacy* means that tasks should be broken down to an adequate level of detail, such that the project team can determine how much time it will take to do them.

It is also important for us to note what the WBS is *not*. First, a WBS is *not* an organization chart for completing a project. This may be confusing because ''org charts'' are visually similar graphics. The WBS is a breakdown of tasks, *not* of titles, roles, or people in an organization. Second, a WBS is (also) *not* a flow chart showing temporal or logical relationships among tasks. In many cases the listing of the tasks will be organized in such a way that a task (e.g., writing the final report) is shown in a different part of the hierarchy than other tasks that must precede it (e.g., all of the design, building, and testing that is being reported). Third and last, a WBS is *not* a listing of all the disciplines or skills that are required to complete the tasks. In many cases the tasks to be completed may require a number of different skills (e.g., electrical engineering and propulsion engineering). Tasks performed by professionals with these skills can be combined into the same part of the hierarchy if that listing of the tasks meets the above criteria of completeness and adequacy.

Figure 10.4 shows another sample WBS, in this case taken from an electrical hardware project. Here the design subtasks are organized in terms of electrical and mechanical design. This may appear to violate our concern with disciplines noted above — until we realize that for this project this is simply a convenient way to organize *the tasks*, which is always our key intent for a WBS.

Figure 10.5 shows yet another sample WBS for some of the engineering tasks of associated with designing a new car. Note that this WBS is not in a graphical form, although it is still hierarchical. While graphical forms may offer clarity, it is perfectly permissible to use a tabular form such as that shown in Figure 10.5. In fact, tables are a common means of collecting the information together once a WBS has been developed into its final form. Note also that this example has the work broken down by the various car components. This is also permissible, so long as the WBS satisfies our concerns about completeness and adequacy.

In the end, then, the WBS is a tool for a project team to use to ensure that they understand all of the tasks that are needed to complete their project. That is why a WBS is so valuable for determining the scope the project.

10.6 LINEAR RESPONSIBILITY CHARTS: KEEPING TRACK OF WHO'S DOING WHAT

Once the tasks that are to be done have been identified in a WBS, a design team has to determine whether or not it has the people, the human resources, to accomplish those tasks. The team also has to decide who will take responsibility for each task. This can be done by building a *linear responsibility chart* (LRC). The LRC lists the tasks to be managed and accounted for and matches them to any or all of the project participants. Figure 10.6 shows a simplified LRC corresponding with many of the tasks in the beverage container design WBS of Figure 10.3. In addition to all of the top-level tasks, the subtasks associated with several of the lower levels are also given. In practice, it is advisable to show all of the top-level tasks, as well as those subtasks that may require management attention. Because of the evolving nature of design projects, less experienced project managers may want to assign responsibility only for the top-level tasks and the next set of tasks that require the team's attention. This allows the team's roles and responsibilities to develop with experience, along with the project.

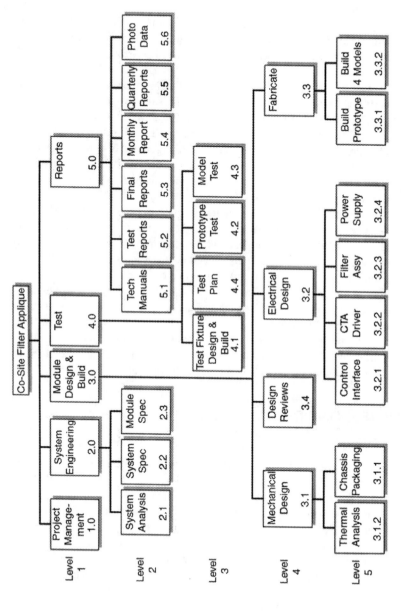

FIGURE 10.4 A work breakdown structure (WBS) for a hardware development project. Note that at the lowest level (level 5), the electrical design tasks have been broken down to the level of the components to be designed. At this stage, the designer of the power supply might follow another WBS similar to Figure 10.3. (After Kezsbom, Schilling, and Edward 1989.)

```
                    PRIMAVERA PROJECT PLANNER

  Date 08JAN98    -----WORK BREAKDOWN STRUCTURE-----

  ENGR - Active Projects for the Fiscal Year

  Structure : xxx.xxx.xx.x

  WBS Code          Title

  94 All Projects
     94E  All Engineering Projects
        94E.101  Project E101
           94 E.101.A  General
           94E.101.A7
           94E.101.B   Air Bag
           94E.101.C   Mechanical Release System
           94E.101.D   Electrical Systems
           94E.101.E   Interior Dashboard
           94E.101.F   Structural Door System
  94E.102 Retrofit Automobile Plant
           94E.102.A   Enclosure
           94E.102.B   Structural System
           94E.102.C   Mechanical System
           94E.102.D   Electrical System
           94E.102.E   Estimating
           94E.102.F   Specifications
           94E.102.G   General
     94I  All Installation Projects
        94I.101  Tooling & Equipment Installation
           94I.101.A   Structural Slab
           94I.101.B   Piping
           94I.101.C   Equipment
           94I.101.D   Electricity
           94I.101.E   Interior Finishes
           94I.101.F   Ventilation & Plumbing
           94I.101.G   General
```

FIGURE 10.5 A work breakdown structure (WBS) for the engineering projects of an automotive firm. This nongraphical WBS organizes the firm's activities according to systems for the autos and overall factory installation projects. The level of detail is not very high, and presumably the firm would have many supporting WBSs for all of these projects.

As we can see in Figure 10.6, there is a row for each task, within which the role, if any, of each project participant is given. These roles do not necessarily mean assuming the primary responsibility. Indeed, most of the participants will play some sort of supporting role for many of the tasks, such as reviewing, consulting, or working at the direction of whoever is responsible. A column is assigned to each of the participants; this allows them to scan down the chart to determine their responsibilities to the project. For example, we see that the client (or the client's designated liaison) will be called upon to give final approval to the objectives tree, the test protocol, the selected design, and the final report. The liaison must also be consulted during some of the pre-design activities, and will be asked to review various intermediate work products. The client's research director

Linear Responsibility Chart	Team Member #1	Team Member #2	Team Member #3	Team Member #4	Team Member #5	Director of Design	Client Liaison	Client Research Director	Outside Consultant
1.0 Understand Customer Requirements	1								
1.1 Clarify Problem Statement	1	2	2	2	2		3	4	
1.2 Conduct Research	1	2		2	2		4	4	4
1.3 Develop Objectives Tree	1								4
1.3.1 Draft Objectives Tree			2	2		5	5	3	4
1.3.2 Review w/ Client	1		2			5	5	3	4
1.3.3 Revise Objectives Tree	1		2	2		6		4	
2.0 Analyze Function Requirements	2	2	1	2	2	5	4	3	3
3.0 Generate Alternatives				1					
4.0 Evaluate Alternatives	5	1	2	2	2				
4.1 Weigh Objectives	1	2				5	6		
4.2 Develop Test Protocol	5	1			2	5	4	3	3
4.3 Conduct Tests		1	2		2			5	3
4.4 Report Test Results	5	2	2		1	5	5	5	5
5.0 Select Preferred Design	1	2			2	5	6	4	4
6.0 Document Design Results		1							
6.1 Design Specifications	1			2		6			
6.2 Draft Final Report	5	1		2		5	5		4
6.3 Design Review w/ Client	1	2		2		5	3	4	3
6.4 Final Report	5	1		2	2	5	6	4	4
7.0 Project Management	1								
7.1 Weekly Meetings	1	2	2	2	2				
7.2 Develop Project Plan	1	2	2	2					
7.3 Track Progress	1					5			
7.4 Progress Reports	1						5		
Key:									
1 = Primary responsibility									
2 = Support/work									
3 = Must be consulted									
4 = May be consulted									
5 = Review									
6 = Final Approval									

FIGURE 10.6 A linear responsibility chart (LRC) for the beverage container design project. Each participant in the project can read down his column and determine his responsibilities over the entire project. Alternatively, the Project Manager can read across a row and determine who is involved with each task.

is somewhat of a resource to the team, and may be consulted at different points of the project, but *must* be consulted regarding the test protocol. The team's boss, the director of design, has asserted the right to be kept informed at a number of points in the project, most notably in terms of design reviews. The project also has access to one or more outside experts, who are generally available for consultation and who must review the design and certain other (unspecified) documents.

We also note from the LRC in Figure 10.6 that the team leader does not always have the primary responsibility for the project. It is often the case in team projects that the team leader will not be responsible for tasks that are outside of her technical area of expertise, although she may want to specify a review or support role in order to remain informed. Sharing responsibility this way is sometimes quite difficult both for team leaders and for teams. Sharing responsibility, which is strongly tied to the storming phase of group formation, requires practice. Therefore, the LRC can be used to make this part of team formation more explicit to the team, and to allow the team to reach consensus on who will be doing what in the project.

LRCs can ensure that the workload is distributed fairly and equitably.

The LRC can also be used to let outside stakeholders in a project understand what they are expected to do. In the beverage container example, the client's research director clearly has an important role to play in the safe conduct of the testing phase. It is very important that this person know early on what is expected, and to be allowed to plan accordingly. Similarly, the outside experts may need to allocate time to ensure availability, and the director of design may need to make resources available to pay for the experts' time.

It should be clear by now that the LRC can be a very important document for translating the "what" of the WBS into the "who" of responsibility. At the same time, we might be tempted to use the LRC to avoid admitting that the team doesn't know something. For example, if every task has every team member assigned in a support or work role, it should raise serious doubts in our minds about whether we really do understand those roles. Similarly, if a team leader claims primary responsibility for all the tasks, her team will certainly be tempted to consider the LRC as little more than a power grab or a mirror of the team leader's insecurities. It is also important that a team understand that it may be necessary to revisit roles as the project unfolds, especially if the team is relatively inexperienced or the project is initially ambiguous.

10.7 SCHEDULES AND OTHER TIME MANAGEMENT TOOLS: KEEPING TRACK OF TIME

Scheduling and similar time management tools help us identify in advance those things that will really mess up our project if they aren't done on time. Three primary scheduling tools are frequently used in project management: a calendar, an activity network, and a Gantt chart. A team calendar is likely the most familiar as it performs many of the same functions as personal calendars or diaries: It simply maps project deadlines or due dates onto a conventional calendar.

The activity network and the Gantt chart are more powerful and, consequently, possibly more useful for large-scale projects. Both are graphical representations of the logical relationships between tasks and the time frames in which they are to be done. Indeed, most software programs for project management use the same information to generate both activity networks and Gantt charts. It is outside the scope of most conceptual design projects conducted by students to develop Gantt charts and activity networks, so we will not consider them further.

Team calendars should be reviewed weekly (at least!).

A team calendar is simply a mapping of deadlines onto a conventional desk or wall calendar. Such deadlines will certainly include externally imposed ones, such as commitments to clients (or to professors for academic projects), but should also include team-generated deadlines for the tasks developed in the WBS. In this sense, the team calendar is really an agreement by the team to assign the resources and time necessary to meet the deadlines shown on the calendar. Figure 10.7 shows a team calendar for a student design team that is seeking to complete its project by the end of April, an externally imposed deadline. Note that the calendar includes several deadlines over which the team probably has no control, such as when the final report is due and when in-class presentation of results is to be done. It also includes routine or recurring activities, such as Tuesday night team meetings. Finally, it includes some deadlines that the team has committed to realizing, such as completing a prototype by 5 p.m. on April 2.

Several points should be kept in mind in setting up a team calendar. First, the idea of a team calendar implies that the deadlines are all understood and agreed to by everyone on the team. As such, the calendar becomes a document that can — and should — be reviewed at every team meeting. Second, the team calendar should allow times that are at least consistent with the time estimates generated in the WBS. If a task was determined to take two weeks to complete, there is little point in allowing it only one week on the team calendar. A final point to note is that the team calendar, while easily understood by members of the team, *cannot by itself capture the relationship between activities*. For example, in Figure 10.7 we see that building the prototype precedes proof-of-concept testing *only* because the team chose to put it that way. For many artifacts a proof of concept may actually precede building a final prototype. The team calendar cannot address this sort of problem, nor can it "remember" team decisions of this sort. For this reason, a team calendar is most useful for small projects or in cases where it is supplemented with other project management tools (such as those we discuss in the following sections).

10.8 BUDGETS: FOLLOW THE MONEY

Budgets are difficult but essential tools for project management. They permit teams to identify the financial and other resources required, and to match those requirements to the available resources. Budgets also require teams to account for how they are spending project monies. Finally, budgets serve to formalize the support of the larger organization from which the team is drawn.

Many of the engineering economics concepts that we will discuss in Chapter 11 are relevant for budgets, so we will defer much of our discussion of these concepts at this time. Further, we usually don't need large, complex budgets for doing the sort of design projects that are likely to be done in academic or similar settings. (Remember, as we have said before, we are concerned with the *budget for doing the designs*, not with the budget for making the designed objects.) Thus, design project budgets normally include research expenses, materials for prototypes, and support expenses related to the project.

We will limit our budget discussions to the above categories of costs, that is, to materials, travel, and incidental expenses. This means that in attempting to budget for a design project, it is necessary at the outset to try to determine what sorts of solutions are

	March		Design Team			May	

March
S M T W T F S
1 2 3 4 5 6
7 8 9 10 11 12 13
14 15 16 17 18 19 20
21 22 23 24 25 26 27
28 29 30 31

Design Team

April

May
S M T W T F S
1
2 3 4 5 6 7 8
9 10 11 12 3 14 15
16 17 18 19 20 21 22
23 24 25 26 27 28 29
30 31

Sun	Mon	Tue	Wed	Thu	Fri	Sat
				1	2 5:00PM Prototype Built	3
4	5	6 7:00–8:15PM Team Meeting	7	8	9 11:00AM Proof of Concept Due	10
11	12 11:00AM Rough Outline Due	13 7:00–8:15PM Team Meeting	14	15	16 5:00PM Topic Stce Outline Due	17
18	19 11:00AM Prsntion Outline Due	20 7:00–8:15PM Team Meeting	21 11:00AM Slides Due	22	23 5:00PM Draft Final Report Due	24
25	26 10:00–11:00AM Present Results	27 7:00–8:15PM Team Meeting	28	29	30 5:00PM Final Report Due	

FIGURE 10.7 A team calendar for a student design project. Note that externally imposed deadlines, team commitments, and recurring meetings are all included on the calendar. It is usually better to make the team calendar "too complete" than it is to leave out potentially important milestones or deadlines.

possible. This is not to say that we have determined the solution, rather that we should consider resource needs earlier than might really be desirable. One effect of this is that design projects often try to establish *not to exceed* budgets that set limits on what can be expended by specifying the highest cost that might come about. The danger in this approach is that if it is done this way routinely, for all projects in an organization, resources may be set aside for design projects that will never be used.

As a final note, it is important to properly value the time invested in a design project by each and every member of a design team. This is important even in student design projects done in design courses. (In fact, there is a tendency to undervalue this very scarce resource just because we did not call out the time in the budget.) One way to place a value on a team member's time is to adapt the "algorithm" that employers use to "bill out" the time of an engineer who is working on a project. Most firms charge between *two and four times* an employee's direct compensation when they bill a client for that employee's time. That multiplier covers fringe benefits, overhead costs, supervision, and profit. If we were to bill student time at a minimum wage rate of only $8.00 per hour, a team of four students working ten hours each week on a project for ten weeks would be billed out by a design firm at $6,400–12,800 for the entire project. Put in the simplest terms, time is a valuable, scarce, and irreplaceable resource — don't waste it!

10.9   TOOLS FOR MONITORING AND CONTROLLING: MEASURING OUR PROGRESS

We have now developed a plan, a schedule, and a budget. How do we track our team's performance relative to the plan? This is an important question, but it can be very hard to answer. The manager of a construction project can go out and see if a task has been done by the date planned. In a design project, however, monitoring and control are more subtle and, in some ways, more difficult. Therefore, it is essential that the members of a team agree upon a process for monitoring their joint progress before the project gets very far underway.

There are a number of techniques and tools available to monitor projects, but these often involve team members filling out time sheets, punching clocks, or other accounting tools. For smaller design projects, especially academic projects, these kinds of tools may not be very effective. We will therefore present a simplified version of the *percent-complete matrix* (PCM) that is widely used in industry to relate the extent of work done on the parts of a project to the status of the overall project.

The goal of the PCM is to use the information in the WBS and the budget to determine the overall status of the project. Constructing a PCM requires only that we know the cost of each item or area of interest, and the percent of the total cost corresponding to that item. Then the PCM allows us to input the percent of the work on that task or work item and, by summing over all the items in the project, we can calculate the total percent of the project completed. In general, the method is best suited to cases where there some clear method of calculating progress is available. If, for example, the foundation work of a building project constitutes 25% of the total expected costs of a project, then we have completed at least 12.5% of the project when we have completed one-half of the foundation work. A manager can periodically update progress in each of the general areas in the WBS to determine overall project progress.

In some cases a physical measure can serve as a proxy for progress, such as yards of concrete poured compared to the total volume called for in the plan, or tons of steel erected compared to budgeted totals. While this approach has significant appeal in standard projects, design projects are generally more concerned with progress relative to allowed time than to available budget, and physical measures are not likely to be available. One alternative is to use the estimated duration of the team's activities instead of dollar budgets, together with a simple rule for tracking progress. One simple rule is that a team can immediately claim 33% progress for an activity if work on the activity has begun. However, the team gains no additional progress on the task until the activity is completed. The team receives the remaining 67% credit for the task only on activity completion. In no case is a team given more than 100% credit for a task, no matter how long that activity really takes. Further, the team gets full credit when the work is done, regardless of how long it actually took. (This convention clearly places a premium on careful and accurate decomposition of the work in the WBS.)

Time is the scarcest resource for most design projects — it must be monitored and controlled!

Consider Figure 10.8, in which we show a modified PCM for the beverage container design project. Each of the tasks used in the activity network has been included, except

Percent Complete Matrix

Task	Planned Duration (days)	Percent of Total	Status (see key)	Credit (days)
Start Project	0	0%	2	0.0
Clarify Problem Statement	3	3%	2	3.0
Conduct Research	10	11%	2	10.0
Draft Objectives Tree	2	2%	2	2.0
Review OT	1	1%	2	1.0
Revise OT	2	2%	2	2.0
Analyze Functions	10	11%	1	3.3
Generate Alternatives	10	11%	1	3.3
Develop Weighted Objectives	10	11%	2	10.0
Develop Test Protocol	8	9%	1	2.6
Conduct Tests	20	21%	0	0.0
Report Test Results	5	5%	0	0.0
Select Among Alternatives	3	3%	0	0.0
Document Design Process	10	11%	0	0.0
End Project	0	0%	0	0.0
Total Days Budgeted	94	100%		39.6%

Key: 0 = Not Started, No Credit, 1 = In Process 1/3 Credit, 2 = Completed, Full Credit

FIGURE 10.8 A percent-complete matrix (PCM) for the beverage container design project. Each activity and its share of the overall project is given. The team is given 33% credit when an activity is begun and the balance upon completion. Unless the tasks have been broken down sufficiently, this method can be misleading, but for small projects it does provide a reasonable approximation of progress.

for the summary tasks, such as understanding customer requirements and evaluating alternatives. (The detailed breakouts for these have been included instead.) The PCM shows the planned or budgeted duration for each task, the percent of the total project that the task accounts for, and its status. In those cases where a task has either been started or completed, credit toward the overall project is given. In cases where there is no progress, no credit is given. Three observations are in order. First, the project manager or team leader could give a more exact percentage of completion than the 0, 33%, 100% used in this example, or could do so for selected tasks. Remember that it is the team that chooses the values for the simple, standard rule. Second, the team can compare the progress achieved thus far with the overall time allocated for the project. For example, if the design project were in the fourth week of a ten-week project, the PCM would seem to indicate that the project is more or less on plan. If the team was in the eighth week, this PCM would be cause for alarm. Third and last, we note that if the team has done a good job of determining the nature and duration of the tasks required to finish the project, this PCM and this method will allow them to monitor their work. If they have not, this method is simply an illusion.

10.10   MANAGING THE DANBURY ARM SUPPORT PROJECT

One of our illustrative examples is the student project to design a means to support and stabilize a child's arm while painting or writing. In this section, we look at some interesting and creative approaches that students have used to present their management processes, and use them to reinforce some points made above. A comment is probably appropriate at the outset. These student projects are of very limited duration, and the teams follow the formal design processes discussed throughout this book. It is also the second time they have been through the process, so they are quite familiar with the requirements. They are, however, under considerable pressure to complete the project on time. Because of this, they are allowed some latitude to find their own approach to using the management tools. This leads to some very creative approaches to managing a small project, and also some errors relative to the specific tools. Correcting these mistakes can a useful way of drawing attention to management issues at a time when the team's focus is on the project itself rather than the tools.

Figure 10.9 is an example of a team charter for a Danbury team. Notice that the charter explicitly recognizes that the goals of the team, the engineering faculty, and the client are not necessarily the same and so must all be stated. The charter also clarifies the team's obligation to place the safety of the student at Danbury as its highest priority. Notice that budgetary and other limits are clarified, which is straightforward in this case, but could be much more involved in many circumstances.

Both the teams we have been following introduced their own version of a work breakdown structure, combining other elements into the document in a way that both increased the information content but also introduced some problems. Figure 10.10, for example uses a clever graphical approach to organize the project into short time blocks that allow the team to understand what they need to do and the time frame within which they must act. This can only be effective if the team understands their tasks very well. (They also used the graphic to highlight key time constraints and deadlines in an inset.)

DANBURY SCHOOL ARM SUPPORT PROJECT

TEAM CHARTER

This charter documents key information regarding a project to be conducted by students at Harvey Mudd College (HMC) on behalf of a student at the Danbury School. The faculty advisor will be Professor _____, the liaison for Danbury will be Mr./Ms. _____. The project manager/team leader will be selected at the outset of the project by and from among the members of the student team.

The team agrees to abide by all restrictions and regulations of Danbury School when on site, to place the safety of the client as its highest priority, and to work in accord with HMC's Honor Code.

Goals

The project is assigned as part of HMC's introductory design course, and the students understand that they are expected to work to accomplish HMC's goals for the course, Danbury's goals for the project, and the needs of a particular user.

The goals of the course E4 are to:
1. develop an understanding and experience of the conceptual design process;
2. give the students experience in team dynamics; and
3. enable students to learn to manage small design projects.

The goals of the team are to:
1. satisfy the client's needs and meet E4 course requirements;
2. learn about engineering and engineering design; and
3. have fun while doing a good thing for someone.

The goals of the Danbury school are to:
1. improve Jessica's learning environment and quality of life;
2. increase public awareness about the conditions faced by students like Jessica.

Deliverables

The following deliverables will be completed by the last Friday before final exam week:
1. a prototype arm stabilization device which has been designed, built and tested for Jessica's use;
2. final design documentation for Danbury School and for the E4 teaching team; and
3. a public presentation of the team's design process and results.

Resource Limits

The team is expected to work an average of 10 hours per week per team member. HMC will provide not more than $125 to the team for the purpose of purchasing supplies and materials for the project.

Other Restrictions or Information

The team will hold weekly meetings with their faculty advisor, and will meet regularly with the sponsors at Danbury.

FIGURE 10.9 A team charter for the Danbury arm support project. Note that the document sets forth the goals and all of the relevant parties, and it shows the deliverables, time limits, and resource restrictions.

Because they understand their tasks well, they are able to meet our concern that the WBS go to a level of understanding such that we can know what is required for success. On the other hand, the graphic is not a proper WBS — the tasks are not organized hierarchically, but rather sequentially. This would become a particular problem if the team encounters problems with some area, or if a team member is not skilled in an area or responsibility.

In Figure 10.11, we see a text-based WBS of sorts. Once again, the team is using time as the organizing principle, but in this case, they are simply using particular months.

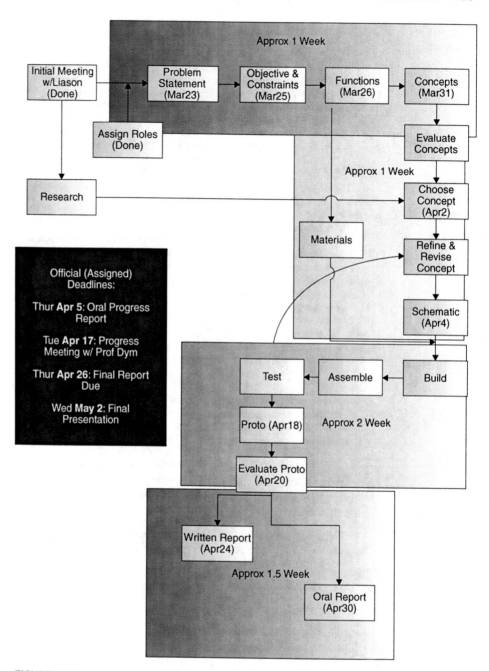

FIGURE 10.10   A management graphic for one of the Danbury student teams. The team was concerned with its ability to complete the project within the time limits and so opted to organize the project into time-based elements that correspond to analysis, design alternative generation, building and testing, and reporting.

Work Breakdown Structure/Schedule

March Activities:

3/21	Meeting with client to ask questions and gather information
3/26	Revise the problem statement
3/26	Identify objectives and constraints
3/26	Devise metrics for evaluating design options
3/26	Compile preliminary research
3/28	Compile completed research
3/28	List functions and means and create a morph chart

April Activities

4/2	Create several design options and apply metrics to choose a final design
4/4	Deliver oral progress report
4/4	Acquire necessary materials to build prototype of chosen design
4/8	Refine chosen prototype design
4/8	Begin building preliminary prototype of structure and piston design
4/15	Complete preliminary prototype of structure and piston design
4/16	Take final measurements at Danbury
4/16	Begin building full prototype
4/18	Complete outline of written report
4/19	Preliminary fit test at Danbury
4/20	Modified structure and fit test at Danbury–Preliminary damping test
4/23	Complete building of the prototype
4/24	Complete testing of the prototype
4/25	Complete written report
4/26	Begin preparing oral presentation

May Activities:

5/1	Complete presentation
5/2	Deliver presentation

FIGURE 10.11 Another management document for a different Danbury student team. Notice that this team has organized the WBS around the expected dates, essentially converting the WBS to a team calendar. This is a potentially risky approach for many reasons. How else might they have arranged their WBS?

This is a very risky strategy for several reasons. First, the use of dates alone suggests that the various tasks have some similarity in scope, difficulty, or perhaps importance. We can see the problem when we are treating generating alternatives and selecting one as being on the same level as a scheduled progress meeting with the sponsor. Secondly, if a due date slips from one time period to the next, it is likely the team will not continue to track it. In this way, the WBS may lead the team to overlook key tasks until they are nearly out of time. Finally, the schedule the team is following is necessarily somewhat arbitrary at the outset, so the use of calendar dates as the primary scheme undermines the team's ability to use the WBS as a scoping document.

There are many ways the team could have chosen to organize their WBS. For example, the team in Figure 10.10 could very easily adapt theirs to one based on the design process given in an earlier beverage container example. Similarly, the second team could have organized their WBS around early, middle, and late activities without tying themselves to a particular set of specific dates.

Finally, it is important to see that managers often modify tools to suit their own situation. This is what the Danbury teams have tried to do, and they were encouraged in their creative application of the tools. They might, however, have made their own tasks easier had they also used them more correctly.

10.11   NOTES

Section 10.2: The definition of projects is from (Meredith and Mantel 1995).

Section 10.3: The underlying conceptual model for the 3S approach is from (Oberlander 1993). The example in Figure 10.3 is adapted from (Kezsbom, Schilling, and Edward 1989).

Section 10.6: The text-based WBS example shown in Figure 10.4 modifies a sample set included in the software Primavera Project Planner, Release 2.0.

Section 10.7: Linear responsibility charts are explained further in most introductions to project management, for example (Meredith and Mantel 1995).

Section 10.9: The use of the standard form of the percent-complete matrix is given in Oberlander (1993). The form as modified draws upon a method given in (CIIP 1986).

Section 10.10: The WBSs of Figures 10.10 and 10.11 are from (Attarian et al. 2007) and (Best et al. 2007).

10.12   EXERCISES

10.1   Explain the differences between managing design projects and managing the implementations of design projects. For example, consider the differences between designing a highway interchange and building that interchange.

10.2   Develop a work breakdown structure (WBS) and a linear responsibility chart (LRC) for an on-campus benefit to raise money for the homeless.

10.3   Develop a work breakdown structure (WBS) and a linear responsibility chart (LRC) for an on-campus meeting to be used by the women in the cooperative cited in Exercise 3.5 as they arrive at a nearby airport and travel to your campus. (Hint: You are not allowed to do the courteous thing and pick them up at the airport.)

10.4   Develop a work breakdown structure (WBS) and a linear responsibility chart (LRC) for a project to design a robot that will be entered into a national collegiate competition.

10.5   Develop a schedule and a budget for the on-campus benefit of Exercise 10.2.

CHAPTER 11

DESIGNING FOR...

What are future directions in design research and practice?

A **CENTRAL THEME** of this book is that engineering design is usually done by teams rather than individuals. This idea reflects the recent experience of engineers in industrial settings throughout the world. Design teams usually include not only engineers, but also manufacturing experts (who may be industrial engineers), marketing and sales professionals, reliability experts, cost accountants, lawyers, and so on. Such teams are concerned with understanding and optimizing the product under development for its *entire life*, including its design, development, manufacturing, marketing, distribution, use, and, eventually, disposal. Concern with all of these areas and with their impact on the design process has come to be known as *concurrent engineering*. While it is beyond the scope of many smaller projects (e.g., the student design examples described earlier) to apply concurrent engineering techniques, it is important that engineering designers be aware of concurrent engineering and its implications for their design work.

One of the most important aspects of engineering in the modern commercial setting is the realization that the audience for a good design includes those who will build and maintain the designed artifact. Another important feature is that designs must usually meet economic or cost-related targets. These aspects are part of a more general notion: Engineers have always sought to realize various desirable attributes to some degree in their designs. This is often referred to as "design for *X*," where *X* is an attribute such as manufacturing, maintainability, reliability, or affordability. (Designers and engineers also refer to them with a different name, the *-ilities*, because many of these desirable attributes are expressed as nouns that have an "-ility" suffix.)

As designers, we can use the idea of the life cycle of a product to guide us through some of the *X*'s. Since most products are designed to be built, sold, used, and then disposed of, we will look first at design for manufacturing and assembly, and then at design for affordability, design for reliability and maintainability, and lastly, at design for sustainability. We will see that these and related concepts can be summarized by the idea of *quality*, so we will briefly look at *quality function deployment* (QFD), one approach to design for quality.

11.1 DESIGN FOR MANUFACTURING AND ASSEMBLY: CAN THIS DESIGN BE MADE?

In many cases, a designed artifact will be produced or manufactured in quantity. In recent years, companies have come to learn that the design of a product can have an enormous impact on the cost of producing it, on its resulting quality, and on its other characteristics. Toward this end, globally competitive industries such as the automotive and consumer electronics industries routinely consider how a product is manufactured during the design process. A significant driver of this concern is the large number of products being manufactured, which allows for economies of scale, which we will discuss in Section 11.2. Further, the time it takes to get a product to the consumer, known as the *time to market*, defines a company's ability to shape a market. Design processes that incorporate manufacturing issues can be key elements in speeding products through to commercial production.

11.1.1 Design for manufacturing (DFM)

Design for manufacturing is an iterative process with designers and builders.

Design for manufacturing (DFM) is design based on minimizing the costs of production and/or the time to market for a product, while maintaining an appropriate level of quality. The importance of maintaining an appropriate level of quality cannot be overstated because without an assurance of quality, DFM is reduced to simply producing the lowest-cost product.

DFM begins almost inevitably with the formation of the design team. In commercial settings, design teams committed to DFM tend to be multidisciplinary, including engineers, manufacturing managers, logistics specialists, cost accountants, and marketing and sales professionals. Each brings particular interests and experience to a design project, but all must move beyond their primary expertise to focus on the project itself. In many world-class companies, such multidisciplinary teams have become the de facto standard of the modern design organization.

Manufacturing and design tend to interact iteratively during product development. That is, the design team learns either of a problem in producing a proposed design or of an opportunity to reduce design cost or timing, as a result of which the team reconsiders its design. Similarly, a design team may be able to suggest alternative production approaches that lead manufacturing specialists to restructure processes. In order to achieve fruitful and synergistic interaction between the manufacturing and design processes, it is important that DFM be considered in each and every one of the design phases, including the early conceptual design stages.

One basic methodology for DFM consists of six steps:

1. estimate the manufacturing costs for a given design alternative;
2. reduce the costs of components;
3. reduce the costs of assembly;
4. reduce the costs of supporting production;
5. consider the effects of DFM on other objectives; and
6. if the results are not acceptable, revise the design once again.

This approach clearly depends upon an understanding of all the objectives of the design; otherwise the iteration called for in Step 6 cannot occur meaningfully. An understanding of the economics of production (some of which are discussed in Section 11.2) is also required. In addition to these, however, there are engineering and process decisions that can directly influence the cost of producing a product. Some processes for shaping and forming metal, for example, cost much more than others and are called for only to meet particular engineering needs. Similarly, some types of electronic circuits can be made with high-volume, high-speed production machines, while others require hand assembly. Some design choices that require higher costs for small production runs may actually be less expensive if the design can also be used for another, higher-volume purpose. In each of these instances, a successful design can be completed only by combining deep knowledge of manufacturing techniques with deep design experience.

11.1.2   Design for assembly (DFA)

Design for assembly is a related, but formally different brand of design for X. Assembly refers to the way in which the various parts, components, and subsystems are joined, attached, or otherwise grouped together to form the final product. Assembly can be characterized as consisting of a set of processes by which the assembler (1) handles parts or components (i.e., retrieves and positions them appropriately relative to each other), and (2) inserts (or mates or combines) the parts into a finished subsystem or system. For example, assembling a ball-point pen might require that the ink cartridge be inserted into the tube that forms the handgrip, and that caps be attached to each end. This assembly process can be done in a number of ways, and the designer needs to consider approaches that will make it possible for the manufacturer to reduce the costs of assembly while maintaining high quality in the finished product. Clearly then, assembly is a key aspect of manufacturing and must be considered either as part of design for manufacturing or as a separate, yet strongly related design task.

Because of its central place in manufacturing, a great deal of thought has been put into development of guidelines and techniques for making assembly more effective and efficient. Some of the approaches typically considered are:

1. *Limiting the number of components to the fewest that are essential to the working of the finished product.* Among other things, this implies that the designer will differentiate between parts that could be eliminated by combining other parts and those that must be distinct as a matter of necessity. The usual issues for this are to identify:
   - parts that must move relative to one another;
   - parts that must be made of different materials (for strength, for example, or insulation); and
   - parts that must be separated in order for assembly to proceed.
2. *Using standard fasteners and/or integrating fasteners into the product itself* Using standard fasteners also allows an assembler to develop standard routines for component assembly, including automation. Reducing the number and type of fasteners allows the assembler to construct a product without having to retrieve as many

components and parts. The designer should also consider that fasteners tend to induce stress concentrations and may thus cause reliability concerns of the sort we will discuss in Section 11.3.

3. *Designing the product to have a base component on which other components can be located,* and designing for the assembly to proceed with as little motion of the base component as possible. This guideline enables an assembler (whether human or machine) to work to a fixed reference point in the assembly process and to minimize the degree to which the assembler must reset reference points.

4. *Designing the product to have components that facilitate retrieval and assembly* This may include elements of detailed design that, for example, reduce the tendency of parts and subassemblies to become tangled with one another, or designing parts that are symmetric, so that once retrieved they can be assembled without turning to a preferred end or orientation.

5. *Designing the product and its component parts to maximize accessibility, during both manufacturing and subsequent repair and maintenance* While it is important that the components be efficient in their use of space, the designer must balance this need with the ability of an assembler or repairer to gain access to and manipulate parts, both for initial fabrication and later replacement.

Consider the number of parts in a design, and how they will be assembled.

While these guidelines and heuristics represent only a small set of the design considerations that make up design for assembly, they do provide a starting point for thinking about both DFA and DFM.

11.1.3 The Bill of Materials (BOM)

Effective design for manufacturing also requires a deep understanding of production processes, among the most important of which are ways to plan and control inventories. A common inventory planning technique is Materials Requirements Planning (MRP). It utilizes the assembly drawings (discussed in Chapter 6) to develop a bill of materials (BOM) and an assembly chart, called a "gozinto" chart, which shows the order in which the parts on the BOM are put together. The BOM is a list of all of the parts, including the quantities of each part required to assemble a designed object. We might think of the BOM as being a recipe that specifies (1) all of the ingredients that are needed, (2) the precise quantities needed to make a specified lot size, and, when paired with the "gozinto" chart, (3) the process for putting the ingredients together.

When a company has determined the size and timing of its production schedule, the BOM is used to determine the size and timing of inventory orders. (Most companies now use *just-in-time* delivery of parts as they try to avoid carrying large inventories of parts that are paid for but not generating revenues until after they are assembled and shipped.) The importance of the assembly drawings and the BOM in managing the production process cannot be overstated. To be effective, not only must the design team develop accurate methods of reporting their design, but the entire organization must be committed to the discipline that any design changes, or *engineering change orders*, will be reported accurately and thoroughly to *all* the affected parties. In Section 11.2, we will see that the BOM is also useful in estimating some of the costs of producing the designed artifact.

A final point to note is that manufacturing concerns include both logistics and distribution, so that these elements have also become an important part of design for manufacturing. One of the major changes in how business is done today is that companies now work toward forging links among the suppliers of materials needed to make a product, the fabricators who manufacture that product, and the channels needed to efficiently distribute the finished product. This set of related activities, often referred to as the *supply chain*, requires a designer to understand elements of the entire product cycle. It is beyond our scope to explore the role of supply-chain management in design, except to note that in many industries, successful designers understand not only their own production and manufacturing processes, but also those of their suppliers and their customers. This requirement for integrated understanding of commercial processes will surely increase in the future.

11.2   DESIGN FOR AFFORDABILITY: HOW MUCH DOES THIS DESIGN COST?

To be able to afford something is, according to the dictionary, to be able to bear the cost of something or to pay the price for it. In the design context, whether or not we can afford something is an issue that will be faced by the client (e.g., can I afford to make this product? or perhaps, can I afford not to make it?), the manufacturer (e.g., can I afford to make this at a given price?), and the user (e.g., can I afford to buy this product?). Thus, affordability is really about expressing an important dimension of the object or system being designed in terms that all stakeholders recognize and understand, money. This dimension is typically covered by the field known as *engineering economics*. Engineering and economics have been closely linked for almost as long as the two fields have existed. Indeed, economists recognize that engineers were the developers of a number of important elements of economic theory. For example, what economists call utility theory and price discrimination were both first articulated by the 19th-century engineer Jules Dupuit, and location theory was developed by a civil engineer named Arthur M. Wellington. In fact, Wellington is credited with the definition of engineering as "the art of doing that well with one dollar which any bungler can do with two." This linking of engineering and economics should come as no surprise to the designer, since it is a rare project for which money really is "no object."

Virtually all engineering decisions have economic elements.

Engineering economics is concerned with understanding the economic or financial implications of engineering decisions, including choosing among alternatives (e.g., cost-benefit analysis), deciding if or when to replace machines or other systems (replacement analysis), and predicting the full costs of devices over the period of time that they will be owned and used (life cycle analysis). These topics can easily fill entire courses in an engineering curriculum and are well beyond our current scope. However, there are some topics sufficiently important to a designer that they must be briefly introduced, perhaps the most important of which is the *time value of money*. A second important topic is *cost estimation*. Without at least a rudimentary knowledge of these, design and engineering teams are likely to make good choices about designs only by luck.

11.2.1   The time value of money

If someone were to offer us $100 today or $100 one year from now, we would almost certainly prefer to take the money *now*. Having the money sooner offers a number of advantages, including providing us with the ability to invest or otherwise use the money during the intervening year; eliminating the risk that the money might not be available next year; and eliminating the risk that inflation might reduce the purchasing power of the money during the next year. This simple example highlights one of the most important concepts in engineering economics, namely, the *time value of money*: Money obtained sooner is more valuable than money obtained later, and money spent sooner is more costly than money spent later.

As indicated, the time value of money captures the effects of both foregone opportunities, referred to as *opportunity costs*, and *risk*. An opportunity cost is a measure of how much the deferred money could have earned in the intervening time. The risk captures both the risk that the money will be worth less (because of inflation) and the risk that the money will simply have become unavailable during the intervening time. Economists and financial professionals bundle together the extent of these risks and their associated lost opportunities in the *discount rate*. A discount rate acts much like an interest rate on a savings account or credit card, except that it views the money as being worth less to us today because of the risks and missed opportunities. The interest rate on a savings account measures how much a bank is willing to pay for the privilege of using our money in the coming year. An interest rate on a credit card measures how much we must pay the card issuer for the privilege of using their money; these rates vary, with higher charges assigned to customers with poorer credit ratings or limited credit histories. Interest calculations thus typically show dollar amounts increasing from a given time onward or forward. Discount rates typically work in reverse, showing the value today of money that will be available at some point in the future.

A dollar today is worth more than the promise of a dollar tomorrow.

Measuring risks and opportunity costs can be a complex process, but as engineers we should remember that design decisions and choices made today will translate into streams of prospective "financial events" that will occur at different times in the future. Some of these financial events are *costs* we will incur (e.g., for manufacturing or distribution), and some are *benefits* (e.g., revenues from sales) we will derive. The more immediate the costs and benefits, the more impact they have on decisions made by clients, users, and designers.

How do we distinguish between "$100 today" and "$100 a year from now" in a rational and consistent way? The answer is that, given a discount rate and a set of future financial events or cash flows, both toward us and away from us, we translate all of our events into a common time frame, either a current time or a future one. Consider again our choice between getting $100 today or $100 next year. If the annual discount rate is 10%, then we would expect to need $110 dollars in a year in order compensate us for not having $100 today (or to purchase then what we could buy today with $100). Then we would be getting the same amount of money if we accept $100 now or $110 a year from now.

We can also work this the other way, asking how much we need today to have the equivalent of $100 a year from now. Next year's $100 is worth the same amount as some amount of money, X, plus the 10% that X could earn in the coming year. That is, $1.10 \cdot \$X = \$100$. Solving for X shows that taking the $100 in a year is equivalent to getting and investing about $91 today. We won't go through all of the various formulas and arithmetic associated with the time value of money, except to point out that we can carry out discounting calculations as far into the future as we want. The principle is the same. Thus, $100 promised for two years from now would be worth even less than $100 were today or even $100 promised for next year, since it wouldn't have accounted for being able to use the $100 for both years, or our being able to use the $10 earned in the first year.

Economists have developed standard approaches to discounting money and to determining the present value of future dollars, whether costs or benefits, and vice versa. Application of these formulas can become quite involved when inflation or unusual timing issues are involved, but virtually all such analysis is based on the relationship

$$PV = FV\left(\frac{1}{1+r}\right)^t \qquad (11.1)$$

where PV is the present value of the costs or benefits, FV is the future value, r is the discount rate, and t is the time period over which a cost is incurred or a benefit is realized. Consider again, the decision concerning the worth of $100 next year. In this case, the future value, FV, is $100, the time period, t, is 1 year, and the discount rate, r, is 10% per year, or 0.10. If we substitute these values into eq. (11.1), the present value, PV, will be the $91 found above. In other words, an offer of $100 a year from now would be equivalent to an offer of $9 less today. The ability to translate future costs into present equivalent values, known as *discounting*, is very important in some design projects and may affect how we might choose among designs. We turn to that topic now.

11.2.2   The time value of money affects design choices

Imagine for a moment that we are asked to choose between two alternative vehicle designs for a transit agency. Design alternative A has a significantly higher initial purchase price, but design alternative B has higher operating costs over the life of the vehicle. In this case we have to reconcile the different costs at different points in the lifetimes of the two choices. One obvious question is, How big are those cost differences and when do they occur? If design B can be bought for much less money than design A, its higher operating costs may not be as important because they are incurred so much later than the immediate initial saving. If, on the other hand, the operating costs of B are both much higher and occur relatively early in the vehicle's life, then the immediate savings gained by buying design B may be illusory. Making a rational choice in such a case requires that we understand how to properly analyze the time value of money (cf. Section 11.1.1) because we have to compare all of the costs in equivalent dollar values.

Now, imagine further that our designs have not only different purchase prices and different operating and maintenance costs, but also different expected lifetimes. This means that we have to adjust all of the costs in a way that makes the values equivalent over the same time frame. Engineering economists have developed a methodology for doing that, although we will not provide a detailed description here. Called *equivalent uniform annual costs* (EUAC), it essentially treats all of the alternatives as though they are replaced with a one-for-one swap whenever they wear out. EUAC then transforms the resulting infinite series of replacements into a series of annual payments. The point to be

Designs are purchased today and over their lifetimes.

drawn here is simply that the series of future costs and benefits of all design alternatives must be considered over the lifetime of each of the alternatives and then translated into a format that allows us to fairly compare those alternatives. The essential wisdom is that it is insufficient to look only at the initial purchase costs of design alternatives as a way of finding out what designs really cost. A true cost analysis requires that we consider the entire life cycle of a design.

11.2.3 Estimating costs

We took it for granted in the previous section that the cost of the final design is known, as are the operating and maintenance costs over the life of the device. In practice cost estimation is not usually so simple. It requires skills and experience, and it can easily consume an entire text. However, several points are relevant during conceptual design.

It is easy to say that the costs of a design typically include labor, materials, overhead, and profits for various stakeholders. However, this simple statement masks the complexity of detailing or structuring the cost of all but the simplest of artifacts. In many cases, estimating the production and distribution costs of a design is extremely difficult. Here we limit ourselves to describing only the principal elements that make up the cost categories listed just above.

Labor costs include payments to the employees who construct the artifact, as well as to support personnel who perform necessary but often invisible tasks, such as answering the phone, filling orders, packing and shipping the product, etc. Labor costs also include a variety of *indirect costs* that are less evident because they are not payments made directly to employees. Such indirect costs are called *fringe benefits* because they are typically payments made to third parties on behalf of the employees. The fringe benefits include health and life insurance, retirement benefits, employers' contributions to Social Security, and other mandated payroll taxes. These indirect costs of labor are often neglected or overlooked by designers estimating the cost of a design, yet for many companies they are as much as 50% of direct labor payments or wages.

In Section 11.1 we discussed the importance of the bill of materials (BOM) in controlling inventories and managing the manufacturing of items. The BOM is also useful for estimating the material costs associated with various designs. *Materials* include those directly used in building the device, as well as intermediate materials and inventories used in ways that may not be obvious. For example, some inventory is wasted during manufacturing, while other inventory may be classified as being part of work-in-progress. The BOM provides a guideline for the number and type of parts that make up the device or object. The BOM is particularly useful since it is developed directly from the assembly drawings, and so it reflects the designer's final intent.

We must exercise care when we use a BOM to estimate costs since both labor and materials are subject to *economies of scale*, the idea that the *unit cost* or production cost per (single) item can often be reduced by making many identical copies, rather than just making a few "originals." The genius of Henry Ford's assembly line, wherein he came up with a way to make millions of copies of his cars, is a reflection of the economies of scale. Ford lowered his unit costs and sold all of the cars that he did because many more people could afford to buy them. Of course, Ford realized such economies of scale by developing new technologies, but that's still another story about how engineering and economics interact.

Costs incurred by a manufacturer that cannot be directly assigned to a single product are termed *overhead*. If, for example, a device is made in a factory that also produces twenty other products, the cost of the building, the machines, the janitorial staff, the electricity, etc., must somehow be shared or distributed among all of the 21 items. If each product was priced to ignore these overhead costs, the company would soon find itself unable to pay either for the building or for the services necessary to maintain it. Other elements of overhead include the salaries of executives, who are presumably using some share of their time to supervise each of the company's activities, and personnel and related costs of needed business functions such as accounting, billing, and advertising. While there are accounting standards that define cost categories and their attributes, precise estimates of overhead costs vary greatly with the structure and practices of the company in question. One company may have only a small number of products and a very lean organization, with most of its costs directly attributed to the products made and sold, and only a small percentage allocated to overhead. In other organizations the overhead can be equal to or greater than the labor costs that are directly assignable to one or more products. In many universities and colleges, for example, the overhead rate associated with research (which pays for laboratories, support staff, college presidents, and other essentials) runs as high as 65% of the researchers' salaries and benefits. The key point is that estimating the costs of producing a design requires careful consultation with clients or their suppliers.

Cost estimates produced during the conceptual stage of a design project are often quite inaccurate when compared with those made for detailed designs. In heavy construction projects, for example, an accuracy of $\pm 35\%$ is considered acceptable for initial estimates. However, this tolerance of inaccuracy should not be taken as a license to be sloppy or casual in early cost estimating.

In practice, each of the engineering disciplines has its unique approaches to cost estimating, and these approaches will often be captured by some useful heuristics or *rules of thumb*, or general guidelines, that are most relevant at the conceptual design stage. In civil engineering, for example, the R. S. Means Cost Guide provides cost estimates per square foot for the various elements in different kinds of construction projects. The Richardson's Manual offers similar information for chemical plant and petroleum refinery projects. On the other hand, costs per square inch may be more relevant for printed circuit board designs. In each of the disciplines, we should carefully consult with experienced professionals to estimate costs successfully, even at the more general levels we need to make conceptual design choices.

Finally on cost estimation, we want to highlight the distinction between the cost of designing an artifact, on the one hand, and the cost of manufacturing and distributing it, on the other. In many cases, the cost of design is a relatively small part of the final project cost, such as in the case of a dam or other large structure. Notwithstanding that, however, most clients expect that a design team will correctly estimate its own costs and budget them accurately. Thus, even when costing out the design activity, an effective design team will seek to understand and control its costs accurately.

11.2.4 Costing and pricing

Finally, it should be noted that while costing is an important element in the *profitability* of a design, it is generally *not* a key factor in the *pricing* of the artifact. This seeming contradiction can be easily explained by noting that gross profits (i.e., profits before taxes and other considerations) are simply the net of revenues minus costs. As such, costs are an important element in the profit equation. Revenues, on the other hand, are determined by the price charged for an item multiplied by the number of items sold. For most profit-maximizing firms, prices are not set on the basis of costs, but rather in terms of what the market is willing to pay. Some examples will illustrate this.

While costs affect profitability, prices are based on value, not costs.

Consider a high-quality graphite tennis racket. It may command a price of hundreds of dollars when initially introduced. It is clear upon inspection, however, that the costs of making such a racket are nowhere near such figures. The materials for the racket may cost just a few dollars, the labor is virtually negligible, and the costs of technological developments are relatively modest when *amortized* (spread out) over the many thousands of sales of the racket. The distribution costs are clearly no different for high-end rackets than for the $10 rackets found at discount stores. However, since there is an obvious demand for such expensive rackets, their prices are set high because there are customers willing to pay high prices. Indeed, the role of marketing professionals on the design team usually includes identifying design attributes that will cause consumers to pay a premium for a designed product.

This example also serves to highlight an aspect of *reliability*. A manufacturer can offer a virtually lifetime replacement guarantee if the costs of manufacture are far below the selling the price. Thus, the high-quality service of certain brands reflects the disparity between their price and their cost structure.

A similar example can be found in the airline industry. Here the provider of the service faces essentially the same costs regardless of whether a flight is almost full or almost empty. This explains why airlines are willing to offer certain deeply discounted fares at some times, and almost no discounts at others (such as holidays). It also explains why the airlines are willing to invest heavily in modeling and tracking the wide variety of fare options they make available.

In some industries a convention has arisen to compensate designers or even providers of certain products on a "cost-plus" basis. For example, most large public works projects, such as highways or dams, are built on the basis of the costs of the contractor or designer plus an additional percentage as a profit allowance. While this is common practice in some cases, the norm in the private sector is to select prices to maximize profits, not to simply tack on a profit factor.

The point is that an engineering designer must control the costs of the design to ensure that the objectives of the client and users are realized. Beyond that, however, the ultimate profitability of an artifact may turn out to be beyond a designer's control, as in the cases of pet rocks and beanie babies.

11.3 DESIGN FOR RELIABILITY: HOW LONG WILL THIS DESIGN WORK?

Most of us have a personal, visceral understanding of reliability and unreliability as a consequence of our own experience with everyday objects. We say that the family car is unreliable, or that a good friend is very reliable — someone we can count on. While such informal assessments are acceptable in our personal lives, we need greater understanding and accuracy when we are functioning as engineering designers. Thus, we now describe how engineers approach reliability, along with its sister concept, maintainability.

11.3.1 Reliability

Reliability is the probability that an item will function under stated conditions for a stated measure of usage.

To an engineer, reliability can be defined as "the probability that an item will perform its function under stated conditions of use and maintenance for a stated measure of a variate (time, distance, etc.)." This definition has a number of elements that warrant further comment. The first is that we can properly measure the reliability of a component or system *only* under the assumption that it has been or will be used under some particular set of usage and maintenance conditions. The second point is that the appropriate measure of use of the design, called the *variate*, may be something other than time. For example, the variate for a vehicle would be miles, while for a piece of vibrating machinery the variate would be the number of cycles of operation. Third, we must examine reliability in the context of the functions discussed in Chapter 4, which emphasizes the care we should take in developing and defining the functions that a design must perform. Finally, note that reliability is treated as a probability, and hence can be characterized by a distribution. In mathematical terms, this means that we can express our expectations of how reliable, safe, or successful a product or a system to be in terms of a cumulative distribution function or a probability density function.

In practice, our use of a probabilistic definition enables us to consider reliability in the context of the opposite of success, that is, in terms of *failure*. In other words, we can frame our consideration of reliability in terms of the probability that a unit will fail to perform its functions under stated conditions within a specified window of time. This requires us to consider carefully what we mean by failure. British Standard 4778 defines a failure as "the termination of the ability of an item to perform a required function." This definition, while helpful at some level, does not capture some important subtleties that we, as designers, must keep in mind: It doesn't capture the many kinds of failures that can afflict a complex device or system, their degree of severity, their timing, or their effect on the performance of the overall system.

For example, we find it useful to distinguish between *when* a system fails and *how* it fails. If the item fails when in use, the failure can be characterized as an *in-service failure*. If the item fails, but the consequences are not detectable until some other activity takes place, we refer to that as an *incidental failure*. A *catastrophic failure* occurs when a failure of some function is such that the entire system in which the item is embedded fails. For example, if our car breaks down while we're on a trip and needs a repair in order for us to complete the trip, we would call that an in-service failure. An incidental failure might be some part that our favorite mechanic suggests we replace during the routine servicing of our car. A catastrophic, accident-causing failure might follow from the failure of a critical part of the car while we are driving at freeway speeds. Each type of failure has its own consequences for the users of the designed artifact, and so must be considered carefully by designers.

We often specify reliability in part by using measures such as the mean time between failures (MTBF), or miles per in-service failure, or some other variate or metric. However, we should note that framing the definition of reliability in terms of probabilities gives us some insight into the limitations inherent in such measures. Consider the two

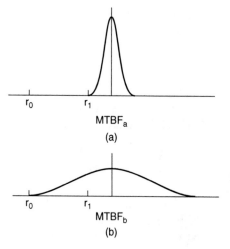

MTBF$_a$

(a)

MTBF$_b$

(b)

FIGURE 11.1 Failure distributions (also called probability density functions) for two different components. Note that both curves have the same value of MTBF, but that the dispersions of possible failures differ markedly. The second design (sketch b) would be viewed as less reliable because more failures would during the early life of the component (i.e., during the time interval $t_0 \leq t \leq t_1$).

failure distributions shown in Figure 11.1. These two reliability probability distributions have the same mean (or average), that is, MTBF$_a$ = MTBF$_b$, but they have very different degrees of dispersion (typically measured as the variance or standard deviation) about that mean. If we are not concerned with both mean and variance, we may wind up choosing a design alternative that is seemingly better in terms of MTBF, but is much worse in terms of variance. That is, we may even choose a design for which the MTBF is acceptable, but for which the number of early failures is unacceptably high.

One of the most important reliability issues for a designer is how the various parts of the design come together and what the impact is likely to be if any one part does fail. Consider, for example, the conceptual sketch of the *series system* design shown in Figure 11.2. It is a chain of parts or elements, the failure of any one of which would break the chain, which in turn will cause the system to fail. Just as a chain is no stronger than its weakest link, a series system is no more reliable than its most unreliable part. In fact, the reliability — or probability that the system will function as designed — of a series system whose individual parts have reliability (or probability of successful performance) $R_i(t)$ is given by:

$$R_S(t) = R_1(t) \bullet R_2(t) \bullet \cdots \bullet R_n(t)$$

or

$$R_S(t) = \prod_{i=1}^{n} R_i(t) \tag{11.2}$$

| 0.99 | 0.99 | 0.70 |

FIGURE 11.2 This is a simple example of a *series system*. Each of the elements in the system has a given reliability. The reliability of the system as a whole can be no higher than that of any one of the parts because the failure of any one part will cause the system to cease operating. What is the reliability of this system as calculated with eq. (11.2)?

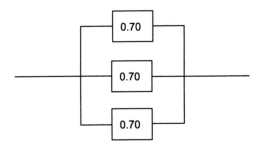

FIGURE 11.3 This is a simple example of a *parallel system.* Note that every one of the components must fail in order for the system to cease working. While such a system has high reliability, it is also quite expensive. Most designers seek to incorporate such redundancy when necessary, but look for other solutions wherever possible. How does the reliability of this parallel system, as calculated with eq. (11.3), compare with that of the series system in Figure 11.2?

where $R_S(t)$ is the reliability of the entire series system, and $\prod_{i=1}^{n}$ is the product function. We see from eq. (11.2) that the overall reliability of a series system is equal to the product of all of the individual reliabilities of the elements or parts within the system. This means that if any one component has low reliability, such as the proverbial weak link, then the entire system will have low reliability and the chain will break.

Designers have long understood that redundancy is important for dealing with the weakest-link phenomenon. A *redundant system* is one in which some or all of the parts have backups or replacement parts that can substitute for them in the event of failure. Consider the conceptual sketch of the *parallel system* of three parts or elements shown in Figure 11.3. In this simple case, each of the components must fail in order for the system to fail. The reliability $R_P(t)$ of this entire parallel system is given by:

$$R_P(t) = 1 - [(1 - R_1(t)) \bullet (1 - R_2(t)) \bullet \cdots \bullet (1 - R_n(t))]$$

or

$$R_P(t) = 1 - \prod_{i=1}^{n}[1 - R_i(t)] \tag{11.3}$$

We see from eq. (11.3) that the reliability of this parallel system (i.e., the probability that the parallel system will operate successfully) is now such that if any one of the elements functions, the system will still function.

Parallel systems have obvious advantages in terms of reliability, since all of the redundant or duplicate parts must fail in order for the system to fail. Parallel systems are also more expensive, since many duplicate parts or elements are included only for contingent use, that is, they are used only if another part fails. For this reason, we must carefully weigh the consequences of failure of a part against failure of the system, along with costs attendant to reducing the likelihood of a failure. In most cases designers will opt for some level of redundancy, while allowing other components to stand alone. For example, a car usually has two headlights, in part so that if one fails the car can continue to operate safely at night. The same car will usually have one radio, since its failure is unlikely to be catastrophic. The mathematics of combining series and parallel systems is beyond our scope, but we clearly have to learn and use them to design systems that have any impact on the safety of users.

Redundancy usually increases both reliability and costs.

Designers can consider modes of failure and develop estimates of reliability only if they truly know how components might fail. Such knowledge is gained by performing experiments, analyzing the statistics of prior failures, or carefully modeling the underlying physical phenomena. Designers lacking deep experience in understanding component failure should consult experienced engineers and designers, users, and the client in order to ascertain that an appropriate level of reliability is being designed. Often the experiences of others allow a designer to answer reliability questions without performing a full set of experiments. For example, the appropriateness of different kinds of materials for various designs can be discussed with materials engineers, while properties such as tensile strength and fatigue life are documented in the engineering literature.

11.3.2   Maintainability

Our understanding of reliability also leads us to conclude that many of the systems that we design do fail if they are used without being maintained, and that they may need some amount of repair even when they are properly maintained. This fact of life leads engineers to consider how best to design things so that the necessary maintenance can be performed effectively and efficiently. Maintainability can be defined as "the probability that a failed component or system will be restored or repaired to a specific condition within a period of time when maintenance is performed within prescribed procedures." As with our definition of reliability, we can learn from this definition.

First, maintainability depends upon a prior specification of the condition of the part or device, and on any maintenance or repair actions, which are part of the designer's responsibilities. Second, maintainability is concerned with the time needed to return a failed unit to service. As such, the same concerns regarding inappropriate use of a single measure, (e.g., the mean time to repair (MTTR)), come into play.

Designing for maintainability requires that the designer take an active role in setting goals for maintenance, such as times to repair, and in determining the specifications for maintenance and repair activities in order to realize these goals. This can take a number of forms, including:

- selecting parts that are easily accessed and repaired;
- providing redundancy so that systems can be operated while maintenance continues;
- specifying preventive or predictive maintenance procedures; and
- indicating the number and type of spare parts that should be held in inventories in order to reduce downtime when systems fail.

There are costs and consequences in each of these design choices. For example, a system designed to have increased redundancy in order to limit downtime while maintenance is ongoing, as is the case for an air traffic control system, will have very large attendant capital costs. Similarly, the cost of carrying inventories of spare parts can be quite high, especially if failures are rare. One strategy that has been increasingly adopted in many industries is to work toward making parts standard and components modular. Then spare parts inventories can be used more flexibly and efficiently, and components or subassemblies can be easily accessed and replaced. Any removed subassemblies can be repaired while the repaired system has been returned to service.

If high maintainability has been established as a significant design objective, design teams must take active steps in the design process to meet that goal. As such, a design team should ask itself what maintenance actions (e.g., service) reduce failures, what elements of the design support early detection of problems or failures (e.g., inspection), and what elements ease the return of failed items to use (e.g., repair). While no one would intentionally design systems to make maintenance more difficult, the world is fraught with examples in which it is difficult to believe otherwise, including a new car in which the owner would have to remove the dashboard just to change a fuse!

11.4 DESIGN FOR SUSTAINABILITY: WHAT ABOUT THE ENVIRONMENT?

Engineers have an ethical obligation to consider the environmental consequences of the things they design.

Many people have come to hold negative views of technology and engineered systems because of the realization that one generation's progress may produce an environmental nightmare for the next. There are certainly enough examples of short-sighted projects (such as irrigation systems that created deserts or flood-control schemes that eliminated rivers entirely) that responsible engineers can feel at least some anxiety about what *their* best ideas might eventually produce. The engineering profession has come to appreciate these concerns over the past several decades, and has incorporated environmental responsibility directly into the ethical obligations of engineers. The American Society of Civil Engineers, for example, specifically directs engineers to "strive to comply with the principles of sustainable development"; the American Society of Mechanical Engineers Code of Ethics includes Canon 8, "Engineers shall consider environmental impact in the performance of their duties." A number of tools to understand the environmental effects are being introduced into engineering design in order to help with these issues and obligations. Some even take on the force of law, such as the need for environmental impact statements for certain projects. Environmental life cycle assessments (LCAs) are increasingly used, so we note some of the key environmental challenges facing designers and describe LCAs.

11.4.1 Environmental issues and design

Environmental concerns relevant to design can be organized in any number of ways. Transportation engineering texts, for example, concern themselves with the impacts of engineered systems on water and air quality, while electrical engineering texts consider the effects of power generation and transmission, or focus on the particular environmental effects of some of the solvents and other chemicals associated producing chips or printed circuit boards. A more general approach is to think in terms of particular aspects of the environment and then consider the likely short- and long-term consequences of design alternatives.

We can often characterize the environmental implications of a design in terms of the effects on air quality, water quality, energy consumption, and waste generation. In each case, we need to address both short-term issues, which will likely come up as part of the economic effects, and long-term issues, which may not. Unfortunately, experience shows that the long-term effects of our design choices often completely overwhelm short-term benefits.

Air quality almost immediately springs to mind when we list environmental concerns related to design. Urban areas have tremendous smog problems, small towns often have an industry with a large smokestack, and even national forests are experiencing loss of habitat due to acid rain and other air-quality issues. It is important to realize that these enormous problems often begin with relatively small emissions from various steps in the production of everyday objects. Each mile we drive in a standard internal combustion engine powered car adds a tiny bit of particulate matter, nitrous oxide, and carbon monoxide to the atmosphere. In addition, refining the fuel, smelting the steel, and curing the rubber for the tires add further emissions to the air. Less obvious but similar air-quality problems result from the production of everyday materials in paper bags and plastic toys. In other words, designers concerned about the environment must consider both the manufacture of the product and its use.

Environmentally conscious engineers should also concern themselves with issues of *water quality* and *water consumption*. We take the availability of clean water for granted. In fact, however, many of the world's major bodies of water are already under stress from overuse and pollution. As with air quality, this is a direct result of the multiple uses made of our water supplies. Many states have experienced severe droughts in the recent years, and in the southwestern United States, water is becoming the single biggest environmental constraint on further growth. Effective designers must consider and calculate the water requirements for producing and using their designs. Estimating changes to water resulting from particular designs is of great significance. These can include changes in water temperature (which for large processes can affect fish and other parts of the ecosystem) and the addition of chemicals, particularly hazardous or long-lived compounds.

The production and use of designed systems needs *energy*. However, the energy demands of a system can be much higher than designers realize, or may come from sources that are particularly problematic environmentally. Several years ago, California faced an energy crisis that led to sporadic blackouts. Design choices about common household appliances such as refrigerators do affect an increasingly energy-starved world. The variety of sizes, shapes, and levels of efficiency of refrigerators highlights the many design choices made by engineers and product design teams. Beneath the surface of such devices, however, there are further design choices made by engineers while generating and selecting alternatives. The principal energy consumer in a refrigerator is the compressor, which can be made more energy-efficient by judicious selection of components. Within the refrigerator walls, the use of insulation materials has a tremendous effect on how well cold temperatures are retained. Even door designs and their placement affect how much energy a refrigerator consumes. Designers must approach such projects systematically, applying all of the skills and techniques learned in their engineering science courses, and accounting for the consequences of their design choices.

Products must be disposed of after fulfilling their useful life. In some cases, perfectly good designs become serious disposal problems. For example, consider the wooden railroad tie used to secure and stabilize train tracks and distribute the loads into the underlying ballast. Properly maintained and supported, ties treated with creosote typically last more than 30 years, even under heavy loads and demanding weather conditions. Not surprisingly, most railroads use such ties. At the end of their lives, however, the same chemical treatment that made them last so long creates a major disposal problem. Improperly disposed of, the chemicals can leach into water supplies, making them harmful to living things. The ties also emit highly noxious, even toxic, fumes when burned. Thus, managing the *waste streams* associated with products and systems has become an important consideration in contemporary design. A great solution to one problem has become a problem in itself. The railroad industry has sponsored a number of research projects to explore ways to reuse, recycle, or at least better dispose of used ties, but the results remain to be seen.

Sometimes the market fails to support the planned post-consumer disposal, even for products designed to be recyclable or reusable. Recycling is certainly the intended use of many paper and plastic products, for example, but many cities have found it difficult to successfully dispose of recycled paper and so are forced to place it in landfills. Battery companies have tried to develop recycling facilities to capture and control metals and other dangerous waste products, but the small and omnipresent nature of batteries has made this very difficult.

11.4.2   Global warming

Among the most pressing concerns facing us are the effects of climate change, also known as global warming. There is overwhelming evidence that the average annual temperatures on the planet are rising, and a very strong consensus in the scientific community that human activity is responsible for some or all of this increase. The consequences of even modest increases in global temperature are likely to be catastrophic for some regions, such as polar ice caps, which are melting at surprising rates, and for some species that depend on particular climate conditions (such as polar bears). Engineers have a special obligation to involve themselves in finding ways to address global warming, both because they have played a key role in responsible technologies and because they have skills that may help moderate climate changes.

One of the most important elements of global climate change is the extent to which carbon is emitted into the atmosphere, introducing what are referred to as "greenhouse gases." Many technologies emit carbon in ways that can surprise us, and when these technologies are used extensively the effects can be very significant. Airplane engines, for example, emit very large volumes of carbon as a combustion byproduct. Indeed, a pound of grapes flown from Chile to the U.S. results in six pounds of carbon being emitted into the atmosphere. Aircraft designers are working very hard to find ways to reduce carbon emissions from engines, but much work remains to be done.

Designing to reduce carbon emissions often begins with the measurement of the "carbon footprint" associated with producing the technology. The designer attempts to measure or estimate all the greenhouse gases emitted in all the processes to produce the product in question. This is still a very new analysis technique, and standards and methods are in flux at this time, but responsible engineers will certainly be expected to understand and apply these techniques when designing for sustainability. As measuring the carbon footprint of technologies becomes better understood, the methods will certainly find their way into life-cycle assessment, an important approach described in the next section.

11.4.3   Environmental life-cycle assessment

Life-cycle assessment (LCA) was developed to help understand, analyze and document the full range of environmental effects of design, manufacture, transport, sale, use and disposal of products. Depending on the nature of the LCA and the product, such analysis begins with the acquisition and processing of raw materials (such as petroleum drilling and refining for plastic products or foresting and processing of railroad ties), and continues until the product has been reused, recycled, or placed in a landfill. LCA has three essential steps:

- *Inventory analysis* lists all inputs (raw materials and energy) and outputs (products, wastes, and energy), as well as any intermediate outputs.
- *Impact analysis* lists all of the effects on the environment of each item identified in the inventory analysis, and quantifies or qualitatively describes the consequences (e.g., adverse health effects, impacts on ecosystems, or resource depletion).
- *Improvement analysis* lists, measures, and evaluates the needs and opportunities to address adverse effects found in the first two steps.

Obviously, one of the keys in LCA is the setting of assessment boundaries. Another is the determination of appropriate measures and data sources for conducting the LCA. Designers cannot expect to find good, consistent data for all elements in the LCA, and so they must reconcile information from multiple sources. Because of differing boundaries, data sources and reconciliation techniques, analysts may produce different figures for the overall effects of a product, even when acting in good faith. To address this issue, it is particularly important to list all assumptions made and to document all data sources used.

Currently, LCA is still in its earliest stages of development as a tool for engineering designers (and others concerned with the environmental effects of technologies). Notwithstanding its youth, however, LCA is already a useful conceptual model for design, and is likely to become increasingly important for the evaluation of engineered systems.

11.5 DESIGN FOR QUALITY: BUILDING A HOUSE OF QUALITY

Quality unites almost all of the elements of conceptual design.

In a certain sense, all the previous X's we have looked at can be considered dimensions of *design for quality*. Quality itself has been defined in a variety of ways, some very brief, some very complicated. One of our favorites is also one of the simplest: *Quality* is "fitness for use," that is, it is a measure of how well a product or service meets its required or desired specifications. By this definition, much of the problem definition activities discussed in Chapters 2, 3 and 4 are aimed at determining what a "quality" design requires. A design will generally be considered a high-quality design if it satisfies all constraints, is fully functional within the desired performance specifications, and meets the objectives as well as or better than alternative designs. In that sense, all of the work we have done in conceptual design is directed to design for quality.

Having said that, however, it is often difficult to tie together all of the elements of a good design. This difficulty can come from problems at the design level, which is our primary focus, or at the implementation level, as in manufacturing or distributing seemingly good products. Designers and manufacturers have developed a variety of tools and techniques to address this difficulty and improve the quality of their products. These include process improvement techniques such as *flowcharting* and *statistical process control* (SPC), external comparisons such as *benchmarking* against other high-quality products, and improving the distribution and delivery of finished goods, known as *supply chain management*.

One of the most important tools used by many designers is *quality function deployment* (QFD). QFD uses — and is also referred to — as the *house of quality*, a matrix that combines information about stakeholders, desirable characteristics of designed products, current designs, performance metrics, and tradeoffs. Figure 11.4 shows the general structure of a house of quality and illustrates how the house metaphor developed. The *Who* in the figure refers to the stakeholders in the design process, that is, client(s), users, and other affected parties. The *Whats* correspond to the desirable attributes that the design must have, to objectives and, in some cases, to functions. The *Now* in the house are existing products or designs. They are typically found as a result of research that has been conducted during the problem definition phase, ands they will be used for benchmarking proposed designs. The *Hows* in the house of quality refer to the metrics and specifications used to measure how well an objective or function has been met. In some versions of the house of quality, some of the functions are placed in the *Hows* part of the matrix, particularly if qualitative measures are being used. *How muches* refers to the goals or targets for the *Whats*. In each of the remaining sections, the relationships, values, or tradeoffs among these elements are displayed. For example, relationships among ways to realize the desired attributes are exposed in the roof of the house. Thus a device made more reliable with redundant systems will also be more costly, so we might put a minus sign in the box corresponding to these measures.

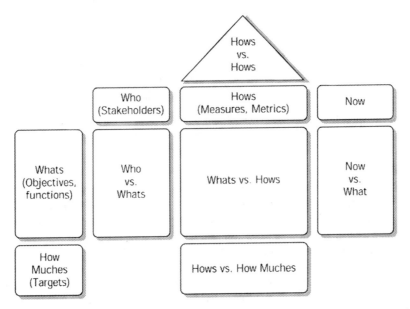

FIGURE 11.4 An elementary abstraction of a *House of Quality* that displays and relates stakeholder interests, desired design attributes, measures and metrics, targets, and current products. The 'house' reports values for these quantities and helps designers explore relationships among them. (After (Ullman 1997).)

Consider the example in Figure 11.5. This simple house of quality is used to explore a housing for a laptop computer. Many people have recently begun to use laptops both while traveling and in the office. A computer maker might want to explore the design space for computer housings that satisfy both office and laptop computers. The

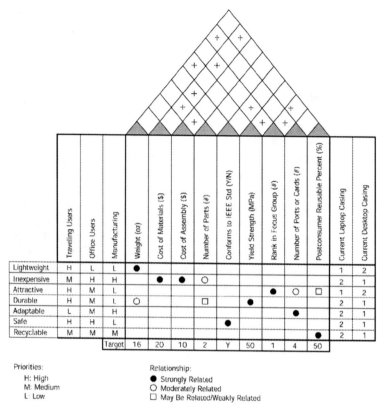

	Traveling Users	Office Users	Manufacturing	Weight (oz)	Cost of Materials ($)	Cost of Assembly ($)	Number of Parts (#)	Conforms to IEEE Std (Y/N)	Yield Strength (MPa)	Rank in Focus Group (#)	Number of Ports or Cards (#)	Postconsumer Reusable Percent (%)	Current Laptop Casing	Current Desktop Casing
Lightweight	H	L	L	●									1	2
Inexpensive	M	H	H		●	●	O						2	1
Attractive	H	M	L							●	O	□	1	2
Durable	H	M	L	O			□		●				2	1
Adaptable	L	M	H								●		2	1
Safe	H	H	L					●					2	1
Recyclable	M	M	M									●	2	1
Target				16	20	10	2	Y	50	1	4	50		

Priorities:
  H: High
  M: Medium
  L: Low

Relationship:
  ● Strongly Related
  O Moderately Related
  □ May Be Related/Weakly Related

FIGURE 11.5 A first-draft house of quality for the design of the housing of a computer that would be used as a laptop and in the office. Note that different users may have different priorities, and the roof of the house helps identify tradeoffs among various measures and attributes.

stakeholders include traveling users, office users, and the manufacturer's production group. In the *Who* vs. *Whats* section, we see that travelers place a high priority on characteristics such as lightweight and durable, while office users are more concerned with cost and adaptability. We might imagine two existing designs, one a standard laptop case and the other a standard desktop/tower casing. The *Whats* vs. *Hows* section shows the relationship between the various metrics and the attributes of a "good" design. Notice, for example, that cost of raw materials and cost of assembly are both strongly related to inexpensive, while number of parts is only modestly so. Similarly, the number of cards and ports that can be accepted is also modestly related to inexpensive, since these require additional assembly work or more parts. *Now* vs. *What* is the result of benchmarking the two existing design choices, and highlights the possibility that a "universal" housing might be able to satisfy more users in total if it can address the shortcomings of either design. Finally, the roof of the house shows some of the relationships and tradeoffs that designers will need to consider. Making the case lighter, for example, is likely to trade off negatively with resistance to forces. Increasing the number of parts is likely to result in higher costs to assemble, as noted in Section 11.1.

This simple example shows that the house of quality can help tie together many of the concepts we have considered throughout this book. An important question is when should QFD be introduced into a design process. Virtually all proponents of the house of quality acknowledge that it requires a good deal of time and effort, just as the processes we have presented do. Our own experience suggests that it is a useful way to plan to gather information, to organize that information as it becomes available, and to foster and enhance discussion within the design team and with stakeholders. However, as with some of the other tools, it should not be taken as an algorithm that produces a hard-and-fast decision, as it cannot produce better results than the materials on which it works!

11.6 NOTES

Section 11.1: This section draws heavily upon (Pahl and Beitz 1996) and (Ulrich and Eppinger 1995). In particular, the six-step process is a direct extension of a five-step approach in (Ulrich and Eppinger 1995), with the iteration made explicit. Our discussion of DFA is adapted from (Dixon and Poli 1995) and (Ullman 1997); rules of assembly are cited in many places but are generally derived from (Boothroyd and Dewhurst 1989). The analogy of the BOM to a recipe is taken from (Schroeder 1993).

Section 11.2: Our review of basic concepts derives from (Riggs and West 1986), but there are any number of excellent texts in engineering economics that can be used to go further with these topics. The cost estimating material draws upon classnotes prepared by our colleague Donald Remer for his cost estimating course and on (Oberlander 1993). A view of engineering costing that is based more on accounting may be found in (Riggs 1994). The relationship between pricing and costs is discussed in (Nagle 1987) and (Philips 1985).

Section 11.3: The definition of reliability comes from U.S. Military Standards Handbook 217B (MIL-STD-217B 1970) as quoted in (Carter 1986). The failure discussion draws heavily from (Little 1991). There are a number of formal treatments of reliability and the associated mathematics, including (Ebeling 1997) and (Lewis 1987). The definition of maintainability is from (Ebeling 1997). The distinction between service, inspection, and repair is from (Pahl and Beitz 1996).

Section 11.4: Codes of ethics for engineers are discussed further in Chapter 12. Sections 11.4.1 and 11.4.3 draw heavily on (Rubin 2001), which also includes a very instructive example of LCA written by Cliff Davidson. The figure for carbon emissions from transporting grapes comes from (McKibbon 2007). Methodologies for calculating carbon footprints are given in (Weidmann and Minx 2007)

Section 11.5: The definition of quality is from (Juran 1979). The standard reference to the house of quality is (Hauser and Clausing 1988), and a number of modifications and extensions have been offered since. The generalized diagram is an adaptation based on (Ullman 1997) who devotes an entire chapter to a more detailed methodology for developing a house of quality.

11.7 EXERCISES

11.1 If you were asked to design a product for recyclability, how would you determine what that meant? In addition, what sorts of questions should you be prepared to ask and answer?

11.2 How might DFA considerations differ for products made in large volume (e.g., the portable electric guitar) and those made in very small quantities (e.g., a greenhouse)?

11.3 Your design team has produced two alternative designs for city buses. Alternative *A* has an initial cost of $100,000, estimated annual operating costs of $10,000, and will require a $50,000 overhaul after five years. Alternative *B* has an initial cost of $150,000, estimated annual operating costs of $5,000, and will not require an overhaul after five years. Both alternatives will last ten years. If all other vehicle performance characteristics are the same, determine which bus is preferable using a discount rate of 10%.

11.4 Would the decision reached in Exercise 11.3 change if the discount rate was 20%? What would happen if, instead, the discount rate was 15%? How do the resulting cost figures influence your assessments of the given cost estimates?

11.5 Your design team has produced two alternative designs for greenhouses in a developing country. Alternative A has an initial cost of $200 and will last two years. Alternative B has an initial cost of $1,000 and will last ten years. All other things being equal, determine which greenhouse is more economical with a discount rate of 10%. Which other other factors can influence this decision?

11.6 What is the reliability of the system portrayed in Figure 11.2?

11.7 What is the reliability of the system portrayed in Figure 11.3? How does this result compare with that of Exercise 11.6? Why?

11.8 On what basis would you choose between a single system, all of whose parts are redundant, and two copies of a system that has no redundancy?

11.9 What are the factors to consider in an environmental analysis of the beverage container design problem? How might an environmental life-cycle assessment help in addressing some of these questions?

11.10 Draw the house of quality for the beverage container design problem used in the earlier chapters.

CHAPTER 12

ETHICS IN DESIGN

Is design really just a technical matter?

DESIGN IS fundamentally a human endeavor. It involves the interactions among members of a design team, the relationships between designers, clients, and manufacturers, and the ways that purchasers of designed devices use them in their lives. In many cases, design affects the lives of people who were not part of the designer-client-user triangle we discussed in Chapter 1. Since design touches so many facets of people's daily lives, we must consider how people interact with each other and how they are acted upon by the designs we create. To design means to *accept responsibility for creating designs for people*. That is, design is not done in a vacuum; *design is a social activity*. Designers are influenced by the social milieu in which they work, and society is influenced by the products of design. Therefore, ethics and ethical behavior must be considered in our examination of how designs are created and used.

12.1 ETHICS: UNDERSTANDING OBLIGATIONS

Words like ethics, morals, obligations, and duty are used in a variety of ways, including seemingly contradictory or unclear ones. As we did with many of the engineering terms earlier in the book, we start this discussion with some dictionary definitions. First, the word *ethics*:

- **ethics 1** the discipline dealing with what is good and bad and with moral duty and obligation **2 a:** a set of moral principles or values **b:** a theory or system of moral values **c:** the principles of conduct governing an individual or group

And, since it is referenced so often in the definition of ethics, the word *moral*:

- **moral 1 a:** of or relating to principles of right or wrong in behavior **b:** expressing or teaching a conception of right behavior

Besides defining a discipline or field of study, these definitions define ethics as a set of guiding principles or a system that people can use to help them behave well. Most of us learn right and wrong from our parents, or perhaps as a set of beliefs from one of the religious traditions that emphasize faith in God (e.g., Christianity, Judaism, and Islam) or those that stress faith in a *right path* (e.g., Buddhism, Confucianism, and Taoism). Either way, virtually all of us have a deep connection with notions such as honesty and integrity, and about the injunction to treat others as we would want to be treated ourselves.

If we already know these things, why do we need another, external set of rules? If we don't, and the law doesn't keep us in line, what is the use of a set of ethical principles? The truth is that the kinds of lessons we learn at home, in school, and in our religious forums may not provide enough explicit guidance about many of the situations we face in life, especially in our professional lives. In addition, given the diversity and complexity of our society, we are likely better off to have some standards of professional behavior that are universally agreed upon, across all of our traditions and individual upbringings. (Our dependence on laws and lawyers would be significantly and happily diminished if every-one's individually-learned lessons were sufficient!)

Ethics are principles of conduct for both individuals and groups.

Our professional lives are even more complicated because our responsibilities often involve obligations to many stakeholders, some of whom are obvious (e.g., clients, users, the immediately surrounding public) and some of whom are not (e.g., some government agencies, professional societies). We will elaborate these obligations further in Sections 12.2–12.5, but we note now that these obligations often conflict. For example, a client may want one thing, while a group of people affected by a design may want something altogether different. Further, those affected by a design may not even know how they are being affected until *after* the design is complete and has been implemented. In addition, it is also interesting to point out that our concern for mediating between conflicting ethical obligations is similar to our concern in conceptual design: that we correctly assess the relative importance of the client's objectives when they compete. Remember that there is no formula or algorithm to apply because the priorities we place on objectives are subjec-tive in nature — as is our personal assessment of the relative importance that we attach to our conflicting obligations. However, professional ethics and their expression in associ-ated codes of ethics provide a means of reconciling such competing obligations, and we will discuss them in Section 12.2.

Consider the well known case in which group of engineers tried unsuccessfully to delay the launch of the space shuttle *Challenger* on January 28, 1986. While severe doubt was expressed by some engineers about the safety of the *Challenger*'s O-rings because of the cold weather before the flight, the upper management of Morton-Thiokol, the com-pany that made the *Challenger*'s booster rockets, and NASA approved the launch. These managers determined that their concerns about Morton-Thiokol's image and the stature and visibility of NASA's shuttle program outweighed the judgments of the engineers closest to the booster design. The Morton-Thiokol engineers ultimately publicized the dismissal of their recommendation not to launch by engaging in *whistleblowing*, wherein someone "blows the whistle" in order to stop a faulty decision made within a company, agency, or some other institution.

Whistleblowing is not new or unique. Another famous case is that of an industrial engineer, Ernest Fitzgerald, who blew the whistle on major cost overruns in the procure-ment of the Air Force's giant C–5A cargo plane. The Air Force was so displeased with Fitzgerald's actions that it took bureaucratic actions to keep him from further work on the plane: It "lost" his Civil Service tenure and then reconstructed that part of the bureaucracy in which had Fitzgerald worked so as to eliminate his position! After an arduous and expensive legal battle, Fitzgerald earned a substantial settlement for wrongful termination and was reinstated in his position.

While such stories are to some extent discouraging, they also show heroic behavior under trying circumstances. More to the point, these examples show how ''doing what's right'' can be perceived quite differently within an organization. An engineer may well be faced with just the kind of clash of obligations that lies at the crux of any discussion of engineering ethics. If that happens, to whom can the designer or engineer turn to for help? While part of the answer lies in the roots of the engineer's personal understanding of ethics, another part of the answer lies in the support of immediate professional colleagues and peers, as these have been seen to be very effective in both righting the perceived wrong and in sustaining the whistleblower. One of the primary sources of insight and guidance, however, is the professional engineering societies and their codes of ethics.

12.2 CODES OF ETHICS: WHAT ARE OUR PROFESSIONAL OBLIGATIONS?

Imagine you are a mining engineer who has been engaged by the owner of a mine to design a new shaft extension. As part of that design task you survey the mine and find that part of it runs under someone else's property. Are you obligated to simply complete the survey and the design for the mine owner who engaged you and who is paying you, and then go on to your next professional engagement?

You may suspect that the mine owner hasn't notified the landowner that his mineral rights are being excavated out from underneath him. Should you do something about that? If so, what? Further, what compels you to do something? Is it personal morality? Is there a law? How are you responsible, and to whom?

The chain of questions just started can easily be lengthened, and the situation made more complicated. For example, what if the mine was the only mine in town and its owner controlled your livelihood, and those of many town residents? Or, if you find out that the mine runs under, or perilously close to, an elementary school, does that change things?

This story highlights some of the many actors and obligations that could arise in an engineering project. In fact, scenarios such as this occurred late in the 19th century and early in the 20th, and it was just such situations that provided part of the impetus for the formation of professional societies and the development of codes of ethics by these societies as a form of protection for their individual members.

Over time, the professional societies also undertook other kinds of activities, including promulgating standards for design endeavors, and providing forums for reporting research and innovations in practice. But the situation remains that professional engineering societies continue to play a leading role in setting ethical standards for designers and engineers. These ethical standards clearly speak to the various and often conflicting obligations that an engineer must meet. The societies also provide mechanisms for helping engineers deal with and resolve conflicting obligations, and, when asked, they provide the means for investigating and evaluating ethical behavior.

Most professional engineering societies have published codes of ethics. We show the codes of ethics of the American Society of Civil Engineers (ASCE) in Figure 12.1

ASCE CODE OF ETHICS

Fundamental Principle

Engineers uphold and advance the integrity, honor and dignity of the engineering profession by:

1. using their knowledge and skill for the enhancement of human welfare and the environment;
2. being honest and impartial and serving with fidelity the public, their employers and clients;
3. striving to increase the competence and prestige of the engineering profession; and
4. supporting the professional and technical societies of their disciplines.

Fundamental Canons

1. Engineers shall hold paramount the safety, health and welfare of the public and shall strive to comply with the principles of sustainable development [1] in the performance of their professional duties.
2. Engineers shall perform services only in areas of their competence.
3. Engineers shall issue public statements only in an objective and truthful manner.
4. Engineers shall act in professional matters for each employer or client as faithful agents or trustees, and shall avoid conflicts of interest.
5. Engineers shall build their professional reputation on the merit of their services and shall not compete unfairly with others.
6. Engineers shall act in such a manner as to uphold and enhance the honor, integrity, and dignity of the engineering profession and shall act with zero-tolerance for bribery, fraud, and corruption.
7. Engineers shall continue their professional development throughout their careers, and shall provide opportunities for the professional development of those engineers under their supervision.

[1] In November 1996, the ASCE Board of Direction adopted the following definition of Sustainable Development: "Sustainable Development is the challenge of meeting human needs for natural resources, industrial products, energy, food, transportation, shelter, and effective waste management while conserving and protecting environmental quality and the natural resource base essential for future development."

FIGURE 12.1   The code of ethics of the American Society of Civil Engineers (ASCE), as modified July 2006. It is similar, although not identical, to the code adopted by the IEEE that is displayed in Figure 12.2.

and of the Institute of Electronics and Electrical Engineers (IEEE) in Figure 12.2. While both codes emphasize integrity and honesty, they do appear to value certain kinds of behavior differently. For example, the ASCE code enjoins its members from competing unfairly with others, a subject not mentioned by the IEEE. Similarly, the IEEE specifically calls for its members to ''fairly treat all persons regardless of such factors as race, religion, gender...''. There are other differences as well in the styles of language. The ASCE presents a set of injunctions about what engineers ''shall'' do, while the IEEE code is phrased as a set of commitments to undertake certain behaviors.

IEEE CODE OF ETHICS

We, the members of the IEEE, in recognition of the importance of our technologies in affecting the quality of life throughout the world, and in accepting a personal obligation to our profession, its members and the communities we serve, do hereby commit ourselves to the highest ethical and professional conduct and agree:

1. to accept responsibility in making decisions consistent with the safety, health and welfare of the public, and to disclose promptly factors that might endanger the public or the environment;

2. to avoid real or perceived conflicts of interest whenever possible, and to disclose them to affected parties when they do exist;

3. to be honest and realistic in stating claims or estimates based on available data;

4. to reject bribery in all its forms;

5. to improve the understanding of technology, its appropriate application, and potential consequences;

6. to maintain and improve our technical competence and to undertake technological tasks for others only if qualified by training or experience, or after full disclosure of pertinent limitations;

7. to seek, accept, and offer honest criticism of technical work, to acknowledge and correct errors, and to credit properly the contributions of others;

8. to treat fairly all persons regardless of such factors as race, religion, gender, disability, age, or national origin;

9. to avoid injuring others, their property, reputation, or employment by false or malicious action;

10. to assist colleagues and co-workers in their professional development and to support them in following this code of ethics.

FIGURE 12.2 The code of ethics of the Institute of Electronics and Electrical Engineers (IEEE), dated February 2006. How does the IEEE code of ethics differ from that adopted by the ASCE displayed in Figure 12.1?

Notwithstanding these differences, both codes of ethics set out guidelines or standards of how to behave with respect to: clients (e.g., ASCE's ''as faithful agents or trustees''); the profession (e.g., IEEE's ''assist colleagues and co-workers in their professional development''); the law (e.g., IEEE's ''reject bribery in all its forms''); and the public (e.g., ASCE's ''shall issue public statements only in an objective and truthful manner''). Perhaps most noteworthy, both place a primary concern on protection of the health, safety and welfare of the public. We will return to this paramount principle in Section 12.5.

One of an engineer's obligations is to adhere to a code of ethics.

The codes of ethics, along with the interpretations and guidance offered by the societies, lay out rules of the road for dealing with conflicting obligations, including the task of assessing whether these conflicts are ''only'' of perception or of a ''real'' and potentially damaging nature.

There are some points to make regarding the professional societies and their codes. First, the differences in the codes reflect different styles of engineering practice in the various disciplines much more than differences in their views of the importance of ethics.

For example, most civil engineers who are not employed by a government agency work in small companies that are people-intensive, rather than capital-intensive. These firms obtain much of their work through public, competitive bidding. Electrical engineers, on the other hand, more often than not work for large corporations that sell products more than they do services, one result of which is that they have significant manufacturing operations and are capital-intensive. Such different practices produce different cultures and, hence, different statements of ethical standards.

A second point is that the professional societies, notwithstanding their promulgation of codes of ethics, have not always been seen as active and visible protectors of whistle-blowers and other professionals who raise concerns about specific engineering or design instances. This situation is improving, steadily if slowly, but many engineers still find it difficult to look to their societies, and especially their local branch sections, for first-line assistance and support in times of need. Of course, as we all make ethical behavior a higher priority, the need for such support will lessen, and its ready availability will surely increase.

Finally, we should note that the codes of ethics adopted by the professional societies we are describing are those found in the United States and Canada, which are not necessarily the same as those in other parts of the world. In countries with a strongly Islamic culture and government, for example, the codes of ethics often reflect an alignment between religious values and professional practice that is alien to our traditions of separating church and state. Similarly, the code of ethics of the *Verein Deutscher Ingenieure* (VDI), or Association of German Engineers, reflect the historical need of German engineers to reflect upon and respond to the willingness of many of their colleagues to work in support of the Nazis during the 1930s and 1940s. It is important for us, as professional engineers, understand and respond to the culture in which we work, while remaining true to our own values.

12.3 OBLIGATIONS MAY START WITH THE CLIENT...

Let us consider in greater depth our various obligations to a client or to an employer. As designers or engineers, we owe our client or employer a professional effort at solving a design problem, by which we mean being technically competent, conscientious, and thorough, and that we should undertake technical tasks only if we are properly ''qualified by training or experience.'' We must avoid any conflicts of interest, and disclose any that may exist. And we must serve our employer by being ''honest and impartial'' and by ''serving with fidelity...'' Most of these obligations are clearly delineated in codes of ethics (e.g., compare the quotes with Figures 12.1 and 12.2), but there's at least one curious obligation on this list: What does it mean to serve with ''fidelity''?

A thesaurus would tell us that fidelity has several synonyms, including constancy, fealty, allegiance, and loyalty. Thus, one implication we can draw from the ASCE code of ethics is that we should be loyal to our employer or our client. This suggests that one of our obligations is to look out for the best interests of our client or employer, and to maintain a clear picture of those interests as we do our design work. But loyalty is a very touchy matter; it is not a simple, one-dimensional attribute. In fact, clients and companies earn the loyalty of their consultants and staffs in at least two ways. One, called *agency-loyalty*, derives from the nature of any contracts between the designer and her client (e.g., "work for hire") or between the designer and her employer (e.g., a "hired worker"). As it is dictated by contract, agency-loyalty is clearly obligatory for the designer. The second kind of loyalty, *identification-loyalty*, is more likely to be seen as optional. It stems from the engineer identifying with the client or company because she admires its goals or sees its behavior as mirroring her own values. To the extent that identification-loyalty is optional, it will be earned by clients and companies only if they reciprocate by demonstrating loyalty to their own staff designers.

Agency-loyalty provides one reason to maintain a "design notebook" to document design work. As we have noted before, keeping such a record is good design practice because it is very useful for recapitulating our thinking as we move through different stages of the design process and for real-time tracking. A dated design notebook also provides a legal basis for documenting how new, patentable ideas were developed. Such documentation is essential to an employer or client if a patent application is in any way challenged. Further, as is typically specified in contracts and employment agreements, the intellectual work done in the creation of a design is itself the intellectual property of the client or the employer. A client or an employer may share the rights to that intellectual property with its creators, but the fundamental decisions about the ownership of the property generally belong to the client or owner. It is important for a designer to keep that in mind, and also to document any separate, private work that she is doing, just to avoid any confusion about who owns any particular piece of design work.

Since identification-loyalty is optional, it provides fertile ground for clashes of obligations because other loyalties have the space to make themselves felt here. As we will discuss further in Section 12.6, modern codes of ethics normally articulate some form of obligation to the health and welfare of the public. For example, the ASCE code of ethics (Figure 12.1) suggests that civil engineers work toward enhancing both human welfare and the environment, and that they "shall hold paramount the safety, health, and welfare of the public . . . ." Similarly, the IEEE code (Figure 12.2) suggests that its members commit to "making engineering decisions consistent with the safety, health and welfare of the public . . . ." These are clear calls to engineers to identify other loyalties to which they should feel allegiance. There is little doubt that it was just such divided loyalties that emerged in the cases of whistleblowing discussed above.

Engineers must be loyal to their clients, to other stakeholders, And to themselves.

In the case of the explosion of the *Challenger*, those who argued against its launching felt that lives would be endangered. They placed a higher value on the lives being risked than they did on the loyalties being demanded by Morton-Thiokol (i.e., to secure its place as a government contractor) and by NASA (i.e., to successfully argue for the shuttle program before Congress and the public). Similar conflicting loyalties emerge for engineers as toxic waste sites are cleaned up under the Super Fund program of the Environmental Protection Agency (EPA). In many instances, employees felt that they needed to look out for their own companies, sometimes because they felt the companies should not be penalized for doing what was once legal, in others because they were pressured to do so by peers and bosses and by the possible loss of their jobs. There are also cases in which engineers were apparently willing, or at least able, to rank their loyalties to their companies first, to the point where falsified emission test data were reported to the government (by engineers and managers at the Ford Motor Company) or parts known to be faulty were delivered to the Air Force (by engineers and managers at the B. F. Goodrich Company).

An apparent disloyalty to a company or an organization may sometimes be, in a longer term, an act of greater, successfully merged loyalties. When the Ford Pinto was initially being designed, for example, some of its engineers wanted to perform crash tests that were not then called for in the relevant U.S. Department of Transportation regulations. The managers charged with developing the Pinto felt that such tests could not benefit the program and, in fact, might only prove to be a burden. Why run a test that is not required, only to risk failing that test? The designers who proposed the tests were seen as disloyal to Ford and to the Pinto program. In fact, what happened was that the placement of the drive train and the gas tank resulted in fiery crashes, lives lost, and major public relations and financial headaches for Ford. Clearly, Ford would have been better off in the long run to have conducted the tests, so the engineers who proposed them could be said to have been looking out for the company's long-term interests.

If there is one point that emerges from the discussion thus far, it is that ethical issues do not arise from a *single* obligation. Indeed, were issues so easily categorized, choices would vanish and ethical conflicts would not be a problem. In fact, as we have already discussed, the very existence of professional codes of ethics both testifies to the reality of conflicting obligations and at the same time provides guidance to mediating among those conflicts.

12.4 . . . BUT WHAT ABOUT THE PUBLIC AND THE PROFESSION?

We now tell a story that shows that when people behave responsibly, things can turn out well even in bad situations. In fact, to start with the ending, the protagonist-hero of this story said, "In return for getting a [professional engineering] license and being regarded with respect, you're supposed to be self-sacrificing and look beyond the interests of yourself and your client to society as a whole. And the most wonderful part of my story is that when I did it nothing bad happened."

Our hero is William J. LeMessurier (pronouncd "LeMeasure") of Cambridge, Massachusetts, one of the most highly regarded structural engineers and designers in the world. He served as the structural consultant to a noted architect, Hugh Stubbins, Jr., for the design of a new headquarters building for Citicorp in New York City. Completed in 1978, the 59-story Citicorp Center is one of the most dramatic and interesting skyscrapers in a city filled with some of the world's great buildings (see Figure 12.3). In many ways, LeMessurier's conceptual design for Citicorp resembles other striking skyscrapers in that it used the *tube* concept in which a building is designed as a tall, hollow tube that has a comparatively rigid or stiff tube wall. (In structural engineering terminology, the tube's main lateral stability elements are located at the outer perimeter and tied together at the corners.) Fazlur Kahn's John Hancock Center in Chicago is a similar design (see Figure 12.4). The outer "tube" or "main lateral stability elements" are the multistory diagonal elements that are joined to large columns at the corners. Kahn's design benefited from a deliberate architectural decision to expose the tube's details, perhaps to illustrate the famous dictum that *form follows function* (which is routinely and wrongly attributed to Frank Lloyd Wright, but was pronounced by Louis Sullivan, the noted Chicago architect who was also Wright's mentor).

FIGURE 12.3 One view of the 59-story Citicorp Center, designed by architect Hugh Stubbins, Jr., with William J. LeMessurier serving as structural consultant. One of this building's notable features is that it rests on four massive columns that are placed at the midpoints of the building's sides, rather than at the corners. This enabled the architects to include under the Citicorp's sheltering canopy a new building for St. Peter's Church. (Photo by Clive L. Dym.)

FIGURE 12.4 The 102-story John Hancock Center, designed by the architectural firm of Skidmore, Owings and Merrill, with Fazlur Kahn serving as structural engineer. Note how the exposed diagonal and column elements make up the tube that is the building's underlying conceptual design. (Photo by Clive L. Dym.)

LeMessurier's Citicorp design was innovative in several ways. One, not visible from the outside, was the inclusion of a large mass, floating on a sheet of oil, within the triangular roof structure. It was added as a damper to reduce or damp out any oscillations the building might undergo due to wind forces. Another innovation was LeMessurier's adaptation of the tube concept to an unusual situation. The land on which the Citicorp Center was built had belonged to St. Peter's Church, with the church occupying an old (dating from 1905) and decaying Gothic building on the lot's west side. When St. Peter's sold the building lot to Citicorp, it also negotiated that a new church be erected "under" the Citicorp skyscraper. In order to manage this, LeMessurier moved the "corners" of the building to the midpoints of each side (see Figure 12.5). This enabled the creation of a large space for the new church because the office tower itself was then cantilevered out over the church, at a height of some nine stories. Looking at the sidewalls of the tube — and here we have to peel off the building's skin because architect Stubbins did not want the structures exposed as they were in the Hancock tower — we see that the wall's

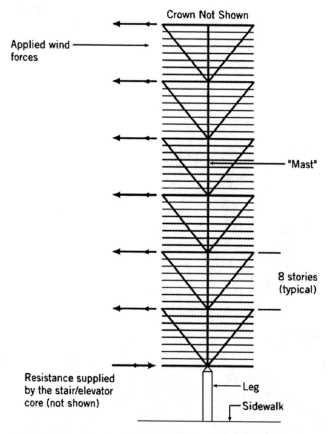

FIGURE 12.5 A sketch of LeMessurier's Citicorp design. Here the tube is made up of (unexposed) diagonal elements, organized as rigid triangles and connected to the four columns at the midpoints of the sides of the building. (Adapted from *Civil Engineering*.)

rigidity comes from large triangles, made up of diagonal and horizontal elements, that all connect at the midpoints of the sides. Thus, LeMessurier's triangles serve the same purpose as Kahn's large X-frames.

The ethics problem arose soon after the building was completed and occupied. LeMessurier received a call from an engineering student in New Jersey who was told by a professor that the building's columns had been put in the wrong place. In fact, LeMessurier was very proud of his idea of placing the columns at the midpoints. He explained to the student how the 48 diagonal braces he had superposed on the midspan columns added great stiffness to the building's tube framework, particularly with respect to wind forces. The student's questions sufficiently intrigued LeMessurier that he reviewed his original design and calculations to see just how strong the wind bracing system would be. He found himself looking at a case that was not examined under then-current practice and building codes. Practice at the time called for wind force effects to be calculated when the wind flow hit a side of a building dead on, that is, normal to the building faces. However, the calculation of the effect of a *quartering wind*, under which the wind hits a building on a 45-degree diagonal and the resulting wind pressure is then distributed over the two immediately adjacent faces (see Figure 12.6), had not been called for previously. A quartering wind on the Citicorp Center leaves some diagonals unstressed and others doubly loaded, with calculated strain increases of 40%. Normally, even this increase in strain (and stress) would not have been a problem because of the basic assumptions under which the entire system was designed.

However, to LeMessurier's further discomfort, he found out just a few weeks later that the actual connections in the finished diagonal bracing system were not the high-strength welds he had stipulated. Rather, the connections were bolted because Bethlehem Steel, the steel fabricator, had determined and suggested to LeMessurier's New York office that bolts would be more than strong enough and, at the same time, significantly cheaper. The choice of bolts was sound and quite correct professionally. However, to

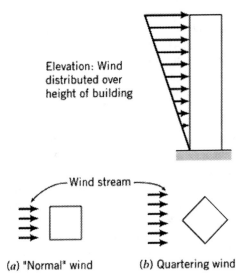

Elevation: Wind distributed over height of building

(a) "Normal" wind   (b) Quartering wind   — Wind stream —

FIGURE 12.6 A sketch of how wind forces are seen on buildings. (a) This is the standard case of wind streams being normal or perpendicular to the building faces. It was the only one called for in design codes at the time the Citicorp center was designed. (b) This is the case of quartering winds, in which case the wind stream comes along a 45-degree diagonal and thus simultaneously applies pressure along two faces at once.

LeMessurier the bolts meant that the margin of safety against forces due to a quartering wind — which, again, structural engineers were not then called upon to consider — was not as large as he would have liked. (It is interesting to note that while New York City's building code did not then require that quartering winds be considered in building design, Boston's code did — and had since the 1950s!)

Spurred by his new calculations and the news about the bolts, and by hearing of some other detailed design assumptions made by engineers in his New York office, LeMessurier retreated to the privacy of his summer home on an island in Maine to carefully review all of the calculations and changes and their implications. After doing a member-by-member calculation of the forces and reviewing the weather statistics for New York City, LeMessurier determined that statistically, once every sixteen years the Citicorp Center would be subjected to winds that could produce a catastrophic failure. Thus, in terminology used by meteorologists to describe both winds and floods, the Citicorp Center would fail in a sixteen-year storm — when it had supposedly been designed to withstand a fifty-year storm. So, what was LeMessurier to do?

In fact, LeMessurier considered several options, reportedly including driving into a freeway bridge abutment at high speed. He also considered remaining silent, as he tried to reassure himself that his innovative rooftop mass damper actually reduced the probabilities of such failure to the fifty-year level. On the other hand, if the power went out, the mass damper wouldn't be there to help. So, what did LeMessurier actually do?

He first tried to contact the architect, Hugh Stubbins, who was away on a trip. He then called Stubbins' lawyer, after which he talked first with his own insurance carrier and then with the principal officers of Citicorp, one of whom had studied engineering before choosing to become a banker. While some early consideration was given to evacuating the building, especially since hurricane season was just over the horizon, it was decided instead that all of the connections at risk should be redesigned and retroactively fixed. Steel plate "Band-Aids," two inches thick, would be welded onto each of 200 bolted connections. However, there were some very interesting implementation problems, only some of which we mention here. The building's occupants had to be informed without being alarmed, because the repair work would go on every night for two months or more. The public had to be informed why the bank's new flagship headquarters suddenly needed immediate modifications. (In fact, the entire process was open and attentive to public concerns.) Skilled structural welders, who were in short supply, had to be found, as did an adequate supply of the right grade of steel plate. Discreet, even secret, evacuation plans had to be put in place, just in case an unexpectedly high wind did show up while the repairs were being made. And New York City's Building Commissioner and the Department of Buildings and its inspectors had to be brought into the loop because they were central to resolving the problem. They had to be informed about the problem and its proposed solution, and they had to agree to inspect that solution. All in all, a dazzling array of concerns, institutions, and, of course, personalities.

In the end, the steel Band-Aids were applied and the entire business was completed professionally, with no finger-pointing and no public assignments of blame. LeMessurier, who had thought his career might precipitously end, came away with still greater stature, occasioned by his willingness to face up to the problem candidly and propose a realistic, carefully crafted solution. In the words of one of the engineers involved in implementing LeMessurier's solution, "It wasn't a case of 'We caught you, you skunk.' It started with a guy who stood up and said, 'I got a problem, I made the problem, let's fix the problem.' If you're gonna kill a guy like LeMessurier, why should anybody ever talk?''

As we have said before, this is a case where everyone involved behaved well. In fact, it is to everyone's credit that all of the actors behaved to a very high standard of professionalism and understanding. It is, therefore, a case that we can study with pleasure, particularly as engineers. It also is a case that could have gone other ways, so we will close our discussion by posing a few questions that you, the reader, might face were you in LeMessurier's position:

- Would you have "blown the whistle," or not?
- What would you have done if you determined that the revised probability of failure was higher (i.e., worse) than for the original design, but still within the range permitted by code?
- What would you have done if your insurance carrier had said to "keep quiet"?
- What would you have done if the building owner, or the city, had said to "keep quiet"?
- Who should pay for the repair?

12.5   ON ENGINEERING PRACTICE AND THE WELFARE OF THE PUBLIC

It is easy to imagine a scenario where we are asked to design a product that we think needn't be made, or perhaps even shouldn't be made. In Chapter 4, for example, we referred to the design of a cigarette lighter, which we also thought of as an igniter of leafy matter. While this example seems trivial, even nonsensical, it points to another facet of divided loyalties. It suggests that designing cigarette lighters might somehow be morally troubling. In the United States today, there are many people who would consider designing cigarette lighters and cigarette-making machinery as being at least "politically incorrect," and perhaps even morally wrong. On the other hand, isn't it up to individuals to choose to smoke or not? If a product is legal, shouldn't we allow ourselves to design it without feeling uncomfortable?

One glaring instance along these lines emerges from the design of the large-scale ovens and associated specialized buildings made in Germany in the 1930s and 1940s. Still another might be the design of nuclear weapons in the U.S. and in the Soviet Union since the late 1940s (and today in a seemingly growing number of developing countries!). The technologies are different, and while some engineers and physicists were excited by the intellectual challenge of designing devices to harness nuclear fission, it's hard to imagine oven designers having similar feelings. So, were these sets of designers merely being loyal to their clients, their governments and to their societies? If so, were they being loyal toward "human welfare and the environment"?

Not everything that can be designed should be designed. Recall that the codes of ethics we discussed in Section 12.2 place the health, safety, and welfare of the public in the first or paramount position. Historically, most engineers and the professional societies have focused almost entirely on the health and safety aspects of these phrases. In a manner similar to the medical profession's admonition "first, do no harm," engineers are certainly committed to ensuring that the things we design are not willfully dangerous, and that the process of design is rigorous, thorough, and honest about potential risks to the public. Unfortunately, the "welfare of the public" phrase has not always been so deeply explored or considered. Some philosophers of technology have challenged engineers to consider these issues more carefully, and we will do so briefly in the next section.

12.5.1   Ethical behavior and "the good life"

For most of us, concern with human welfare begins with meeting fundamental human needs, such as ensuring adequate food, water, and shelter. While starting from basic needs, and perhaps taking a cue from contemporary economics, we often extend these concerns so that "more" and "better" become synonyms. They certainly are the same for those around the world who live in abject poverty. It is an open question whether or not "more" and "better" are the same in the developed world. Clearly, when we speak of the "welfare of the public," we are no longer in the land of the purely technical: The welfare of the public is implicitly about what constitutes "the good life." The depletion of key resources, the degradation of our environment, and the changes in our global climate should give us pause about what we mean by "the good life."

Deciding what constitutes the welfare of the public means addressing social and political concerns. And just as with our concerns about balancing competing objectives and reconciling conflicting obligations, we must be aware that our social and political concerns are also subjective, and not objective. For example, the ASCE has issued Guidelines to Practice that suggest that in order to adhere to the canons of its code of ethics (Figure 12.1) engineers should "recognize that the lives, safety, health and welfare of the general public are dependent upon engineering judgments, decisions and practices incorporated into structures, machines, products, processes and devices." *Judgment, decision,* and *practice* are *political words*: they signify far more than just technical or scientific aspects. In the first place, the context in which engineers practice is always political because judgments are made about desirable ends and are exercised in the contexts of relationships in which power is not equally or necessarily fairly distributed. Secondly, as this guideline to the canon recognizes, the judgments, decisions, and practices of engineers are embodied in the structures, machines, processes and devices they make — and these, in turn, are implicated in the wellbeing of society more generally. Thus, the judgments and decisions engineers make in practicing their vocation are never "just technical." This implies that the practice of engineering has much in common with other professional practices (such as law and medicine), and that engineers bear a special responsibility to practice their profession, at least in part, as if they were also practicing politics in a societal environment.

While most of us would immediately grant the societal nature of the context in which engineering is practiced, many engineers shrink from its political aspects because the word "politics" has acquired a pejorative flavor, that is, it carries unfortunate baggage. For example, many working engineers regularly offer "politics" as something that stands in the way of getting the real job done, how rivals seek competitive advantage, or how winners placate losers either following or in advance of their loss. In this sense, when a practicing engineer is forced by circumstance to do politics, she understands herself to be doing something other than engineering, something that she must tolerate but which is not intrinsic to her practice. Engineering is engineering, but *that* is just ''politics.'' However, as the author Langdon Winner has pointed out, there is a point ''where ethics finds its limits and politics begins . . . when we move beyond questions of individual conduct to consider the nature of human collectivities and our membership in them.'' It is precisely at this meeting point where ethics and politics overlap that many professional engineers prefer not to go. But aren't such design criteria as efficiency, effectiveness, and economy subjective in nature? Are they not situated historically? Don't they reflect socially constructed standards of ''goodness''?

In light of this, it is unfortunate that so many of us regard politics in a diminished and pejorative sense, a view held not only by engineers but also by a huge share of the general population. On a small scale, we see politics as the incessant jockeying for position by the people and groups we encounter daily in our professional and personal lives. On a larger scale, we are too often subjected to a harmful version of politics as a media spectacle of partisan competition for office, a game marked by negative campaigning and character assassination rather than a discourse about what might constitute a good and just society.

Fortunately, we need not limit our imagination to the impoverished meaning attached to politics in the current climate. A more robust conceptualization might more readily suggest positive ways in which our work as engineers can be understood as political. At its core, politics concerns judgment, in particular judgment about good ends, and about just means for reaching toward those ends. Politics is about making judgments about what we ought to do, and how we ought to do it. In making these judgments, we set out what we think we should be and, in carrying them out, we become what we are. Judgment of this sort is also political by virtue of its public character: Political judgment is carried out *in* public, *by* public persons, and in a manner that accounts for or addresses the public nature and implications of whatever is being considered. Viewed as public judgment about public matters that have public implications, politics transcends the private calculation of private interests for private advantage that characterizes so much of what we label as politics today.

Thus, engineers must exercise political judgment before, and while, doing design because of the politics that arise *after* design, once a device or system is released for public use. Our society seems culturally disposed to view technology either as completely neutral or, where some substantive character is assigned to technology, as primarily oriented toward achieving the ''goods'' of progress, prosperity, variety and convenience. However, the argument that technologies are simply neutral or primarily progressive is difficult to sustain: We can drive a car here or there, or not at all, but a city organized to make automobile traffic efficient is radically different than a city made for pedestrians, and it closes down as many options as it opens up. Thus, when standing in the middle of a parking lot beside a street with eight lanes but no sidewalks, we do not always understand this as progress. In a sense, technology is much like other fundamental principles. An American would never say that the principle of freedom of expression is ''neutral,'' or devoid of substantive content that distinguishes it from its opposite, simply because people can use freedom of expression to say they hate their government as easily as they can use it to say they love it. Despite the fact that freedom of expression can be exercised in a variety of ways, it is far from neutral: It embodies and structures a particular way of life; it establishes political relationships, permission and prohibitions; and it is bound up with the distribution of political power.

The same can be said of technology. Technology is implicated in particular possibilities for social organization and relationships, in the establishment and enforcement of permissions and prohibitions, and in the distribution of economic, social, and political power. This insight has been expressed most clearly by Andrew Feenberg in his book *Questioning Technology*:

> *Technology is power in modern societies, a greater power in many domains than the political system itself. The masters of technical systems, corporate and military leaders, physicians and engineers, have far more control over patterns of urban growth, the design of dwellings and transportation systems, the selection of innovations, our experience as employees, patients and consumers, than all the electoral institutions of our society put together. But, if this is true, technology should be considered as a new kind of legislation, not so very different from other public decisions. The technical codes that shape our lives reflect particular social interests to which we have delegated the power to decide where and how we live, what kinds of food we eat, how we communicate, are entertained, healed and so on.*

12.5.2   Engineering publics

One of the more challenging questions facing engineers who take the welfare of the public seriously is "Who, or what, is the public?" When an engineer begins a project to design a water treatment facility, she needs to be attentive to the complexity of the public whose wellbeing she has committed to serve by subscribing to the ethical obligations that define her profession. Engineers can go a long way towards satisfying this need simply by taking it seriously, and by questioning the ways in which the public and its interest are often mischaracterized. The public and its interest cannot be taken for granted. The public is not simply an object out there waiting to be served or observed, it is comprised of real people who must be recognized and activated through ongoing engagement. Given the considerable diversity with which contemporary American society is blessed, this means that attending to the public interest really means integrating into engineering practice an ongoing, good-faith effort to engage the interests of multiple *publics*.

Engineers must also resist the temptation to accept the various surrogates that often stand in for publics today. The public sphere is not simply a market, and publics are not the same as shareholders or stakeholders, audiences, clients, or consumers. A market is a mechanism for the self-interested exchange of commodities and the calculation of price; it is not a forum for public-interest deliberation concerning human goods that may, in fact, be priceless. Shareholders are those who have something to gain in a transaction, and stakeholders are those who have something to lose. In other words, they have interests, but their interests are typically private, not public. Audiences watch and listen, typically quite passively; publics, on the other hand, engage, act, and express themselves. Clients make demands, and consumers make choices from among the alternatives they are offered; publics express their interests by making claims, telling stories, and asking questions. The citizens that comprise these publics don't just choose between available alternatives, but they also imagine alternatives to the choices they have been offered. In each of these examples, what distinguishes the public sphere as a space of encounter and a public as a social form is a characteristic mode of address between people. In other words, publics are not simply aggregations of isolated, individual subjects and their private interests and preferences. They are, instead, social bodies whose identity and interests are constructed through deliberate and dynamic encounters that take a variety of forms, including dialogue, debate, narrative, celebration, and conflict, to name but a few.

As engineers, we should realize that what we design creates publics. Publics are most often created simply by the act of addressing them. The engineer who proposes a highway to a group of homeowners may find to his dismay that he has created a public (and energized its opposition to the project) without ever intending to do so. But creating a public need not be a negative or obstructing thing for the designer. While no one would argue that the Internet is a perfect vehicle for communication, it is hard to deny the social networks and, in some cases, thoughtful publics that it has enabled or empowered.

There are many ways of addressing and, therefore, initiating a public. Laws, literature, speeches, the drawing of political boundaries, and architecture are obvious examples. Each of these modes of address initiates a public, which then goes about the work of shaping and expressing itself to itself, as well as to others, and, in so doing, achieves its definition and independence. In designing and making things, engineers are also making publics. When an engineer designs a bridge, or a weapon, a windmill or a computer network, she is enacting an address that initiates the self-organization of a public. Obviously, the characteristics of the thing's design — both what is present and what is absent, as well as options taken and rejected — will have an important influence on the shape and concerns of the public that is created. The engineer's attention to the public and its welfare, then, must start right here, in imagining, thinking about, and ultimately engaging with the potentially multiple publics arising out of the thing she is designing.

12.6   ETHICS: ALWAYS A PART OF ENGINEERING PRACTICE

Ultimately, ethics is necessarily an issue that is intensely personal. Returning to our question about designing cigarette lighters, the matter always resolves to, should *I* be working on this design project? While the professional societies produce and insist upon standards of professional conduct, engineering practice is ultimately done by individual practitioners. There is no way to predict when a serious conflict of obligations and loyalties will arise in our lives. Nor can we know the specific personal and professional circumstances within which such conflicts will be embedded. Nor, unfortunately, is there a single answer to many of the questions posed. If faced with a daunting conflict, we can only hope that we are prepared by our upbringing, our maturity, and our ability to think and reflect about the issues that we have so briefly raised herein.

12.7   NOTES

Section 12.1: (Martin and Schinzinger 1996) and (Glazer and Glazer 1989) are very interesting, useful, and readable books on, respectively, engineering ethics and whistleblowing. Ethics emerges as a major theme in Harr's (1995) tale of a civil lawsuit spawned by inadequate toxic waste cleanup.

Section 12.2: An interesting account of the historical development of the professional societies and codes of ethics is given in (Davis 1992). The issues relating to international codes of ethics are discussed in Little et al. 2007).

Section 12.3: The definitions of agency- and identification-loyalty are derived from (Martin and Schinzinger 1996).

Section 12.4: The Citicorp Center case is adapted from (Morgenstern 1995) and (Goldstein and Rubin 1996). We were greatly helped by William LeMessurier's review of the material.

Section 12.5: This section raises issues that require careful and thoughtful reading. For the sake of brevity, we cite only (Arendt 1963), (Harr 1995), (Feenberg 1990), (Little et al. 2007), and (Winner 1990) among the many sources about the deep and complex issues raised in this section.

12.8 EXERCISES

12.1 Is there a difference between ethics and morals?

12.2 Identify the stakeholders that the design team at HMCI must recognize as it develops its design for the portable electric guitar. Are there obligations to these stakeholders that you should consider and could they conflict with what your client has asked you to do?

12.3 As an engineer testing designs for electronic components, you discover that they fail in a particular location. Subsequent investigation shows that the failures are due to a nearby high-powered radar facility. While you can shield your own designs so that they will work in this environment, you also notice that there is an adjacent nursery school. What actions, if any, should you take?

12.4 You are considering a safety test for a newly designed device. Your supervisor instructs you not to perform this test because the relevant government regulations are silent on this aspect of the design. What actions, if any, should you take?

12.5 As a result of previous experiences as a designer of electronic packaging, you understand a sophisticated heat-treatment process that has not been patented, although it is considered company confidential. In a new job, you are designing beverage containers for BJIC and you believe that this heat-treatment process could be effectively used. Can you use your prior knowledge?

12.6 With reference to Exercise 12.5, suppose your employer is a nonprofit organization that is committed to supplying food to disaster victims. Would that change the actions you might take?

12.7 You are asked to provide a reference for a member of your design team, Jim, in connection with a job application he has filed. You have not been happy with Jim's performance, but you believe that he might do better in a different setting. While you are hopeful that you can replace Jim, you also feel obligated to provide an honest appraisal of Jim's potential. What should you do?

12.8 With reference to Exercise 12.7, would your answer change if you knew that you could not replace Jim?

REFERENCES AND BIBLIOGRAPHY

This combined reference–bibliography contains both the references cited in the notes at the end of each chapter and a sampling of books covering a wide range of issues including design theory, design in different disciplines, product development, project management techniques, optimization theory, applications of artificial intelligence, engineering ethics and the practice of engineering, and more. This list of works is *not* complete—the literatures on design and project management alone are both vast and rapidly expanding. Thus, keep in mind that this listing represents only the tip of a very large iceberg of published work in design and in project management. Some of the works cited are just intellectually interesting, and some are books that students in particular will find useful for project work.

N. Abram, *Measure Twice, Cut Once: Lessons from a Master Carpenter*, Little, Brown, Boston, MA, 1996.

J. L. Adams, *Conceptual Blockbusting: A Guide to Better Ideas*, Stanford Alumni Association, Stanford, CA, 1979.

K. Akiyama, *Function Analysis: Systematic Improvement of Quality and Performance*, Productivity Press, Cambridge, MA, 1991.

C. Alexander, *Notes on the Synthesis of Form*, Harvard University Press, Cambridge, MA, 1964.

Anon., *American National Standard for Ladders—Wood Safety Requirements*, ANSI A14.1–2000, American National Standards Institute (ANSI), Chicago, IL, 2000.

Anon., *Dimensions and Tolerancing*, ANSI Y 14.5M–1994, American National Standards Institute (ANSI), Chicago, Illinois, 1994.

Anon., *Goals and Priorities for Research in Engineering Design*, American Society of Mechanical Engineers, New York, NY, 1986.

Anon., *Improving Engineering Design: Designing for Competitive Advantage*, National Research Council, National Academy Press, Washington, DC, 1991.

Anon., *Managing Projects and Programs*, The Harvard Business Review Book Series, Harvard Business School Press, Cambridge, MA, 1989.

Anon., *Introduction to Application and Interpretation of Geometric Dimensioning and Tolerancing*, Technical Documentation Consultants of America, Ridgecrest, CA, 1996.

E. K. Antonsson and J. Cagan, *Formal Engineering Design Synthesis*, Cambridge University Press, New York, NY, 2001.

H. Arendt, *Eichmann in Jerusalem: A Report on the Banality of Evil*, Viking Press, New York, NY, 1963.

J. S. Arora, *Introduction to Optimum Design*, McGraw-Hill, New York, NY, 1989.

K. J. Arrow, *Social Choice and Individual Values*. 1st ed., John Wiley, New York, 1951.

M. F. Ashby, *Materials Selection in Mechanical Design*, 2nd ed., Butterworth Heinemann, Oxford, England, 1999.

W. Asimow, *Introduction to Design*, Prentice-Hall, Englewood Cliffs, NJ, 1962.

R. Attarian, N. Hasegawa, J. Osgood and A. Lee, *Design and Implementation of an Arm Support Device for a Danbury School Student with Cerebral Palsy*, E4 Project Report, Department of Engineering, Harvey Mudd College, Claremont, CA, 2007.

A. B. Badiru, *Project Management in Manufacturing and High Technology Operations*, John Wiley & Sons, New York, NY, 1996.

K. M. Bartol and D. C. Martin, *Management*, 2nd ed., McGraw-Hill Book Company, New York, NY, 1994.

Barton-Aschman Associates, *North Area Terminal Study*, Technical Report, Barton-Aschman Associates, Evanston, IL, August 1962.

Louis Berger, *Central Artery North Area Project*, Interim Report, Louis Berger & Associates, Cambridge, MA, 1981.

R. Best, M. Honda, J. Karras and A. Kurtis, *Design of Arm Restraint for Student with Cerebral Palsy*, E4 Project Report, Department of Engineering, Harvey Mudd College, Claremont, CA, 2007.

G. Boothroyd and P. Dewhurst, *Product Design for Assembly*, Boothroyd Dewhurst Inc., Wakefield, RI, 1989.

T. Both, G. Breed, C. Stratton and K. V. Horn, *Micro Laryngeal Surgery: An Instrument Stabilizer*, E4 Project Report, Department of Engineering, Harvey Mudd College, Claremont, CA, 2000.

C. L. Bovee, M. J. Houston and J. V. Thill, *Marketing*, 2nd ed., McGraw-Hill Book Company, New York, NY, 1995.

C. L. Bovee, J. V. Thill, M. B. Word and G. P. Dovel, *Management*, McGraw-Hill Book Company, New York, NY, 1993.

E. T. Boyer, F. D. Meyers, F. M. Croft, Jr., M. J. Miller and J. T. Demel, *Technical Graphics*, John Wiley & Sons, New York, NY, 1991.

D. C. Brown, "Design," in S. C. Shapiro (Editor), *Encyclopedia of Artificial Intelligence*, 2nd ed., John Wiley & Sons, New York, NY, 1992.

D. C. Brown and B. Chandrasekaran, *Design Problem Solving*, Pitman, London, and Morgan Kaufmann, Los Altos, CA, 1989.

L. L. Bucciarelli, *Designing Engineers*, MIT Press, Cambridge, MA, 1994.

S. Carlson Skalak, H. Kemser and N. Ter-Minassian, "Defining a Product Development Methodology with Concurrent Engineering for Small Manufacturing Companies," *Journal of Engineering Design*, 8 (4), 305–328, December 1997.

A. D. S. Carter, *Mechanical Reliability*, Macmillan, London, England, 1986.

S. Chan, R. Ellis, M. Hanada and J. Hsu, *Stabilization of Microlaryngeal Surgical Instruments*, E4 Project Report, Department of Engineering, Harvey Mudd College, Claremont, CA, 2000.

J. Corbett, M. Dooner, J. Meleka and C. Pym, *Design for Manufacture: Strategies, Principles and Techniques*, Addison-Wesley, Wokingham, England, 1991.

R. D. Coyne, M. A. Rosenman, A. D. Radford, M. Balachandran and J. S. Gero, *Knowledge-Based Design Systems*, Addison-Wesley, Reading, MA, 1990.

N. Cross, *Engineering Design Methods*, 2nd ed., John Wiley & Sons, Chichester, England, 1994.

M. Davis, "Reflections on the History of Engineering in the United States: A Preface to Engineering Ethics," GTE Lecture at the Center for Academic Ethics & College of Engineering at Wayne State University, November 19, 1992; online at http://ethics.iit.edu/publication/Reflections_on_the_history.pdf, accessed April 15, 2007.

M. L. Dertouzos, R. K. Lester, R. M. Solow and the MIT Commission on Industrial Productivity, *The Making of America: Regaining the Productive Edge*, MIT Press, Cambridge, MA, 1989.

J. R. Dixon, *Design Engineering: Inventiveness, Analysis, and Decision Making*, McGraw-Hill, New York, NY, 1966.

J. R. Dixon, "Engineering Design Science: The State of Education," *Mechanical Engineering*, 113 (2), February 1991.

J. R. Dixon, "Engineering Design Science: New Goals for Education," *Mechanical Engineering*, 113 (3), March 1991.

J. R. Dixon and C. Poli, *Engineering Design and Design for Manufacturing*, Field Stone Publishers, Conway, MA, 1995.

C. L. Dym (Editor), *Applications of Knowledge-Based Systems to Engineering Analysis and Design*, American Society of Mechanical Engineers, New York, NY, 1985.

C. L. Dym (Editor), *Computing Futures in Engineering Design*, Harvey Mudd College, Claremont, CA, 1997.

C. L. Dym (Editor), *Designing Design Education for the 21st Century*, Harvey Mudd College, Claremont, CA, 1999.

C. L. Dym, *E4 (Engineering Projects) Handbook*, Department of Engineering, Harvey Mudd College, Claremont, CA, Spring 1993.

C. L. Dym, *Engineering Design: A Synthesis of Views*, Cambridge University Press, New York, NY, 1994a.

C. L. Dym, Letter to the Editor, *Mechanical Engineering, 114* (8), August 1992.

C. L. Dym, "The Role of Symbolic Representation in Engineering Education," *IEEE Transactions on Education, 35* (2), March 1993.

C. L. Dym, "Teaching Design to Freshmen: Style and Content," *Journal of Engineering Education, 83* (4), 303–310, October 1994b.

C. L. Dym, *Structural Modeling and Analysis*, Cambridge University Press, New York, 1997.

C. L. Dym, *Principles of Mathematical Modeling*, 2nd Edition, Elsevier Academic Press, New York, NY, 2004.

C. L. Dym, "Basic Elements of Mathematical Modeling," in P. Fishwick (Editor), *CRC Handbook of Dynamic System Modeling*, CRC Press, Boca Raton, Florida, pp. 5.1–5.20, 2007.

C. L. Dym, A. M. Agogino, D. D. Frey, O. Eris and L. J. Leifer, "Engineering Design Thinking, Teaching and Learning," *Journal of Engineering Education, 94* (1), 103–120, January 2005.

C. L. Dym and R. E. Levitt, *Knowledge-Based Systems in Engineering*, McGraw-Hill, New York, NY, 1991.

C. L. Dym and L. Winner (Editors), *Social Dimensions of Engineering Design*, Harvey Mudd College, Claremont, CA, 2001.

C. L. Dym, W. H. Wood and M. J. Scott, "Rank Ordering Engineering Designs: Pairwise Comparison Charts and Borda Counts," *Research in Engineering Design, 13* (4), 236–242, 2003.

C. E. Ebeling, *An Introduction to Reliability and Maintainability Engineering*, McGraw-Hill, New York, NY, 1997.

D. L. Edel, Jr. (Editor), *Introduction to Creative Design*, Prentice-Hall, Englewood Cliffs, NJ, 1967.

K. S. Edwards, Jr. and R. B. McKee, *Fundamentals of Mechanical Component Design*, McGraw-Hill, New York, NY, 1991.

K. A. Ericsson and H. A. Simon, *Protocol Analysis: Verbal Reports as Data*, MIT Press, Cambridge, MA, 1984.

A. Ertas and J. C. Jones, *The Engineering Design Process*, 2nd ed., John Wiley & Sons, New York, NY, 1996.

D. L. Evans (Coordinator), "Special Issue: Integrating Design Throughout the Curriculum," *Engineering Education, 80* (5), 1990.

J. H. Faupel, *Engineering Design*, John Wiley & Sons, New York, NY, 1964.

L. Feagan, T. Galvani, S. Kelley and M. Ong, *Device for Microlaryngeal Instrument Stabilization*, E4 Project Report, Department of Engineering, Harvey Mudd College, Claremont, CA, 2000.

A. Feenberg, *Questioning Technology*, Routledge, New York, 1999.

R. L. Fox, *Optimization Methods for Engineering Design*, Addison-Wesley, Reading, MA, 1971.

J. Fortune and G. Peters, *Learning From Failure—The Systems Approach*, John Wiley & Sons, Chichester, UK, 1995.

M. E. French, *Conceptual Design for Engineers*, 2nd ed., Design Council Books, London, England, 1985.

M. E. French, *Form, Structure and Mechanism*, MacMillan, London, England, 1992.

D. C. Gause and G. M. Weinberg, *Exploring Requirements: Quality Before Design*, Dorset House Publishing, New York, NY, 1989.

J. S. Gero (Editor), *Design Optimization*, Academic Press, Orlando, FL, 1985.

J. S. Gero (Editor), *Proceedings of AI in Design '92*, Kluwer Academic Publishers, Dordrecht, The Netherlands, 1992.

J. S. Gero (Editor), *Proceedings of AI in Design '94*, Kluwer Academic Publishers, Dordrecht, The Netherlands, 1994.

J. S. Gero (Editor), *Proceedings of AI in Design '96*, Kluwer Academic Publishers, Dordrecht, The Netherlands, 1996.

M. P. Glazer and P. M. Glazer, *The Whistleblowers: Exposing Corruption in Government and Industry*, Basic Books, New York, NY, 1989.

G. L. Glegg, *The Design of Design*, Cambridge University Press, Cambridge, England, 1969.

G. L. Glegg, *The Science of Design*, Cambridge University Press, Cambridge, England, 1973.

G. L. Glegg, *The Selection of Design*, Cambridge University Press, Cambridge, England, 1972.

T. J. Glover, *Pocket Ref*, Sequoia Publishing, Littleton, CO, 1993.

B. A. Goetsch, J. A. Nelson and W. S. Chalk, *Technical Drawing*, 4th ed., Delmar Publishers, Albany, New York, 2000.

S. H. Goldstein and R. A. Rubin, "Engineering Ethics," *Civil Engineering*, October 1996.

P. Graham (Editor), *Mary Parker Follett—Prophet of Management: A Celebration of Writings From the 1920s*, Harvard Business School Press, Boston, MA, 1996.

P. Gutierrez, J. Kimball, B. Maul, A. Thurston and J. Walker, *Design of a Chicken Coop*, E4 Project Report, Department of Engineering, Harvey Mudd College, Claremont, CA, 1997.

C. Hales, *Managing Engineering Design*, Longman Scientific & Technical, Harlow, England, 1993.

J. Harr, *A Civil Action*, Vintage Books, New York, NY, 1995.

B. Hartmann, B. Hulse, S. Jayaweera, A. Lamb, B. Massey and R. Minneman, *Design of a "Building Block" Analog Computer*, E4 Project Report, Department of Engineering, Harvey Mudd College, Claremont, CA, 1993.

J. R. Hauser and D. Clausing, "The House of Quality," *Harvard Business Review*, 63–73, May-June 1988.

S. I. Hayakawa, *Language in Thought and Action*, 4th Edition, Harcourt Brace Jovanovich, San Diego, CA, 1978.

R. T. Hays, "Value Management," in W. K. Hodson (Editor), *Maynard's Industrial Engineering Handbook*, 4th ed., McGraw-Hill Book Company, New York, NY, 1992.

G. H. Hazelrigg, *Systems Engineering: An Approach to Information-Based Design*, Prentice Hall, Upper Saddle River, NJ, 1996.

G. H. Hazelrigg, "Validation of Engineering Design Alternative Selection Methods," unpublished manuscript, courtesy of the author, 2001

J. Heskett, *Industrial Design*, Thames and Hudson, London, 1980.

R. S. House, *The Human Side of Project Management*, Addison-Wesley, Reading, MA, 1988.

V. Hubka, M. M. Andreasen and W. E. Eder, *Practical Studies in Systematic Design*, Butterworths, London, England, 1988.

V. Hubka and W. E. Eder, *Design Science*, Springer-Verlag, London, England, 1996.

B. Hyman, *Topics in Engineering Design*, Prentice Hall, Englewood Cliffs, NJ, 1998.

D. Jain, G. P. Luth, H. Krawinkler and K. H. Law, *A Formal Approach to Automating Conceptual Structural Design*, Technical Report No. 31, Center for Integrated Facility Engineering, Stanford University, Stanford, CA, 1990.

F. D. Jones, *Ingenious Mechanisms*: Vols. 1–3, The Industrial Press, New York, NY, 1930.

J. C. Jones, *Design Methods*, Wiley-Interscience, Chichester, UK, 1992.

J. Juran, *Quality Control Handbook*, 3rd Edition, McGraw-Hill, New York 1979.

D. Kaminski, "A Method to Avoid the Madness," *The New York Times*, 3 November 1996.

H. Kerzner, *Project Management: A Systems Approach to Planning, Scheduling and Controlling*, Van Nostrand Reinhold, New York, NY, 1992.

D. S. Kezsbom, D. L. Schilling and K. A. Edward, *Dynamic Project Management: A Practical Guide for Managers & Scientists*, John Wiley & Sons, New York, NY, 1989.

A. Kusiak, *Engineering Design: Products, Processes and Systems*, Academic Press, San Diego, CA, 1999.

M. Levy and M. Salvadori, *Why Buildings Fall Down*, Norton, New York, NY, 1992.

E. E. Lewis, *Introduction to Reliability Engineering*, John Wiley & Sons, New York, NY, 1987.

P. Little, *Improving Railroad Car Reliability Using A New Opportunistic Maintenance Heuristic and Other Information System Improvements*, Doctoral Dissertation, Massachusetts Institute of Technology, Cambridge, MA, 1991.

P. Little, R. Hink and D. Barney, "Living Up to the Code: Engineering as Political Judgment," *International Journal of Engineering Education*, Vol. 24, No. 2, pp. 314–327, 2008.

M. W. Martin and R. Schinzinger, *Ethics in Engineering*, 3rd ed., McGraw-Hill Book Company, New York, NY, 1996.

Massachusetts Department of Public Works, *North Terminal*, Draft Environmental Impact Report (Section 4(F) and Section 106 Statements), Massachusetts Department of Public Works, Boston, MA, 1974.

B. McKibben, *Deep Economy: The Wealth of Communities and the Durable Future*, Times Press, New York, 2007.

R. L. Meehan, *Getting Sued and Other Tales of the Engineering Life*, The MIT Press, Cambridge, MA, 1981.

J. R. Meredith and S. J. Mantel, Jr., *Project Management: A Managerial Approach*, John Wiley & Sons, New York, NY, 1995.

F. C. Misch (Editor), *Webster's Ninth New Collegiate Dictionary*, Merriam-Webster, Springfield, MA, 1983.

J. Morgenstern, "The Fifty-nine-story Crisis," *The New Yorker*, 29 May 1995.

T. T. Nagle, *The Strategy and Tactics of Pricing*, Prentice-Hall, Englewood Cliffs, NJ, 1987.

A. Newell and H. A. Simon, *Human Problem Solving*, Prentice-Hall, Englewood Cliffs, NJ, 1972.

R. L. Norton, *Machine Design: An Integrated Approach*, 3rd ed., Prentice Hall, Upper Saddle River, NJ, 2004.

K. N. Otto, "Measurement Methods for Product Evaluation," *Research in Engineering Design*, 7, 86–101, 1995.

K. N. Otto and K. L. Wood, *Product Design: Techniques in Reverse Engineering and New Product Development*, Prentice Hall, Upper Saddle River, NJ, 2001.

G. D. Oberlander, *Project Management for Engineering and Construction*, McGraw-Hill, New York, NY, 1993.

G. Pahl and W. Beitz, *Engineering Design: A Systematic Approach*, 2nd ed., Springer, London, England, 1996.

A. Palladio, *The Four Books of Architecture*, Dover, New York, NY, 1965.

Y. C. Pao, *Elements of Computer-Aided Design and Manufacturing*, John Wiley & Sons, New York, NY, 1984.

P. Y. Papalambros and D. J. Wilde, *Principles of Optimal Design: Modeling and Computation*, 2nd ed., Cambridge University Press, Cambridge, England, 2000.

T. E. Pearsall, *The Elements of Technical Writing*, Allyn & Bacon, Needham Heights, MA, 2001.

H. Petroski, *Design Paradigms*, Cambridge University Press, New York, NY, 1994.

H. Petroski, *Engineers of Dreams*, Alfred A. Knopf, New York, NY, 1995.

H. Petroski, *To Engineer is Human*, St. Martin's Press, New York, NY, 1985.

W. S. Pfeiffer, *Pocket Guide to Technical Writing*, Prentice Hall, Upper Saddle River, NJ, 2001.

L. Phips, *The Economics of Price Discrimination*, Cambridge University Press, Cambridge, England, 1985.

S. Pugh, *Total Design: Integrated Methods for Successful Product Engineering*, Addison-Wesley, Wokingham, England, 1991.

H. E. Riggs, *Financial and Cost Analysis for Engineering and Technology Management*, John Wiley & Sons, New York, NY, 1994.

J. L. Riggs and T. M. West, *Essentials of Engineering Economics*, McGraw-Hill, New York, NY, 1986.

E. S. Rubin, *Introduction to Engineering and the Environment*, McGraw-Hill, New York, NY, 2001.

M. D. Rychener (Editor), *Expert Systems for Engineering Design*, Academic Press, Boston, MA, 1988.

D. G. Saari, *Basic Geometry of Voting*, Springer-Verlag, New York, 1995.

D. G. Saari, "Bad Decisions: Experimental Error or Faulty Decision Procedures," unpublished manuscript, courtesy of the author, 2001a.

D. G. Saari, *Decisions and Elections: Explaining the Unexpected*, Cambridge University Press, New York, 2001b.

M. Salvadori, *Why Buildings Stand Up*, McGraw-Hill, New York, NY, 1980.

A. Samuel, *Make and Test Projects in Engineering Design*, Springer-Verlag, London, UK, 2006.

Y. Saravanos, J. Schauer and C. Wassman, *Sliding Fulcrum Stabilizer*, E4 Project Report, Department of Engineering, Harvey Mudd College, Claremont, CA, 2000.

D. A. Schon, *The Reflective Practitioner*, Basic Books, New York, NY, 1983.

D. Schroeder, "Little Land Bruisers," *Car and Driver*, 96–109, May 1998.

R. G. Schroeder, *Operations Management: Decision Making in the Operations Function*, McGraw-Hill, New York, NY, 1993

M. J. Scott and E. K. Antonsson, "Arrow's Theorem and Engineering Decision Making," *Research in Engineering Design, 11,* 218–228, 1999.

J. J. Shah, "Experimental Investigation of Progressive Idea Generation Techniques in Engineering Design," *Proceedings of the 1998 ASME Design Theory and Methodology Conference,* American Society of Mechanical Engineers, New York, NY, 1998.

S. D. Sheppard and B. H. Tongue, *Statics Analysis and Design of Systems in Equilibrium,* John Wiley, New York, 2005.

J. E. Shigley, C. R. Mischke and R. G. Budynas, *Mechanical Engineering Design,* 7th ed., McGraw Hill, New York, 2006.

H. A. Simon, "Style in Design," in C. M. Eastman (Editor), *Spatial Synthesis in Computer-Aided Building Design,* Applied Science Publishers, London, England, 1975.

H. A. Simon, *The Sciences of the Artificial,* 2nd Edition, MIT Press, Cambridge, MA, 1981.

L. Stauffer, *An Empirical Study on the Process of Mechanical Design,* Thesis, Department of Mechanical Engineering, Oregon State University, Corvallis, OR, 1987.

L. Stauffer, D. G. Ullman and T. G. Dieterich, "Protocol Analysis of Mechanical Engineering Design," in *Proceedings of the 1987 International Conference on Engineering Design.* Boston, MA, 1987.

S. Stevenson and S. Whitmore, *Strategies for Engineering Communication,* John Wiley & Sons, New York, 2002.

G. Stevens, *The Reasoning Architect: Mathematics and Science in Design,* McGraw-Hill, New York, NY, 1990.

G. Stiny and J. Gips, *Algorithmic Aesthetics,* University of California Press, Berkeley, CA, 1978.

N. P. Suh, *Axiomatic Design: Advances and Applications,* Oxford University Press, Oxford, England, 2001.

N. P. Suh, *The Principles of Design,* Oxford University Press, Oxford, England, 1990.

M. C. Thomsett, *The Little Black Book of Project Management,* American Management Association, New York, NY, 1990.

C. Tong and D. Sriram (Editors), *Artificial Intelligence in Engineering Design, Volume I: Design Representation and Models of Routine Design,* Academic Press, Boston, MA, 1992a.

C. Tong and D. Sriram (Editors), *Artificial Intelligence in Engineering Design, Volume II: Models of Innovative Design, Reasoning About Physical Systems, and Reasoning About Geometry,* Academic Press, Boston, MA, 1992b.

C. Tong and D. Sriram (Editors), *Artificial Intelligence in Engineering Design, Volume III: Knowledge Acquisition, Commercial Applications and Integrated Environments,* Academic Press, Boston, MA, 1992c.

B. W. Tuckman, "Developmental Sequences in Small Groups," *Psychological Bulletin, 63,* 384–399, 1965.

E. R. Tufte, *The Visual Display of Quantitative Information,* Graphics Press, Cheshire, CT, 2001.

K. Turabian, *A Manual for Writers of Term Papers, Theses, and Dissertations,* University of Chicago Press, Chicago, 1996.

D. G. Ullman, "A Taxonomy for Mechanical Design," *Research in Engineering Design, 3,* 1992.

D. G. Ullman, *The Mechanical Design Process,* 2nd ed., McGraw-Hill, New York, NY, 1997.

D. G. Ullman and T. G. Dieterich, "Toward Expert CAD," *Computers in Mechanical Engineering, 6* (3), 1987.

D. G. Ullman, T. G. Dieterich and L. Stauffer, "A Model of the Mechanical Design Process Based on Empirical Data," *Artificial Intelligence for Engineering Design, Analysis and Manufacturing, 2* (1), 1988.

D. G. Ullman, S. Wood and D. Craig, "The Importance of Drawing in the Mechanical Design Process," *Computers and Graphics, 14* (2), 1990.

K. T. Ulrich and S. D. Eppinger, *Product Design and Development,* 2nd ed., McGraw-Hill, New York, NY, 2000.

G. N. Vanderplaats, *Numerical Optimization Techniques for Engineering Design,* McGraw-Hill, New York, NY, 1984.

VDI, *VDI–2221: Systematic Approach to the Design of Technical Systems and Products,* Verein Deutscher Ingenieure, VDI-Verlag, Translation of the German Edition 11/1986, 1987.

C. E. Wales, R. A. Stager and T. R. Long, *Guided Engineering Design: Project Book*, West Publishing Company, St. Paul, MN, 1974.

J. Walton, *Engineering Design: From Art to Practice*, West Publishing, St. Paul, MN, 1991.

D. J. Wilde, *Globally Optimal Design*, John Wiley & Sons, New York, NY, 1978.

B. A. Wilson, *Design Dimensioning and Tolerancing*, Goodheart-Wilcox, Tinley Park, IL, 2005.

L. Winner, "Engineering Ethics and Political Imagination," in P.T. Durbin (Editor), *Philosophy of Technology II: Broad and Narrow Interpretations*, Kluwer Academic Publishers, Dordrecht, Netherlands, 1990.

T. T. Woodson, *Introduction to Engineering Design*, McGraw-Hill, New York, NY, 1966.

R. N. Wright, S. J. Fenves and J. R. Harris, *Modeling of Standards: Technical Aids for Their Formulation, Expression and Use*, National Bureau of Standards, Washington, DC, 1980.

C. Zener, *Engineering Design by Geometric Programming*, Wiley-Interscience, New York, NY, 1971.

C. Zozaya-Gorostiza, C. Hendrickson and D. R. Rehak, *Knowledge-Based Process Planning for Construction and Manufacturing*, Academic Press, Boston, MA, 1989.

INDEX

A GUIDE TO WRITING AS AN ENGINEER

THIRD EDITION

David Beer

Department of Electrical and Computer Engineering
University of Texas at Austin

David McMurrey

Austin Community College

WILEY

JOHN WILEY & SONS, INC.

Publisher   Don Fowley
Acquisitions Editor   Michael McDonald
Marketing Manager   Christopher Ruel
Production Manager   Dorothy Sinclair
Production Editor   Sandra Dumas
Senior Designer   Kevin Murphy
Cover Designer   David Levy
Media Editor   Lauren Sapira
Editorial Assistant   Rachael Leblond
Production Management Services   Laserwords Maine
Cover Photo   © IT Stock

This book was set in 10/12 Times Roman by Laserwords Private Limited and printed and bound by Malloy. The cover was printed by Malloy.

The paper in this book was manufactured by a mill whose forest management programs include sustained yield harvesting of its timberlands. Sustained yield harvesting principles ensure that the number of trees cut each year does not exceed the amount of new growth.

This book is printed on acid-free paper. ⊗

Beer, David; McMurrey, David
A Guide to Writing As an Engineer—Third Edition.

ISBN 978-0-470-41701-0

Printed in the United States of America.

10 9 8 7 6 5 4 3 2 1

CONTENTS

1

ENGINEERS AND WRITING

Poor communication skill is the Achilles' heel of many engineers, both young and experienced—and it can even be a career showstopper. In fact, poor communication skills have probably claimed more casualties than corporate downsizing.

> H. T. Roman, "Be a Leader—Mentor Young Engineers,"
> *IEEE USA Today's Engineer*, November 2002.

It is nearly impossible to overstate the benefits of being able to write well. The importance of the written word in storing, sharing, and communicating ideas at all levels of all organizations makes a poor facility with the mechanics of writing a severely career-limiting fault.

> John E. West, *The Only Trait of a Leader: A Field Guide to Success for New Engineers, Scientists, and Technologists*, 2008.
> www.onlytraitofaleader.com

Like a lot of other professionals, many engineers and engineering students dislike writing. After all, don't we go into engineering because we want to work with machines, instruments, and numbers rather than words? Didn't we leave writing behind us when we finished English 101? We may have hoped so, but the fact remains—as the above quotes so bluntly indicate—that to be a successful engineer we must be able to write (and speak) effectively. Even if we could set up our own lab in a vacuum and avoid communication with all others, what good would our ideas and discoveries be if they never got beyond our own mind?

If you personally feel you haven't mastered writing skills in college, the fault probably is not entirely yours. Few engineering colleges offer adequate (if any) courses in technical writing, and many students find what writing skills they did possess are badly rusted from lack of use by the time they graduate with an engineering degree. Ironically, most engineering programs devote less than 5% of their curriculum to communication skills—the very skills that many engineers will use some 20% to 40% of their working time. Even this percentage usually increases with promotion, which is why many young engineers eventually find themselves wishing they had taken more writing courses.

But rather than dwell on the negative, let's look at the needs and opportunities that exist in engineering writing, then see how you can best remove barriers to becoming an efficient and effective writer. You'll soon find that the skills you need to write well are no harder to acquire than many of the technical skills you have already mastered as an engineer or engineering student. First, here are four factors to consider.

1. Engineers write a lot.
2. Engineers write many kinds of documents.
3. A successful engineering career requires strong writing skills.
4. Engineers can learn to write well.

ENGINEERS WRITE A LOT

Many engineers spend over 40% of their work time writing, and usually find the percentage increases as they move up the corporate ladder. It doesn't matter that most of this writing is now sent through electronic mail (email); the need for clear and efficient prose is the same whether it appears on a computer monitor or sheet of paper. Much written material first read on a screen ends up being printed out on paper anyway—and the likelihood of a completely paperless office, lab, engine room (or toilet) still seems pretty remote.

An engineer told us some years ago that while working on the B-1b bomber, he and his colleagues calculated that all the proposals, regulations, manuals, procedures, and memos that the project generated weighed almost as much as the bomber itself. Most large ships carry several tons of maintenance and operations manuals. Two trucks were needed to carry the proposals from Texas to Washington for the ill-fated supercollider project. John Naisbitt estimated in his book *Megatrends* over 25 years ago that some 6,000 to 7,000 scientific articles were being written every day, and even then the amount of recorded scientific and technical information in the world was doubling every five and a half years. Jumping to the present, look what John Bringardner has to say in his short article entitled "Winning the Lawsuit":

Way back in the 20th century, when Ford Motor Company was sued over a faulty ignition switch, its lawyers would gird for the discovery process: a labor-intensive ordeal that involved disgorging thousands of pages of company records. These days, the number of pages commonly involved in commercial litigation discovery has ballooned into the billions. Attorneys on the hunt for a smoking gun now want to see not just the final engineering plans but the emails, drafts, personal data files, and everything else ever produced in the lead-up to the finished product.

Wired Magazine, July 2008, p. 112.

Who generates and transmits—in print, online, graphically, or orally—all this material, together with countless memos, reports, proposals, manuals, and other technical information? Engineers. Perhaps they get some help from a technical editor if their company employs one, and secretaries may play a part in some cases. Nevertheless, the vast body of technical information available in the world today has its genesis in the writing and speaking of engineers, whether they work alone or in teams. Figure 1-1 shows just one response we got when we randomly asked an engineer friend, who works as a software deployment specialist for a large international company, to outline a typical day at his job (our italics indicate where communication skills are called for).

Friday's Schedule *2/15/08*	
7:30	Arrive, *read and reply to several overnight emails.*
8:00	Work on project.
10:30	*Meet with* project manager to *write answer to* department head request.
11:00	*Write up a request* to obtain needed technical support.
11:30	Lunch.
12:00	*Meet with* server group about *submitted application* to fix process problems.
12:20	*Reply to* emails from Sales about prospective customers' *technical questions.*
12:30	*Write to* software vendor about how our product works with their plans.
1:00	*Give presentation to* server hosting group *to explain* what my group is doing.
2:00	*Join* the team *to write up* weekly *progress report.*
2:30	*Write emails to* update customers on the status of solving their problems.
2:45	*Write email reply to* question about *knowledge base article I wrote.*
3:00	*Meet with* group *to discuss project goals* for next four months.
3:30	*Meet with* group to *create presentation of findings* to project management.
4:00	Work on project.
5:00	Leave for day.

Figure 1-1 The working day of a typical engineer calls for plenty of communication skills.

ENGINEERS WRITE MANY KINDS OF DOCUMENTS

As mentioned above, few engineers work in a vacuum. Throughout your career you will interact with a variety of other engineering and nonengineering colleagues, officials, and members of the public. Even if you don't do the actual engineering work, you may have to explain how something was done, should be done, needs to be changed, must be investigated, and so on. The list of all possible engineering situations and contexts in which communication skills are needed is unending. Figure 1-2 identifies just some of the documents you might be involved in producing during your engineering career. (It's worth noting that not all companies label reports by the same name or put them in the same categories as we have. Also, many of these reports would obviously overlap into more than one of the "files" we have somewhat arbitrarily placed them in.)

As we move further into the twenty-first century, electronic communication is rapidly replacing much hard copy. Used for anything from quick pithy notes and

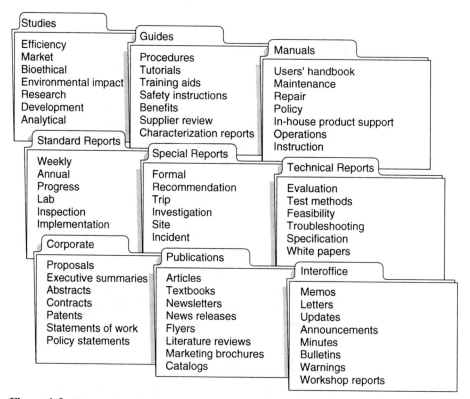

Figure 1-2 Throughout their careers, engineers write many kinds of documents in various contexts and with different purposes and audiences.

memos to complete multivolume documents, email has perhaps become the most popular form of written communication. Yet this fact does not in any way change the need for clarity and organization in engineering writing, and whatever the future holds, solid skills in clear and efficient writing, and the ability to adapt to many different document specifications, will probably be necessary for as long as humans communicate with each other. This probability leads us to our next point.

A SUCCESSFUL ENGINEERING CAREER REQUIRES STRONG WRITING SKILLS

In the engineering field you are rarely judged solely by the quality of your technical expertise or work. People also form opinions of you by what you say and write—and how you say and write it. When you write a memo or report, talk to members of a group, deal with vendors on the phone, or attend meetings, the image others get of you is largely formed by how well you communicate. Even if you work for a large company and don't see a lot of high-level managers, those same managers can still gain an impression of you by the quality of your written reports as well as by what your immediate supervisor tells them. Thus Robert W. Lucky, former Executive Director of AT&T Laboratories and head of research at Telcordia Technologies, and an accomplished writer himself, points out:

> *It is unquestionably true that writing and speaking abilities are essential to the successful engineer. Nearly every engineer who has been unsuccessful in my division had poor communication skills. That does not necessarily mean that they failed because of the lack of these skills, but it does provide strong contributory evidence of the need for good communication. On the contrary, I have seen many quite average engineers be successful because of above-average communication skills.*

rlucky@telcordia.com Accessed August 20, 2008

Moreover, two relatively recent trends are now making communication skills even more vital to the engineering profession. These are *specialization* and *accountability*. Due to the advancement and specialization of technology, engineers are finding it increasingly difficult to communicate with one another. Almost daily, engineering fields once considered unified become progressively fragmented, and it's quite possible for two engineers with similar academic degrees to have large knowledge gaps when it comes to each other's work. In practical terms this means that a fellow engineer may have only a little more understanding of what you are working on than does a layperson. These gaps in knowledge often have to be bridged, but they can't be unless specialists have the skills to communicate clearly and effectively with each other.

In addition to communicating with one another, engineers must also be able to communicate with the public, since engineers and their companies are now being held much more accountable by the public. As the Director of the Center for Engineering Professionalism at Texas Tech University puts it,

> *The expansiveness of technology is such that now, more than ever, society is holding engineering professionals accountable for decisions that affect a full range of daily life activities. Engineers are now responsible for saying: "Can we do it, should we do it, if we do it, can we control it, and are we willing to be accountable for it?" There have been too many "headline type" instances of technology gone astray for it to be otherwise ... Pinto automobiles that burn when hit from the rear, DC-10s that crash when cargo doors don't hold, bridges that collapse, Hyatt Regency walkways that fall, space shuttles that explode on national TV, gas leaks that kill thousands, nuclear plant accidents, computer viruses, oil tanker spills, and on and on.*

> Engineering Ethics Module, Murdough Center for
> Engineering Professionalism, Texas Tech University,
> Lubbock, Texas. www.murdough.ttu.edu/EthicsModule/
> EthicsModule.htm Accessed 2/5/2003

People do want to know *why* a space shuttle crashed (after all, their taxes paid for the mission). They want to know if it really is safe to live near a nuclear reactor or high-power lines. The public—often through the press—wants to know if a plant is environmentally sound or if a project is likely to be worth the tax dollars. Moreover, there is no shortage of lawyers ready to hold engineering firms and projects accountable for their actions. All this means that engineers are being called upon to explain themselves in numerous ways and must now communicate with an increasing variety of people—many of whom are not engineers.

ENGINEERS CAN LEARN TO WRITE WELL

Here are the words of Norman Augustine, former chairman and CEO of Martin Marietta Corporation and also chair of the National Academy of Engineering:

> *Living in a "sound bite" world, engineers must learn to communicate effectively. In my judgment, this remains the greatest shortcoming of most engineers today—particularly insofar as written communication is concerned. It is not sensible to continue to place our candle under a bushel as we too often have in the past. If we put our trust solely in the primacy of logic and technical skills, we*

will lose the contest for the public's attention—and in the end, both the public and the engineer will be the loser.

Norman R. Augustine, in *The Bridge*, The National Academy of Engineering, *24* (3), Fall 1994, p. 13

The danger described above still exists, because writing is not easy for most of us, and just like programming, woodworking, or playing the bagpipes, good writing takes practice. A lot of truth lies in the adage that no one can be a good writer—only a good *rewriter*. If you look at the early drafts of the most famous authors' works you will see various scribbling, additions, deletions, rewordings, and corrections where they have edited their text. So don't expect to produce a masterpiece of writing on your first try. Every initial draft of a document, whether it's a one-page memo or a fifty-page set of procedures, needs to be worked on and improved before being sent to its readers.

As an engineer you have been trained to think logically. In the laboratory or workshop you are concerned with precision and accuracy. From elementary and secondary school you already possess the skills needed for basic written communication, and every day you can see samples of clear writing in newspapers, weekly news magazines, and popular journal articles. Thus you are already in a good position to become an effective writer partly by emulating what you've already been exposed to. All you need is some instruction and practice. This book will give you plenty of the former, and your engineering career will give you many opportunities for the latter. Meanwhile, keep in mind that as an engineering professional you will frequently have to communicate through a variety of documents and mediums, you will certainly enhance your career by being able to do so, and you may even find that it can be fun!

NOISE AND THE COMMUNICATION PROCESS

Have you ever been annoyed by someone talking loudly on a cell phone while you were trying to study or talk to a friend? Or maybe you couldn't enjoy your favorite TV show because someone was using the vacuum cleaner in the next room or the stereo was booming.

In each case, what you were experiencing was noise interfering with the transmission of information. Whenever a message is sent, someone is sending it and someone else is trying to receive it. In communication theory, the sender is the *encoder*, and the receiver is the *decoder*. The message, or *signal*, is sent through a channel, usually speech, writing, or some other conventional set of signs, and anything that prevents the signal from flowing clearly through the channel from the encoder to the decoder is *noise*. Figure 1-3 illustrates this concept. Note how all our actions involving communication are "overshadowed" by the possibility of noise.

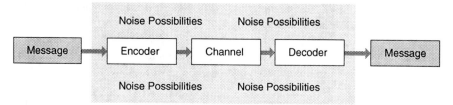

Figure 1-3 In noise-free technical communication, the signal flows from the encoder (writer, speaker) to the decoder (reader, listener) without distortion or ambiguity. When this occurs, the received message is a reliable version of the sent one.

Applying this concept to engineering writing, we can say that anything causing a reader to hesitate in uncertainty, frustration, or even amusement, is noise. Chapter 2 will go into more detail on this, but for now here are just a few simple samples of written noise:

When they bought the machine they werent aware of it's shortcomings.

They were under the allusion that the project could be completed in six weeks.

There was not a sufficient enough number of samples to validate the data.

Our intention is to implement the verification of the reliability of the system in the near future.

In the first sentence two apostrophe problems cause noise. A reader might "trip over" these glitches and momentarily be distracted from the sentence's message (or at least waste time thinking how much smarter he or she is than the writer). The same might be said for the confusion between *allusion* and *illusion* in the second sentence. The third sentence is noisy because of the redundancy and wordiness it contains. Wouldn't you rather just read *There weren't enough samples to validate the data*? The final example is a monument to verbosity. With the noise removed, it simply says: *We want to verify the system's reliability soon.*

It's relatively easy to identify and remove simple noise like this. More challenging is the kind of noise that results from fuzzy and disorganized thinking. Here's a notice posted on a professor's door describing his office hours:

I open most days about 9 or 9:30, occasionally as early as 8, but some days as late as 10 or 10:30. I close about 4 or 4:30, occasionally around 3:30, but sometimes as late as 6 or 6:30. Sometimes in the mornings or afternoons, I'm not here at all, but lately I've been here just about all the time except when I'm somewhere else, but I should be here then, too.

Academic humor, maybe, but it's not hard to find writing in the engineering world that is equally difficult to interpret, as this excerpt from industrial procedures shows:

> If containment is not increasing or it is increasing but MG Press is not trending down and PZR level is not decreasing, the Loss of Offsite Power procedure shall be implemented, starting with step 15, unless NAN-S01 and NAN-S02 are de-energized in which case the Reactor Trip procedure shall be performed. But if the containment THRSP is increasing the Excess Steam Demand procedure shall be implemented when MG Press is trending down and the LIOC procedure shall be implemented when the PZR level is decreasing.

The point isn't just that noise in a written document causes anything from momentary confusion to a complete inability to understand a message. Inevitably, noise costs money—or to put it graphically,

$$NOISE = \$\$\$\$$$

According to engineer Bill Brennan, a senior member of the technical staff at Advanced Micro Devices (AMD) in Austin, Texas, it costs a minimum of $200 to produce one page of an internal technical report and at least five times that much for one page of a technical conference report. Thus, as you learn to reduce noise in your writing, you will become an increasingly valuable asset to your company.

Noise can also occur in spoken communication, of course, as you will see in Chapter 9. For now, maybe you can recall how often you've been distracted by a speaker's monotonous tone, nervous cough, clumsy use of notes, or indecipherable graphics—while you just sat there, a captive audience.

The following chapters contain advice, illustrations, and strategies to help you learn to avoid noise in your communication. Try to keep this concept of noise in mind when you write or edit, whether you are working on a five-sentence memo or a 500-page technical manual. Throughout your school years you may have been reprimanded for "poor writing," "mistakes," "errors to be corrected," "choppy style," and so on, but as an engineer it might be better to think in terms of *noise to be eliminated from the signal*. For efficient and effective communication to take place, the signal-to-noise ratio must be as high as possible. To put it another way, we need to filter as much noise out of our communication as we can.

CONTROLLING THE WRITING SYSTEM

Engineers frequently design, build, and manage systems made up of interconnected parts. Controls have to be built into such systems to guarantee that they function correctly and reliably and that they produce the desired result. The machinery used to mill propeller shafts for large ships must be guided by a control system to ensure that correct tolerances and other specifications are met. If the ATM chews up your card and spits it back out to you in place of the $200 you had hoped for, you'd claim the system is not working right—or that it is out of control. The system is only functioning reliably if the input (your ATM card) produces the desired output (your $200).

What has this got to do with writing? Well, we can view language as a *system* made up of various components such as sounds, words, clauses, sentences, and so on. Whenever we speak or write, we use this system, and like other systems it must be controlled if it is to do its job right. The person who supposedly wrote in an accident report, *Coming home, I drove into the wrong house and collided with a tree I didn't have,* was obviously unable to express what really happened. The input (thought) to the system (language) did not have the desired output (meaning) because the writer was not in control of the system or was not thinking clearly.

In the same way, an instruction like *Pour the concrete when it is above* 40°F indicates a lack of language control since the writer is not clearly stating whether the concrete or the weather must meet the specification of "above 40°F." Thus you might think of language as a system or even a tool you can learn to control so that it will do exactly what you want it to. Learning to control language, namely to write and speak so you get desired results or feedback, is really not much different than training yourself to operate complex machinery or software systems. With some help and effort you can train yourself to eliminate most, if not all, noise that might occur when you transfer information by means of writing and speaking. Figure 1-4 depicts how this works. Note how the end product of your communication is often "feedback," which will give you an indication of how well you are using the language system.

If you get the response (or feedback) you want from your communication, you can be pretty sure you have communicated well. A proposal accepted, a part promptly delivered, a repair quickly made, an applied-for promotion

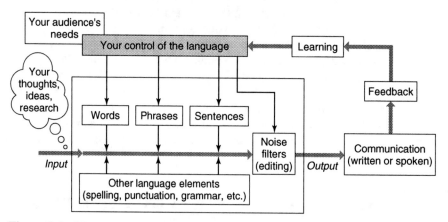

Figure 1-4 The process of communicating can be illustrated as a system with an input and output. How well the input is processed once it is in the system, i.e., how well you convey your information to others, will determine the impact of your message. From the response (feedback) you get, you will learn how to further improve the process.

awarded—these are just a few examples of the payback from effective communication. To put it another way, if you learn to efficiently control the tool you are using (language) so that it's noise-free, you will produce clear and effective written documents that get results.

EXERCISES

1. Ask any professional engineers about the amount and kinds of writing they do on the job. How much of their time is spent writing each day? Is the amount of writing they do related to how long they have been with their company? In what ways do they feel their writing skills have helped (or hindered) them in their careers so far? Do they get any help with their writing from secretaries, peers, or technical writers? What is the attitude of their superiors toward clear writing?

2. Look at the list of technical documents in Figure 1-2. How many are you familiar with? Can you think of examples of some of these documents? When would they likely be important to you as a reader? Can you think of other types of documents not included in Figure 1-2? Ask some engineering friends how many kinds of documents they have worked on, either as individuals or as part of a group.

3. Think of your own engineering major or specialty. List some engineering fields most closely related to yours, some that are marginally related, and some that are only remotely related. What kinds of technical knowledge do you share with people in these fields? At what point is your common knowledge likely to be no longer useful? What problems can you foresee in communicating technical information with engineers in other fields? What problems would you face if you had to talk about your field to a nonengineering audience?

4. As we point out in this chapter, noise is anything that interferes with efficient transmission of information. We've all experienced noise when trying to communicate with another person—and most of us have at times created it. What kinds of noise do you think you create in your written communication? Is it primarily in your spelling, grammar, sentence structure, organization of thoughts, or what? How about in your spoken communication? What kinds of noise sometimes interfere with your receiving and understanding the written or spoken communication of others?

BIBLIOGRAPHY

Cuevas, Vera. "What Companies Want: The 'Whole Engineer.'"*EE Times Online*. www .eetimes.com/salarysurvey/1998/work_companies.html. Accessed February 20, 2003.

Jovin, Ellen. *E-Mail Etiquette for Business Professionals*. New York: Syntaxis Press, 2007.

McMurrey, David A., and Buckley, Joanne. *A Writer's Handbook for Engineers*. Toronto, Ontario: Thomson Engineering, 2008.

Naisbitt, John. *Megatrends*. New York: Warner Books, 1982.

Paradis, James G., and Zimmerman, Muriel L. *The MIT Guide to Science and Engineering Communication*, 2nd ed. Cambridge, MA: MIT Press, 2002.

Pearsall, Thomas E. *The Elements of Technical Writing*, 3rd ed. Boston: Allyn and Bacon, 2008.

2

Eliminating Sporadic Noise in Engineering Writing

I am not a picky person when it comes to spelling and grammar, but when I see a report or memo which has repeated errors I immediately question the ability and dedication of the person who wrote it. Why didn't they take the time and effort to do it right? Most of the successful engineers I know write clear, well-organized memos and reports. Engineers who can't write well are definitely held back from career advancement.

Richard L. Levine, Manager, Bell Northern Research, 1987

There arises from a bad and inapt formation of words, a wonderful obstruction of the mind.

Sir Francis Bacon 1561–1626

Errors that crop up here and there in writing, causing what Bacon calls "a wonderful obstruction of the mind," are often referred to as faulty mechanics by English teachers but can also be thought of as sporadic or intermittent noise. Such noise occurs randomly on a page, rather than affecting the whole document the way a poor choice of font size or confused organization of material might. Of course, enough sporadic noise in a document, such as repeated misspellings or numerous sentence fragments, can easily turn into constant noise. Such noise will give your reader an impression of hastily and carelessly produced work undeserving of the response or feedback you hope for—as is bluntly expressed by an engineering manager in the opening quotation to this chapter.

To help you eliminate intermittent noise, this chapter looks at where it is most likely to occur in spelling, punctuation, sentence structure, and technical usage. We also give some pointers on how to edit your writing in order to remove occasional noise.

SPELLING AND SPELL CHECKERS

You might think electronic spell checkers have eliminated any need to be a careful speller. Unfortunately, this is not the case. With apologies to Shakespeare we took his words "A rose by any other name would smell as sweet" (from *Romeo and Juliet*) and ran them through a spell checker as *A nose by any outer dame wood small as sweat*. No red flags were raised by the program. Nor will spell checkers catch common errors such as confusing *there* for *their*, *to* for *too*, or *it's* for *its*. Some typographical errors simple give you other words that will pass unnoticed, as in this sentence. (Did you see it?) A very slight slip of the finger on the keyboard can make the difference between asking for some forms to be mailed to you or nailed to you. A quick transposition could render a memo *nuclear* rather than simply *unclear*.

At best, the effect of poor spelling on your readers is a sense of annoyance, or at least of having their attention distracted by something other than what you want to communicate. At worst, noise created by spelling glitches can bring readers to a stop and cause them to seriously question your ability as a writer. They might even suspect that a person who is careless with spelling could also be inept in more critical technical matters, as the author of our quote at the top of this chapter implies.

To reduce or eliminate any noise in your writing caused by incorrect spelling, use a spell checker but also have a standard dictionary nearby. A current dictionary is the only resource that can reliably answer questions such as the following:

- Whether there is more than one way to spell a word, or what the accepted plural forms of words such as *appendix* or *matrix* are.
- How words like *well-known* or *so-called* are hyphenated, or whether a computer is *on-line* or *online*.
- Whether it is appropriate to write about *FORTRAN*, *Fortran*, or *fortran*.
- What the difference between British and American spelling or usage might be.
- What the accepted past tense is of recent verbs that have come into technical English such as *input*.

It is especially important for an engineer to use a current dictionary. English is a dynamic language, and the language of science and technology changes even more rapidly as knowledge increases and devices are developed. You won't find words like *software*, *modem*, and *LED* in a dictionary from the 1950s, and since then older words such as *bug, hardware, interface*, and *mouse* have taken on new meanings. Some usage has yet to be decided on: Would a computer shop advertise that it repairs *mice* or *mouses*? Do you send *e-mail*, *E-mail*, or *email*? (As of now all three options are still used, but *email* seems to be winning.)

PUNCTUATION

Would you want to drive on a busy highway or in a city where there were no traffic signs? Controlling the flow of traffic is vital if anyone is to get anywhere. Similarly, within sentences the flow of meaning is controlled by punctuation marks, the conventionally agreed-upon "traffic signals" of written communication. We do the same thing in spoken language by means of pitch, breath pauses, and emphasis. Directing the flow of ideas in writing is not really difficult, and a useful procedure when you're unsure of how to punctuate a sentence is to say it aloud as in normal conversation. Pay careful attention to where you pause naturally within the sentence—that's likely to be where you need some punctuation.

Many detailed guides to punctuation exist, and you may want to look at them if you have a lot of queries in this area. You will also find excellent advice on punctuation in the front or back sections of some standard college dictionaries. Meanwhile, the following suggestions are offered on the most common problems many engineers tend to have with punctuation.

COMMAS

Confusion sometimes exists about commas because frequently their use is optional. *Before we arrived at the meeting we had already decided how to vote* would be written with a comma after *meeting* by some and not by others, because some people tend to be heavy comma users while others go light on them. The question to ask is, Does adding or omitting a comma in a given sentence create noise? In general, if no possible confusion or strain results, the tendency in technical writing is to omit unessential commas.

Often, omitting a comma after introductory words or phrases in a sentence will cause your reader to be momentarily confused—as you would have been if there were no comma after the first word of this sentence. Here are further examples of missing commas causing noise.

Problem:

> After the construction workers finished eating rats emerged to look for the scraps.
>
> In all the containers were in good condition considering the rough journey.
>
> As you can see the efficiency peaks around 10–12%.
>
> If an acoustic horn has a higher throat impedance within a certain frequency range it will act as a filter in that range which is undesirable.

Solution:

> After the construction workers finished eating, rats emerged to look for the scraps.
>
> In all, the containers were in good condition considering the rough journey.
>
> As you can see, the efficiency peaks around 10–12%.
>
> If an acoustic horn has a higher throat impedance within a certain frequency range, it will act as a filter in that range, which is undesirable.

Again, try saying these sentences aloud with their intended meanings. You'll find you put the comma—or pause—where it belongs almost without thinking.

One more point about commas: Most technical editors prefer what is called a "serial comma" when you list words or ideas within a sentence, as in *The serial comma has become practically mandatory in most scientific, technical, and legal writing.* You may have been told that the *and* joining the last two terms replaces the need for a comma, but this is not so in technical writing. See how the serial comma is useful in the following sentences by reading them aloud and noting how you need the pause before the *and*:

> Fresnel's equations determine the reflectance, transmittance, phase, and polarization of a light beam at any angle of incidence.
>
> Tomorrow's engineers will have to be able to manage information overload, communicate skillfully, and employ a computer as an extension of themselves.

A serial comma may also prevent confusion:

> Rathjens, Technobuild, Johnson and Turblex build the best turbines for our purposes.

Unless *Johnson and Turblex* is the name of one company, you will need a serial comma:

> Rathjens, Technobuild, Johnson, and Turblex build the best turbines for our purposes.

SEMICOLONS

Whether we like it or not, the semicolon seems to be disappearing from much engineering writing. Often it is replaced by a comma, which is an error according to traditional punctuation rules. More frequently we simply use a period and start a new sentence, but then a psychological closeness might be lost. Look at these two examples:

> Your program is working well, however mine is a disaster.
>
> Take Professor Hixson's class. You'll find he's a great teacher.

The relationship between these statements could be better stressed by using a semicolon:

> Your program is working well; however, mine is a disaster.
>
> Take Professor Hixson's class; you'll find he's a great teacher.

Perhaps one reason we don't see many semicolons in engineering writing is that fewer and fewer people feel confident using them. Another possibility is that little noise results from using a comma or a period and new sentence, as in the examples above. Note this pair of sentences:

> We wanted to finish the computer program yesterday; however, the network was down all afternoon.
>
> We wanted to finish the computer program yesterday, however, the network was down all afternoon.

Although the first sentence would be considered correct and the second wrong, you will find plenty of examples of the second punctuation around. The main problem in the second sentence is that a reader can't be sure at first whether *however* "belongs" to the first half of the sentence or the second. A semicolon after *yesterday* is really needed to make this clear. If you frequently use words like *however, therefore, namely, consequently,* and *accordingly* to link what could otherwise be two sentences, insert a semicolon before and a comma after them. You'll find this will add a shade of meaning that cannot be achieved otherwise.

Use semicolons to separate a series of short statements listed in a sentence if any one of the statements contains internal punctuation. The semicolon will then divide the larger elements:

> I suggest you choose one social science subject, such as psychology or philosophy; one natural science course, such as chemistry, physics, or biology; and one math class.
>
> The team is made up of Seth Deleery, vice-president of marketing; Nat Beers, director of research; Ruth Ustby, assistant director of training and human relations; and Cate Kanapathy, chief avionics engineer.

COLONS

Colons are used to separate the hour and minute in a time notation and to divide parts of book or article titles:

> This proposal is due on Monday morning at 8:30 sharp.
>
> One of the books recommended for the seminar is *The Limits of Safety: Organization, Accidents, and Nuclear Weapons.*

The most common use of the colon within a sentence, however, is to introduce an informal list:

> For the final exam you will need several items: a pencil, a calculator, and three sheets of graph paper.

You can also use a colon to introduce an illustration or example, as we did in the sentence leading into the above example. Note, however, that in both cases an independent clause—a statement that can stand by itself and have meaning—comes before the colon. You should *not* write the example sentence as

> For the final exam you will need: a pencil, a calculator, and three sheets of graph paper,

because what comes before the colon makes no sense by itself and the colon needlessly interrupts the flow of the sentence. Instead write

> For the final exam you will need a pencil, a calculator, and three sheets of graph paper.

(Note how the same reasoning made us lead into the last two illustrations with no colon after the words "example sentence as" and "Instead write.")

PARENTHESES

Use parentheses to set off facts or references in your writing—almost like a quick interjection in speech:

> Resistor R5 introduces feedback in the circuit (see Figure 5).
>
> This reference book (published in 1993) still contains useful information.

If what you place within parentheses is not a complete sentence, put any required comma or period outside the parentheses:

> Typical indoor levels of radon average 1.5 picocuries per liter (a measure of radioactivity per unit volume of air).
>
> Whenever I design a circuit (like this one), I determine the values of the components in advance.

If your parenthetical material forms a complete sentence, put the period inside the marks:

> I have already calculated the values of the resistors. (R1 is 10.5 KΩ, and R2 is 98 Ω.) The next step is to choose standard values.

Remember, it is best not to use parenthetical material too frequently since these marks force your readers to pause, and are likely to distract them (if only for a brief moment—see what we mean?) from the main intent of your writing.

DASHES

A dash (often mistakenly referred to as a hyphen) will make a sentence seem more emphatic by calling attention to the words set aside or after it: *He was tall, handsome, rich—and stupid.* Since the dash is considered less formal than the other parenthetical punctuation marks (parentheses and commas), you should try to avoid it in very formal writing. If you overuse it, you are in danger of calling wolf too often, and your dashes will lose their effect. With this caution in mind, you may still find dashes helpful for the following purposes:

Emphasis:	Staying up all night to finish a lab project is not so terrible — once in a while.
Summary:	Reading all warnings, wearing safety glasses and hardhats, and avoiding hot materials — all these practices are crucial to sensible workshop procedure.
Insertion:	My opinion — whether you want to hear it or not — is that the drill does not meet the specifications promised by our supplier.

Notice we're talking about the "em" dash here—the dash used between words that practically touches the letters at each end of it, and which we have used in this sentence. The "en" dash is shorter, slightly longer than a hyphen, and used when you cite ranges of numbers: *31–34; $350–400*. Most word processing programs allow you to choose whichever you need.

HYPHENS

Hyphens have been called the most underused punctuation marks in technical writing. Omitting them can sometimes create real noise, as when we read *coop* (an enclosure for poultry or rabbits) but discover that *co-op* was meant. On the other hand, a hyphen sometimes appears where it is unneeded, as in *re-design, sub-question*, or even *un-needed*.

Unfortunately, apart from the general rule that hyphens should be used to divide a word at the end of a line or to join pairs of words acting as a single descriptor—as in *The transistor is a twentieth-century invention*—there is no clear consensus on when to use them. You'll often have to decide for yourself with the help of a recent dictionary, but here are some suggestions:

- Don't hyphenate prefixes such as *pre-, re-, semi-, sub-*, and *non-* unless leaving out a hyphen causes an eyesore or possible confusion. *Preconception* is fine, but *preexisting* needs a hyphen if only for looks. The same might be said of *antiinflationary, ultraadaptable*, or *reengineering*. You may have to distinguish between *recover* (regain) and *re-cover* (to put a new cover on) and the like at times. Again, a good dictionary will help.
- Don't hyphenate compound words before a noun when the first one ends in *ly*. For instance, *early warning system* needs no hyphen since it is clear that *early* modifies *warning*, not *system*. The same applies to *optimally achieved goals, highly sensitive cameras*, and similar constructions.

- Stay alert for sentences in which you can eliminate noise by adding one or more hyphens. As you can see, a hyphen improves the second sentence of each of the following pairs:

We used a 16 key keypad.
We used a 16-key keypad.

We knew Marienet made klystrons would be able to generate a 9.395 GHz microwave.
We knew Marienet-made klystrons would be able to generate a 9.395 GHz microwave.

The equation assumes a one dimensional plane wave propagation inside the horn.
The equation assumes a one-dimensional plane-wave propagation inside the horn.

Research showed the computer aided students improved their grades dramatically.
Research showed the computer-aided students improved their grades dramatically.

With really complex technical terms you may have very little to go on regarding hyphens. For instance, how do you punctuate *direct axis transient open circuit time constant*? The best solution (*direct-axis transient open-circuit time constant*) may only be found in a technical dictionary or by observing what the common practice is among specialists in the field.

EXCLAMATION POINTS

The best advice on the exclamation point is to use it all you want in your novel or personal letters, but avoid it in professional writing except in the case of warnings (*DANGER: Sodium Cyanide is extremely toxic!*). Since engineering documents seek to convey information, any excitement or triumph should be generated by the facts provided in the document rather than by a tagged-on marker. Occasionally, an exclamation mark might even be interpreted by your readers as arrogant or sarcastic:

We soon found that the previous data was unsubstantiated!

After reading your report, I feel you might benefit from our on-site course in technical writing!

Punctuation error!

QUOTATION MARKS

Use quotation marks to set off direct quotations in your text, and put any needed period or comma within them, even if the quoted item is only one word. Although British publishers use different guidelines, the American practice is always to put commas and periods inside quotes, and semicolons and colons outside, as in the following:

The manager stressed to the whole group that the key word was "Preparedness."

"The correct answer is 18.2 Joules," he told me.

We had heard about the "Four-Star Marketing Plan," but no one remembered what it involved.

We left the game right after the band played "The Eyes of Texas"; it was too darn hot and humid to stay any longer.

Sometimes the question of where to put question marks with brief quotations arises. The solution is quite simple: If the question mark applies only to what is within the quotes, it goes inside the final quote marks. No period follows after the quotation marks. If it applies to the whole sentence, it will go outside the final quote marks:

Their manager bluntly asked, "Are we on schedule?"

What is the meaning of the term "antepenultimate"?

If you need to quote material that takes up more than two lines, set it off from your text by a space and indent it from both right and left margins. You might even use a slightly smaller font size, and you should omit the quotation marks, as shown here:

According to the author, specifications should not be written by a single person:

The lead engineer delegates the writing of numerous sections to specialists, who may not be aware of the overall goals of the project, and may have parochial views about certain requirements. The lead engineer is faced with the difficult task of fitting all these pieces together, finding all the places where they may conflict, and adjusting them to be correct and consistent with each other [NAWCTSD Technical Report 93-022, p.11].

The importance of consistency cannot be overstressed in the production of . . .

SENTENCE SENSE

As an engineering writer, your aim is to convey information with a minimum of noise. Thus the only important "rule" of grammar is to eliminate noise so that the readers of your document receive precisely the message you intend. In other words, your signal-to-noise ratio should be as high as possible. This section looks at the grammatical and stylistic areas where noise often seems to occur in engineering writing. Under the heading of "Two Latin Legacies," we also discuss two persistent but outmoded grammar rules you can safely forget.

CONNECTING SUBJECTS TO VERBS

It's unlikely you would write *The machines is broken* without quickly noticing a discrepancy between the subject (*machines*) and the verb (*is*). A problem can occur, however, when several words come between your subject and verb and you forget how you started the sentence. If you are writing in a hurry and leave no time for editing, you might produce something like this:

This <u>combination</u> of electrical components <u>constitute</u> a single-pole RC filter.

A 35 mm <u>film</u> of some high buildings <u>are</u> strongly recommended.

Only <u>one</u> of the pre-1925 high-rise structures <u>were</u> damaged in the quake.

Those plural nouns that follow later (*components, buildings, structures*) can sometimes mislead us into relating the verb to them rather than to the earlier nouns (*combination, film, one*) to which they belong. This danger increases with the length of a sentence and the amount of information intervening between the true subject and verb of a sentence. A good style or grammar program on your word processor may help prevent this from happening, but it is just as well to be alert to the danger.

Sometimes a question arises in engineering writing with units of measurement. Do you write *Twelve ounces of adhesive were added* or *Twelve ounces of adhesive was added*? How about *12 grams of acid was spilled* or *12 grams of acid were spilled*?

The answer is a matter of logic rather than grammar. Even though we're alluding to several ounces or grams here, we "see" them as one unit, and thus the singular verb is preferable. Little or no noise is created, however, if you slip up on this one.

Using *either/or* in a sentence occasionally makes us stop and think. Look at this sentence:

Either the old manual or the recent procedures (is/are?) acceptable.

Which verb should you use? Since a verb is normally controlled by the noun immediately before it, we would write *Either the old manual or the recent procedures* are *acceptable*. Following this practice we could also write

---

Either the recent procedures or the old *manual is* acceptable.

It is best to follow the same rule with *neither/nor*. Thus the following two sentences would be preferred:

Neither the engineers nor their *supervisor was* invited to the planning conference.

Neither the rudder nor the *wings were* badly damaged in the crash.

MODIFIERS

A modifier is a word or group of words whose function is to add meaning to other ideas in a sentence. If you say your company has bought a transceiver, you have certainly conveyed some meaning, but if you say *Our company has bought a TS 840 S transceiver with single sideband capabilities*, you add a lot of meaning to the word *transceiver* by adding some modifiers.

The danger lies in creating noise by misplacing the modifiers in a sentence. Such distortion can produce sentences that don't make sense or that make sense in the wrong way. Misplaced modifiers occur when a reader gets the wrong impression (or no impression) of who is doing what in a sentence. This is frequently because words like "I" or "we" or "the engineers" or some other subject has been omitted. Consider the following:

Jumping briskly into the saddle, the horse galloped across the prairie.

After testing the mechanism, the theory behind it was easily understood.

Once having completed needed modifications and adjustments, the equipment operated correctly and met all specifications.

If we look at these statements logically, we have a horse that rides, a theory that can test a mechanism, and equipment that modifies and adjusts. This is not likely to be what the writer meant. Revising the sentences might result in the following:

> Jumping briskly into the saddle, the outlaw galloped across the prairie.
>
> After testing the mechanism, we easily understood the theory.
>
> Once we had completed needed modifications and adjustments, the equipment operated correctly and met all specifications.

Meanwhile, another problem can crop up if you place a modifier too far from the word or idea it modifies:

> I was ordered to get there as soon as possible by fax.
>
> By the age of 4 his father knew he would be an engineer.

It's not hard to remedy the lack of logic in these sentences and to avoid traveling by fax or having 4-year-old fathers, but sometimes the meaning cannot be extracted, as in the following:

> The tone-detector circuit was too unreliable to be used in our telephone answering device, which was built of analog devices.

The sentence is correct if the telephone answering device is made of analog devices, but much more likely the writer is concerned with the inaccuracies of an analog tone-detector circuit. This is easily fixed:

> The tone-detector circuit, which was built of analog devices, was too unreliable to be used in our telephone answering device.

UNCLEAR PRONOUNS

When you use a pronoun in your writing, it is commonly assumed that you are referring to whatever noun or nouns come just before it in the sentence. Thus, *The promotion was given to Vicky, who really deserved it*, is perfectly clear: The *who* refers to Vicky. Sometimes we get careless though, especially with the pronouns *this* and *that*, their plurals, and *which* and *it*. Look at this example:

We will study the terrain by soil analysis and computer simulation before reaching a decision on whether construction can take place here. This will also enable us to . . .

What does the *This* refer to in the second sentence—study, terrain, analysis, simulation, decision, or construction? According to accepted usage, it should be *construction* since it's the last noun before the pronoun *This*, but that's unlikely to be what the writer meant. The meaning would be much clearer if the second sentence read something like this:

This study will also enable us to . . .

Let's look at another example:

Ambiguous: Back in 1954, three researchers made a series of discoveries about the unknown sources of Barbour's early notebooks. These prompted them to further investigate . . .

Clearer: These discoveries prompted the three to further investigate . . .

PARALLELISM

Parallelism refers to the need for items in a list to share the same grammatical structure. Faulty parallelism creates noise because it violates a sense of logical consistency. Rather than tell someone you *like to jog, wrestling, and play the fiddle*, you would probably say you *like to jog, wrestle, and play the fiddle*, or that you enjoy *jogging, wrestling, and playing the fiddle*. But in longer sentences there is a danger of losing control of this flow.

After a lot of discussion, the team concluded that their alternatives were to call in a consultant, thus increasing the cost of the project, or having three more engineers reassigned to the team.

Note how this sentence reads as if the team's alternatives are (1) to call in a consultant, and (2) having more engineers reassigned—two unparallel statements that can grate on our sense of logical flow. The sentence can be rewritten to state that the alternatives were *to call in a consultant . . . or to have three more engineers reassigned*. See if you can recognize the lack of parallelism in this sentence:

> The back-up system should be efficient, should meet safety specifications, and have complete reliability.

To make this statement parallel, think of the list embedded in it. We are told that the back-up system

1. should be efficient
2. should meet safety specs
3. have complete reliability.

To be consistent, the sentence needs one more *should*—or one less:

> The back-up system should be efficient, should meet safety specifications, and should be completely reliable.
>
> The back-up system should be efficient, meet safety specifications, and be completely reliable.

This might seem like a rather fine point, but since a lack of parallelism can often cause a reader to pause, if only subconsciously, it qualifies as noise when it occurs in a sentence. Keeping parallel structure is even more important when you construct lists, as we will show in Chapter 3 under Use Lists for Some Information.

FRAGMENTS

Sentence fragments are partial statements that create noise because they convey an incomplete unit of information. Here's an example:

> She decided to major in petroleum engineering. Even though it would take five years.

The first sentence makes sense by itself. Try saying the second statement alone, as an independent exclamation, and your listeners will be lost. We must admit, however, that in everyday speech and popular journalism you will find plenty of fragments that seem to cause little or no noise. Look at this example:

> Nearly 60 percent of U.S. households had VCRs by the end of the 1980s. In spite of the microwave oven being the most popular appliance of the decade.

We know what the writer means here, but strictly speaking the second statement is a fragment because it could not stand alone and make sense. The words *In spite of* indicate a contrastive relationship that is clear only in the context of the first statement. It would be more efficient to write the following:

> In spite of the microwave oven being the most popular appliance of the 1980s, nearly 60 percent of U.S. households had VCRs by the end of the decade.

In your formal engineering writing you would do well to avoid incomplete sentences. They can usually be quite easily remedied, as you can see. Here's another example:

> Fragment: Delays in the October shipments have occurred. Due to the strike.
>
> Complete: Delays in the October shipments have occurred due to the strike.
>
> Better: The strike has delayed the October 6 shipments.
>
> or
>
> The October 6 shipments have been delayed by the strike.

ACTIVE OR PASSIVE VOICE?

As indicated in the last pair of sentences, we can use two distinct "voices" in English sentences. The active voice directly states that someone does something, as in *The engineer wrote the report.* The passive voice turns it around to *The report was written by the engineer.* Thus the active voice emphasizes the performer of the action—the engineer, in our example—while the passive emphasizes the recipient of the action, the report.

Many engineering and scientific writers are told to use the passive voice, that is, to leave themselves out of their writing. They might write *It was ascertained that . . .* rather than *We made sure that . . .* , or *The deadline was met* rather than

We met the deadline. Chances are management would rather tell you *It has been decided to terminate your employment* than *We have decided to fire you.* (Perhaps such hedging is necessary at times since it helps conceal responsibility and gives us no one to blame!)

The passive voice is certainly appropriate when writing up your research or describing a process, for example. There are plenty of instances where you don't want the "doer" to get in the way of your description. Also, it's logical to use the passive if the doer of an action is unknown or unimportant, or if what is being done is simply more important than who did it:

Electricity was discovered thousands of years ago.

The bridge was torn down in 1992.

The contaminated material is then taken to a safe environment.

Sometimes the passive will give variety to your writing, even if your inclination is to write predominantly in the active voice:

Computer experts claim that general-purpose processors have unpredictable execution times due to their use of complex architectural features. This conjecture has now been tested by our group and we have found that the architecture really induces little or no unpredictability. Moreover, data gained from our study show how the execution times can be predicted. It was also found that . . .

In spite of the passive's usefulness, however, the natural form of the English sentence is usually the active voice. This form generally tends to be the more efficient one. Look at the following pairs, comparing the first sentence to the second:

Control of the flow is provided by a DJ-12 valve.

A DJ-12 valve controls the flow.

A system for delineating these factors is shown in Figure 5.

Figure 5 shows a system for delineating these factors.

By switching off the motor when it started to vibrate and looking at the tachometer, the resonant frequency was determined.

We determined the resonant frequency by switching off the motor when it started to vibrate and looking at the tachometer.

The passive can become especially burdensome in procedures or instructions:

The button is pressed twice.

vs.

Press the button twice.

Previously entered data in the DataBase is eliminated by the Edit menu being opened and Select All being chosen.

vs.

Eliminate previously entered data in the DataBase by opening the Edit menu and choosing Select All.

Nowadays engineering writers are tending to get away from the rigid use of the passive as they realize there is a lot to be said for using the active voice. Sentences become more vigorous, direct, and efficient in the active form, and by showing that a *person* is involved in the work, you are doing no more than admitting reality. Also, the active voice gives credit where credit is due. If we read in a progress report that *several references were checked out from the library and 25 pages of notes were taken,* are we as impressed by the energy expended as when we read *I checked out several books from the library and took 25 pages of notes?*

One danger of avoiding the active voice is that we can end up saying some pretty awkward things:

Hurrying to complete the project, several wires got soldered together incorrectly.

The supervisor was seen by us, and we were ignored by her.

My darling . . . you are really loved by me. Am I somewhat loved by you?

Perhaps the best policy is to use the active voice in your technical (and everyday) writing if it seems the most natural and efficient way to express yourself, assuming there is no company policy against its use. Don't hesitate to write in the passive, however, if the circumstances seem to call for it or if the specifications for the document you are writing require it.

SEXIST LANGUAGE

Gender, or sex, is now only indicated in English by *she/he, his/hers, her/him,* and by a small group of words describing activities formerly pursued by one sex or the other, such as *mailman, stewardess, chairman,* or *seamstress.* Now of course men might bring the drinks on an airplane and women might deliver the mail, not to mention take an equal place in the engineering workplace. Given this situation, it is unnecessarily restrictive—and to some people offensive—to use gender-specific terms in writing and speech unless there is good reason to do so. The following pairs show how easy it is to reword your sentences or paragraphs to include everyone they should:

Restrictive: *Every engineer should be at his workstation by 9* A.M.

Inclusive: Every engineer should be at his or her workstation by 9 A.M.
or (preferred because less wordy):
Engineers should be at their workstations by 9 A.M.

Restrictive: *An employee can expect a lot of challenges during his career here.*

Inclusive: Employees can expect a lot of challenges during their careers here.

Restrictive: *Every technician must wear safety glasses when he enters the work area.*

Inclusive: Technicians must wear safety glasses when entering the work area.

Most nouns indicating gender in English have already been modified to be inclusive. A recent dictionary can guide you here. One title that still sneaks through, however, especially in organizations traditionally dominated by males, is *chairman.* If the "chairman" is female, is she the chairwoman or chairperson? Both are acceptable, but it's probably simpler to refer to anyone in such a position as *chair:*

Sarah is chair of the new committee on marketing strategy.

or

Sarah is chairing the new committee on marketing strategy.

TWO LATIN LEGACIES

A few grammar rules impressed upon us in the past really do not hold up under careful linguistic or logical inspection. They were based on how Latin works, rather than English. To put it another way, noise rarely occurs when these rules are ignored. Here are the two main ones, together with comments and a caution.

"Never End a Sentence with a Preposition."   In reality a preposition is often the best word to end a sentence with. (A purist might claim we should have just written *... the best word with which to end a sentence*). When an editor criticized Sir Winston Churchill for doing so, Churchill responded with "Young man, this is the kind of nonsense up with which I will not put!" After all, did you find any noise in the opening sentence of this paragraph? Efficient writing sometimes dictates that we end a sentence with a preposition. Compare the following pairs. You can see that in each case the second sample, ending with a preposition, flows better and is more natural:

That's a problem on which we will really have to work.

That's a problem that we will really have to work on.

We must make sure we can find some engineering consultants on whom we can really count.

We must make sure we can find some engineering consultants we can really count on.

"Never Split an Infinitive."   An infinitive is the form of a verb that combines with the word *to*, as in *to go, to work*, or *to think*. Confident writers have dared *to deliberately split* the infinitive whenever doing so was in the best interests of clear writing. Certain TV space adventurers have been venturing *to boldly go* where the rest of us can't for a long time now, and an electrician may find it necessary (and safer) *to entirely separate* the wires in a power line sometimes. But don't overload a split infinitive. If you put too much material between *to* and the rest of the verb, noise or even nonsense might result:

The team has been unable to, except for the lead engineer and one technician who is on temporary assignment with us, master the new program.

Rewrite this as

> Except for the lead engineer and one technician on temporary assignment with us, the team has been unable to master the new program.
>
> or
>
> The team has been unable to master the new program — with the exception of the lead engineer and one technician who is on temporary assignment
> with us.

TRANSITIONS

Transitional words and phrases are signposts that show a reader the way your thinking is going. They help connect ideas, distinguish conditions or exceptions, or point out new directions of thought. Simple words like *therefore, thus, similarly,* and *unfortunately* eliminate ambiguity by helping a reader interpret your information. So if you neglect transitions in your writing, you may create noise, since your reader might miss some important connection. Look at these two sentences:

> *The group's long-range plans for the S-34B project have been extended. The completion date for the project is as originally planned.*

Both sentences are grammatically correct and contain important facts, but can the reader tell how these facts are related? Now notice how the next three illustrations indicate relationships the first example does not:

> The group's long-range plans for the S-34B project have been extended. Nevertheless, the completion date for the project is as originally planned.
>
> The group's long-range plans for the S-34B project have been extended. Unfortunately, the completion date for the project is as originally planned.
>
> Even though the group's long-range plans for the S-34B project have been extended, the completion date for the project is as originally planned.

While facts are important, it is often the *relationships* between the facts that create the whole picture. Thus you should make your transitions and connections as strong as possible. Here are some examples:

To indicate a sequence: *before ... later, first ... second, in addition, additionally, then, next, finally*

Before the project got under way, we felt we could never meet the deadline. Later, it became clear there was a realistic chance of doing so.

To indicate contrast: *but, however, yet, still, nevertheless, although, on the contrary, in contrast, on the other hand*

The GX-40 vehicle scored over 96% in initial dependability testing; nevertheless, the design was scrapped.

To indicate cause and effect: *consequently, therefore, so, thus, hence*

This company has had to downsize lately. Consequently, many of our staff are looking for other positions.

To indicate elaboration: *further, furthermore, for example, moreover, in fact, indeed, certainly, besides*

The automotive airbag has proved to be a major factor in driver survival. Moreover, the bag has generated considerable profits for its producers.

SENTENCE LENGTH

When dealing with highly technical subjects you should rarely write sentences over 20 words long. Technical material can be difficult enough to follow without being presented in lengthy, complex sentences. This difficulty increases if your audience is less familiar with your field than you are. Even nontechnical ideas are hard to grasp in an unnecessarily long-winded sentence:

We finally had a long discussion with the R & D staff but were not able to convince them that they should commit to a specific date for implementation of the design, but instead they responded with a proposal to extend the project, which would result in a lot more work for all of us and a considerable loss of profits for the company.

Nobody wants to be left breathless at the end of a mile-long sentence. If you find your sentences tend to be lengthy, look for ways to break them into two or more separate ones. The readability of your prose will be determined partly by the length of your sentences. On the other hand, too many short sentences may leave your readers feeling like first graders:

> The Kw766XTR is a low-profile desktop scanner. It has outstanding performance. It offers a frequency range of 29–54 and 108–174 MHz. It includes 50 memory channels. The design is sleek. Individual channels can be locked out. They can also be delayed.

Try to vary your style and avoid both lengthy and abrupt sentences. Very short sentences used sparingly, however, can be effective in helping you reinforce a point. Remember this.

TECHNICAL USAGE

USELESS JARGON

In its negative sense, jargon is pure noise since it refers to unintelligible speech or writing. The word derives from a French verb meaning the twittering of birds, and has a lot in common with "gobbledygook," first used to compare the speech of Washington politicians to the gobbling of Texas turkeys. High-tech jargon is sometimes known as technobabble or scispeak. Some people seem to like to sprinkle their writing liberally with such impressive-sounding phrases as *integrated logistical programming, differential heterodyne emission*, or *functional cognitive parameters*. Unfortunately, unless these words hold a precise meaning for both writer and reader, no communication takes place.

Technobabble is so common that with tongue-in-cheek we have created an "electrotechnophrase generator" in Figure 2-1 to help addicts satisfy their habit. Select any three-digit number and read off the corresponding words from the chart below; for example, 2-8-3 gets *differential heterodyne emission*. Readers may have no idea what you mean, but they should be impressed—or afraid to ask for a meaning.

USEFUL JARGON

In another sense, however, jargon is the necessary technical terminology used in specialized fields. A chemist might use the term *deoxyribose* around a group of peers without feeling a need to explain it, just as a geologist could talk about the *Paleozoic* era or *Devonian* period with other geologists. Computer engineers can safely refer to bytes, bauds, and packet switching—among themselves. Communication between experts would be ponderous if not impossible if they had no specialized jargon. Moreover, each year technical language increases greatly as scientific knowledge increases. The recent Eleventh Edition of Merriam-Webster's Collegiate Dictionary (2008) claims to have added some 10,000 new words and senses from various fields of knowledge that were not contained in its previous edition.

Sometimes you will find that common words take on new meanings when used by experts. *Charge, conductor, mole*, and *mud* are just four examples. Typesetters mean something quite different than most of us would when they refer to *widows,*

Column 1	Column 2	Column 3
0. voltaic	integrated	simulation
1. Sholokhov's	semiconductor	algorithm
2. differential	Yagi	attenuator
5. virtual	tracking	parameters
3. Fourier	scaled	emission
4. transient	Q-factor	diode
6. phasor	diffusion	network
7. compound	Doppler	gate
8. thermal	heterodyne	transducer
9. Gaussian	coaxial	magnetron

Figure 2-1   The Electrotechnophrase Generator (courtesy of ECE students at the University of Texas at Austin). Warning: Use only when you want to be sure no one understands you.

orphans, gutters, and *leading.* As engineers, you know and use all sorts of technical jargon. Some you share with practically all engineers, some with those in the same general field of engineering as you—such as chemical, civil, or aerospace—and some technical terms you would use only among peers in highly specialized fields like celestial mechanics or software engineering.

There is only one way to avoid noise when using technical terminology: *Know your audience.* Make certain you are writing or speaking at their level of comprehension, because if you're above their heads, you will be wasting your time and theirs. Explain terms whenever necessary; don't risk confusing readers or completely losing them because they don't know what you are talking about. Definitions within your text, examples, analogies, or a good glossary are all useful tools for the technical writer who must frequently communicate with less technically inclined audiences. These specific tools are discussed more fully in other sections of this book.

ABBREVIATIONS

Abbreviations are necessary in technical communication for the same reason valid technical jargon is: They refer to concepts that would take a great deal of time to spell out fully. It would be time-consuming and boring for a computer expert to read *Computer-Aided Design/Computer-Aided Manufacturing* several times (or hear it in a talk) when *CAD/CAM* would do. However, you will create a lot of noise in your writing if you use abbreviations your readers don't understand. Always spell abbreviations out the first time you use them unless you know this would insult the intelligence of your audience:

Then it goes into the ROM (Read-Only Memory).

To understand our billing process, you first need to know what a British Thermal Unit (BTU) is.

Once you have defined an abbreviation you can normally expect your reader to remember it. The exception to this would be if you are using some highly complicated or unusual abbreviations throughout your document, in which case you may need to remind readers more than once what the abbreviations stand for, or provide a glossary they can refer to.

Initialisms and acronyms.   Abbreviations can be subdivided into *initialisms* and *acronyms*. Initialisms (sometimes called initializations) are formed by taking the first letters from each word of an expression and pronouncing them as initials: *GPA, IBM, LED, UHF*. Acronyms are also created from the first letters or sounds of several words, but are pronounced as words: *AIDS, FORTRAN, NAFTA, NASA, RAM, ROM*. Some acronyms become so commonplace that they are thought of as ordinary words and written in lower case: *bit, laser, pixel, radar, scuba, sonar*.

Don't be surprised if you find a list of both initialisms and acronyms lumped under the title ACRONYMS. Many engineering writers no longer observe the distinction between the two and call any abbreviation an acronym. You probably shouldn't make an issue of it, especially if the writer is your superior.

Two usage pointers:

1. Use the correct form of *a/an* before an initialism. No matter what the first letter is, if it is pronounced with an initial vowel sound (for example, the letter M is pronounced "em"), write *an* before it:

   > *an MTCR (Missile Technology Control Regime)*
   > *an LED readout*
   > *an SRU pin*
   > *an ultrasonic frequency* (but *a UHF receiver*)

   Some abbreviations might fool you. Consider LEM (lunar excursion module) for example. If the custom is to pronounce it as an initialism, L-E-M, then you will have *an* LEM. If it is normally considered an acronym (as one word), you will have *a* LEM.

2. Form the plural of acronyms and initializations by adding a lowercase *s*. Only put an apostrophe between the abbreviation and the *s* if you are indicating a possessive form:

   > *We ordered three CRTs.*
   > *We weren't satisfied with the last CD-ROM's performance.*
   >   *or*
   > *We weren't satisfied with the performance on the last CD-ROM.*

NUMBERS

Engineering means working with numbers a great deal. Frequently, this is where a lot of written noise occurs due to typos, incorrect or inexact numbers, and inconsistencies. Obviously, you can avoid serious noise by making certain any number you write is accurate. You should also give numbers to the necessary degree of precision: Know whether 54.18543 is needed in your report or whether 54.2 will do. Avoid noise from inconsistent use of numbers by following these guidelines:

1. Numbers are expressed as words (twelve) or numerals (12). Cardinal numbers are *one, two, three*, etc. Ordinal numbers are *first, second, third*, etc. Although custom varies, it's a good idea to write the cardinal numbers from one to ten as words and all other numbers as figures.

   two transistors   *232 stainless steel bolts*
   three linear actuators   *12 capacitors*

   However, when more than one number appears in a sentence, write them all the same:

   The IPET has 4000 members and 134 chapters in 6 regions.

   Also, use numerals rather than words when citing time, money, or measurements:

   1 A.M.   *$5.48*   *12.4 m*   *8 ft*

2. Spell out ordinal numbers only if they are single words. Write the rest as numerals plus the last two letters of the ordinal:

   second harmonic   *21st element*
   fourteenth attempt   *73rd cycle*

3. If a number begins a sentence, it's a good idea to spell it out regardless of any other rule.

   Thirty-two computers were manufactured today.

   To avoid writing out a large number at the beginning of a sentence, rewrite the sentence so it doesn't begin with a number:

   Last year, 5198 engines were manufactured in this division.
                                   or
   This division manufactured 5198 engines last year.

Note   You may sometimes see very large numbers written with spaces where you expect commas. Thus 10,354,978 might appear as 10 354 978—to avoid any possible confusion with the practice in some countries of using commas as decimal markers. Decide which method you want to use based on your company's preference and/or your audience.

---

4. Form the plural of a numeral by adding an *s*, with no apostrophe:

   80s   *1920s*

   Make a written number plural by adding *s, es*, or by dropping the *y* and adding *ies:*

   nines   sixes
   fours   nineties

5. Place a zero before the decimal point for numbers less than one. Omit all trailing zeros unless they are needed to indicate precision.

   0.345 cm   *12.00 ft*
   0.5 A   *19.40 tons*

6. Write fractions as numerals when they are joined by a whole number. Connect the whole number and the fraction by a hyphen:

   2-1/2 liters   32-2/3 km

7. Time can be written out when not followed by A.M. or P.M., but you will normally need to be more precise than this. Use numerals to express time in hours and minutes when followed by A.M. and P.M. or when recording data. Universal Time (UTC, from the French for *universal coordinated time*) uses the 24-hour clock.

ten o'clock	*10:41* A.M.	*8:45* P.M.
4 hours 36 minutes	*12 seconds*	*23:41 (=11:41* P.M.*)*

8. When expressing very large or small numbers, use scientific notation. Some numbers are easily read when expressed in either standard or scientific form. Choose the best format and be consistent:

0.0538 m	or	5.38×10^{-2} *m*
8.32×10^{-21} *m/s*	or	*367 345 199 m/s*

UNITS OF MEASUREMENT

Although the public in the United States is still not committed to the metric system, you will find that in general the engineering profession is. Two versions of the metric system exist, but the more modern one, the SI (from French *Système International*), is preferred. The vital rule is to be consistent. Don't mix English and metric units unless you are forced to. Be sure to use the commonly accepted abbreviation or symbol for a unit if you do not write out the complete word, and leave a space between the numeral and the unit.

70 ns	100 dB
12 V	34.62 m
23 e/cm^3	6 Wb/m^2

Many people, including technically trained ones, still think in standard or English units of measurement, so sometimes you may find it advisable to give both referents in your writing. As with many other editorial matters, you can only make this decision after thinking of your readers' needs. When it might be advisable to add "explanatory" units, as with a mixed audience, do so by writing them in parentheses after the primary units:

212°F (100°C)	5.08 cm (2 in)

Make sure you use the correct symbol when referring to units of measurement, and remember that similar symbols may stand for more than one thing. A great deal of noise (or disaster) could result if you confused the following, for example:

°C (degrees Celsius)	C (coulomb — unit of electric charge)
g (gram)	G (gauss — measure of magnetic induction)
m (thousandth)	M (million)
n (nano-)	N newtons
s (second — as in time)	S (siemens — unit of conductance)

Units of measurement derived from a person's name usually are not capitalized, even if the abbreviation for the unit is. Note also that although the name can take a plural form, an *s* is not added to the abbreviation to make it plural:

amperes A	farads F	henrys H	kelvins K
teslas T	volts V	webers Wb	

When working with very large or very small units of measurement, you will need to be familiar with the designated SI expressions and prefixes:

Factor	Prefix	Symbol
10^{24}	yotta	Y
10^{21}	zetta	Z
10^{18}	exa-	E
10^{15}	peta-	P
10^{12}	tera-	T
10^{9}	giga-	G
10^{6}	mega-	M
10^{3}	kilo-	k
10^{2}	hecto-	h
10^{1}	deka-	da
10^{-1}	deci-	d
10^{-2}	centi-	c
10^{-3}	milli-	m
10^{-6}	micro-	μ
10^{-9}	nano-	n
10^{-12}	pico-	p
10^{-15}	femto-	f
10^{-18}	atto-	a
10^{-21}	zepto	z
10^{-24}	yocto	y

A recent dictionary of scientific terms will guide you if you are unsure of the correct spellings or symbols of the units you are using. There is no point using them in your writing, however, if you or your audience don't know what they mean. Symbols and abbreviations are indispensable to an engineer, but use them sparingly when writing for an audience other than your peers. You may sometimes need to define the ones you use, either in your text parenthetically (a brief explanation in parentheses following the term or symbol, like this) or with annotations, as in the following example:

$$P = IE \tag{1}$$

where

P = power, measured in watts

I = current in amperes

E = EMF (electromotive force) in volts

EQUATIONS

It would be hard to do much engineering without equations. They can communicate ideas far more efficiently than words can at times—consider the ideas represented by $E = mc^2$ for example. However, formulas and equations slow down your reader, so use them only when necessary and when certain your audience can follow them.

Many word-processing programs now make it easy to write equations in text, but if you have to write them in longhand do so with care to ensure both accuracy and legibility. An illegible or ambiguous equation is hardly going to communicate data effectively, and an error in an equation could be fatal. In other words, make sure your equations are noise-free.

You should normally center equations on your page and number them sequentially in parentheses to the right for reference. Leave a space between your text and any equation, and between lines of equations. Also, space on both sides of operators such as $=$, $+$, or $-$, as shown in the equations below. If you have more than one equation in your document, try to keep the equal signs and reference numbers parallel throughout:

$$F(x) = \int \log x \, dx \tag{1}$$

$$H(s)(xv_{2)} = X(s)/Y(s) \tag{2}$$

The total harmonic distortion (THD) of voltage at any bus k is defined as

$$THD_k = \frac{\sqrt{\sum_{h=2}^{H} |V_k^h|^2}}{|V_k^l|}, \tag{3}$$

THD can be incorporated into the minimization procedure in [2] by considering a network function that equals the sum of squared THD_ks, or

$$f(I_m) = \sum_{k=1}^{K}(THD_k)^2 = \sum_{k=1}^{K}\left[\frac{\sqrt{\sum_{h=2}^{H}|V_k^h|^2}}{|V_k^l|}\right]^2$$

$$= \sum_{h=2}^{H}\sum_{k=1}^{K}\frac{1}{|V_k^l|^2}|V_k^h|^2. \tag{4}$$

Note that (4) is identical to (2) when $y(h) = 1$ for $h = 2, 3, 4, \ldots, H$, and when

$$b(k) = \frac{1}{|V_K^l|^2}, \quad k = 1, 2, 3, \ldots, K. \tag{5}$$

Since the fundamental frequency voltages are approximately 1.0 pu, the objective function of (4) is a close approximation to that of (1).

Figure 2-2   An example of clear multiline equations. Notice the standard punctuation marks at the appropriate places.

Eventually, you may have to incorporate multiline equations into your technical papers and reports, where they will read (and should be punctuated) just like sentences. An example of such an equation is illustrated in Figure 2-2.

As this example reveals, no material is too complex to be presented clearly in a flowing, natural manner. Punctuation, transitions, accurate grammar, and mechanics are all indispensable tools for conveying highly technical information with a minimum of noise.

EDIT, EDIT, EDIT

If you look at the early handwritten drafts of some of the greatest writers' works, you'll see alterations, additions, deletions, and other squiggles that indicate how much revision went into the draft before it became a finished work. We could all produce better written documents if we always

1. *had* the time to edit our work carefully.

2. *took* the trouble to edit our work carefully.

For an engineer, time is frequently going to be a problem. You can't always find time for a leisurely edit of your work. However, you would be ill-advised to send a first draft of anything of importance to your readers. A quick email note to a friend about lunch isn't worth much concern, but anything more than this, especially if it's going beyond your immediate colleagues, needs at least to be looked over briefly with an editorial eye. How much time you invest in editing should be in direct proportion to the importance of the document. Use all the assistance your word processor will give you, including any spelling, grammar, or readability programs you may have, but don't follow their suggestions blindly. *You* have to be the final arbiter on the clarity and effectiveness of your work—*your* name will be on the document, not your word processor's manufacturer.

COLLABORATIVE PROOFREADING

There is nothing wrong with having a colleague, friend, or spouse look over your writing before you submit it to its intended audience. Two heads are usually better than one for discovering flaws in a piece of writing, and you are no longer in a freshman English class where such help might be considered plagiarism. In industry, experts often cooperate in writing technical reports, proposals, and other documents in the same way that they work together on engineering projects. In fact, most lengthy documents are produced by team effort, where different team members use their particular strengths to ensure that the document is the best it possibly can be.

Collaborative editing can involve something as simple as asking a friend for his or her opinion of the organization, clarity, and mechanics of your work and using those comments to improve your writing where necessary. The more skilled and frank your friend is, the better. With a long document, however, collaborative editing can be done by having different team members check the document for different potential kinds of noise, which is usually better than having everyone searching for whatever they can find. This team approach to editing is fully discussed at the end of Chapter 3, under "Share the Load: Write as a Team." Chapter 3 also gives you several guidelines on how to eliminate noise not just within a sentence but in larger chunks of writing—or even throughout an entire document.

EXERCISES

1. Review some of your own recent writing for problems with spelling, punctuation, or any of the items listed in this chapter under "Sentence Sense." Did you create any noise in your documents by not following these guidelines? How could you use the guidelines as a "quality control" tool when writing in the future?

2. Find what you feel is a good example of technical writing in any field. Analyze it carefully. What makes it effective, noise-free writing? List and give examples of the ways in which the writer has carefully observed many of the guidelines given in this chapter.

3. Look at an article in a professional journal or the first chapter of a textbook and determine who its assumed audience is. Then investigate how the author uses technical terminology. Is it appropriate for the audience? Are explanations or definitions given where they seem

called for? Do you find any examples of unnecessary technical jargon? How might such jargon have been avoided?

4. In technical journals check a number of reports or articles that contain abbreviations, numbers, units of measurement, and equations. Are the authors consistent in the way they write these? Does the way these items are written vary from one report to the next or from one journal to another? In the case of journals, is any information provided on how such things are to be written? Does the journal provide a style guide for writers who might wish to contribute articles?

BIBLIOGRAPHY

The Chicago Manual of Style, 15th ed. Chicago: The University of Chicago Press, 2003.

Hale, Constance, ed. *Wired Style: Principles of English Usage in the Digital Age*, rev. ed. San Francisco: HardWired, 1999.

Pearsall, Thomas E. *The Elements of Technical Writing*, 2nd ed. Boston: Allyn & Bacon, 2001.

Rude, Carolyn. *Technical Editing*, 4th ed. New York: Longman, 2005.

Strunk, William. *The Elements of Style: The Original Edition*. New York: Dover Publications, 2006.

3

GUIDELINES FOR WRITING NOISE-FREE ENGINEERING DOCUMENTS

As every engineer knows, form and content must work together. What is sometimes forgotten is that the relationship of form and content applies to documents as well as to physical phenomena. Without some type of form, be it well or poorly structured, no content can be communicated.... Even the word "in-form-ation" implies that ideas must be structured in some fashion or other.

Susan Stevenson and Steve Whitmore, *Strategies for Engineering Communication* (New York: John Wiley & Sons, 2002), p. 247.

Information isn't a scarce commodity, as a leading economist wrote in the 1970s. Attention is. So, what can you do to sustain your reader's attention?

Bruce Ross-Larson, *Writing for the Information Age* (New York, W. W. Norton & Company, 2002), p. 3.0.

This chapter presents guidelines for producing large sections of noise-free writing, from efficient paragraphs to effective and useful documents. Although different people approach writing tasks in somewhat different ways, these guidelines in general follow the overall process used by successful engineering writers and include important factors you should consider from the time you first face a writing task to the point where you have a final draft you can be proud of. We have also focused on these topics because they represent common problems you as an engineer are likely to face in the course of writing and formatting your documents.

FOCUS ON WHY YOU ARE WRITING

Look at this statement by Ruth Savakinas:

Complex technical writing is likely to be very difficult to read. Readability further decreases when the writer does not define major ideas for the reader and when the written document is not relevant to the reader's experience and interests. These two impediments can be eliminated if you clearly define your purpose and your audience. . . .

From "Ready, Aim—Write!" *IEEE Transactions in Professional Communication, 31*(1), March 1988, p. 5.

What she wrote over twenty years ago still holds: Before starting to write, you should have a good idea of precisely who your audience is and what you want to communicate to them. If these goals aren't first defined in your own mind, you can't really expect your readers to get a clear message. Having this sense of purpose as you write may not guarantee your readers will receive a noise-free message, but writing without a clear goal will almost certainly result in poor communication. Thus, whether you have to write a short memo or a lengthy technical report, you should start with a firm sense of purpose so you can (1) present appropriate supporting data, (2) test its adequacy, and (3) discard anything that is not needed.

Broadly speaking, the purpose of most technical writing is either to present information or to persuade people to act or think in a certain way. Frequently, however, your documents will have to be both informative and persuasive. To fine-tune your sense of purpose before writing, ask yourself the following:

Do I want to

1. **Inform** — provide information without necessarily expecting any action on the part of my reader(s)?

2. **Request** — obtain permission, information, approval, help, or funding?

3. **Instruct** — give information in the form of directions, instructions, procedures, or the like, so my readers will be able to do something?

4. **Propose** — suggest a plan of action or respond to a request for a proposal?

5. **Recommend** — suggest an action or series of actions based on alternative possibilities that I've evaluated?

6. **Persuade** — convince or "sell" my readers, or change their behavior or attitudes based on what I feel to be valid opinion or evidence?

7. **Record** — document for the record how something was researched, carried out, tested, altered, or repaired?

How you write any document should be guided by what you want your audience to do with your information and what they need from the document in order to be able to do it. Thus, your audience plays a defining role in determining how you approach your task. Do they need to be informed, instructed, dissuaded, warned, encouraged, or what? Only a careful analysis of your purpose(s) for writing and the nature of your audience can give you the answers and thus enable you to write to the point.

FOCUS ON YOUR READERS

If you found yourself in a remote region and met people who had never seen anything electronic, you wouldn't hand them your scientific calculator or MP3 player and expect them to use it. First, a great deal of technology transfer would have to take place; you would have to teach your "audience" how to use your gadget (assuming they cared to know). This may seem obvious, but a lot of technical writing fails because writers make inaccurate assumptions regarding the people who read their documents. Engineers often write without taking adequate time initially to consider the nature, needs, interests, levels of expertise, or possible reactions of those who must read their work. Since you will be writing for many different audiences during your career, as Figure 3-1 illustrates, it is well worth taking the time to think about your audience before writing to them.

Audience analysis is not just a question of being polite, thoughtful, or sensitive. Since your goal is to send a clear, noise-free message through your document to your audience, you must consider their abilities and expectations as you plan, write, and revise.

As an engineer, you may find yourself writing to a variety of people either in your immediate group, close by in the company, elsewhere in the company, or outside the company. Sometimes you will write to your professional and technical peers, sometimes to your superiors, and other times to those "below" you. In all

Figure 3-1 You will deal with many different people as your career progresses, so it is best to have a clear picture of who your audience is before beginning to communicate with them.

these writing situations, inadequate audience analysis will inevitably result in noise, since different readers need different kinds of data from you.

No matter who you write to, you write because you expect some kind of resulting action—even if it is only nonphysical "action" such as permission, understanding, or a change of opinion. To get results, your communication must bridge a gap between you and your target audience. In the working world, this gap is likely to be caused by variations in *knowledge*, *ability*, or *interest*. Obviously, the three may overlap, but to determine where you stand before putting any effort into writing, first identify who your audience is and then ask yourself these questions:

Knowledge

- Are my readers engineers in my field of expertise who are seeking technical information and will they be offended or bored by elementary details?
- Are they engineers from a different field who will need some general technical background first?
- Are they managers or supervisors who may be less knowledgeable in my field but who need to make executive decisions based on what I write?
- Are they technicians or others without my expertise and training but with a strong practical knowledge of the field?
- Are they nonexperts from marketing, sales, finance, or other fields who lack engineering or technical backgrounds but who are interested in the subject for nonengineering reasons?
- Are they a mixed audience, such as a panel or committee, made up of experts and lay people?

Ability

- Am I communicating technical information on a level my audience can use?
- Am I using appropriate vocabulary, examples, definitions, and depth of detail?
- Am I expecting more expertise, skill, or action from my audience than I can reasonably expect?

Interest

- Why will my audience want to spend time reading this document?
- Does my document provide the right level of detail and technology to keep my audience's interest without losing them or boring them?
- What is their current attitude likely to be — positive, neutral, or negative?
- Will my document give them the information they want?

The answers to these questions will increase your awareness of the multiple decisions and choices to be made as you plan, write, or revise your document. Remember, in order to deliver a clear message, you should first assess your audience. You need to know who you are writing to and have a clear idea of their technical knowledge, expectations, and attitude toward the subject. If you properly analyze these and address them in your document, you are well on your way to communicating effectively.

SATISFY DOCUMENT SPECIFICATIONS

Before writing, you should be aware of any specifications your document must meet. Many audiences expect documents they receive to be within certain parameters. If management asks for a brief memo, they may be irritated when you overload their circuits with a lengthy, detailed treatise. When a technician requests the specs on a frequency tester, it won't be appreciated if you come up with a flowery prose discussion on the strengths and weaknesses of the equipment. If you respond to an RFP (request for proposal) that calls for a proposal of no more than ten pages but submit something twice that long, chances are your proposal will be eliminated from the competition.

Various document specifications exist. Such specifications may require you to provide sections addressing certain topics in your report, such as experimental problems, environmental impact, decisions reached, budget estimates, and so on. The editors of an engineering journal may put limits on the number of words and the number of graphics your technical paper can include. A word limit is frequently placed on the length of an abstract or summary as well as on other sections of a document. Here is the final requirement for a proposal to obtain a research grant from one large funding organization:

> *Also required is a nontechnical summary (250 words or less) of the research proposed, expressing significance attached to the project and reasons for undertaking it. This summary will be used for public information and should be written in terms that nonscientists can easily understand.*

Many reports have specifications that include requirements not only for their length but also for such matters as headings, spacing, and margin width. Some government agencies, for example, require that the proposals they receive be written in specific formats, in certain fonts, and even with restrictions on how many letters are permitted in each line of text. Here is an example from an RFP for a government research program:

> *Each proposal shall consist of not more than five single-spaced pages plus a cover page, a budget page, a summary page of no more than 300 words, and a page detailing current research funding. All text shall be printed in single-column format on 8-1/2 × 11-inch paper with margins of at least 1 inch on all sides. . . .*

Knowing precisely what is expected of you *before* you begin to write will prevent wasted time and give your document a better chance at success.

GET TO THE POINT

Anyone reading your memos, letters, and reports is likely to be in a hurry. Few engineers have the leisure for "biblical" reading—where one reads from Genesis to Revelation to discover how things turn out. Just as your sentences need to be direct, your documents need to have the most important information at the beginning. This means moving from the general to the specific. Readers would much rather know your key points, complaints, requests, conclusions, or recommendations before they read supporting details. For instance, if you do a series of tests to determine whether some equipment should be replaced, your supervisor will want to know what you have found out and what you recommend. A complete, detailed description of your test procedures may be necessary to support your main points and will likely be verified by others—but it could go unread by those people in management who need only the bottom line.

Where you tell your readers what they most need to know depends on the kind of document you're writing. In a letter it will be in the opening sentences. In a memo you should provide a subject line making more than just a vague reference to the overall topic. Look at these examples:

Vague:	*SUBJECT:*	*Employee safety*
Better:	*SUBJECT:*	*Need for employees to wear hard hats and safety glasses*
Vague:	*SUBJECT:*	*Emergency requisitions*
Better:	*SUBJECT:*	*Recommendations to change the procedures for making emergency requisitions*

Most memos are now sent by email of course, which may limit the number of characters for your subject heading. In this case the challenge is to get as much meaning as possible into a small space and to clearly state your key message in the opening sentences of the memo.

In a longer report your main points should become quickly evident to your reader through an informative title followed by a summary or abstract of your findings, conclusions, recommendations, results, or whatever the important information is. (See the chapters on individual reports and the sections on abstracts and executive summaries in this book.) No matter what kind of document you are producing, however, first determine your audience and purpose, and then give your readers the information they most need in the place they can most efficiently access it—the beginning of the paper, rather than buried somewhere in the middle or at the end.

PROVIDE ACCURATE INFORMATION

Even the clearest writing is useless when the information it conveys is wrong. If you state that an ampere is defined as a coulomb of charges passing a given point in 10 seconds rather than 1 second, you have presented wrong information. If you refer to data in Appendix B of your report when you mean Appendix D, the error could stump your readers and cause them to lose confidence in your report.

Inaccurate references to the work of others also will cause your readers to be highly suspicious of the reliability of your entire report—and even of your honesty as a writer (see Chapter 11). Inaccurate directions in a set of instructions or procedures might be frustrating at best, disastrous at worst. Considerable problems have resulted when engineers gave measurements in standard units that were assumed to be metric by others. Another kind of inaccuracy might be a claim that is true sometimes but not under all conditions, for example, that water always boils at 100° C. What about purity and variations in atmospheric pressure?

There is also a great difference between fact and opinion. A *fact* is a dependable statement about external reality that can be verified by others. An *opinion* expresses a feeling or impression that may not be readily verifiable by others. The danger comes when opinions are stated as facts. Note the difference between these two:

Fact:	*The NR-48 tool features multiple programmable transmitters and a five-station receiver array.*
Opinion:	*The NR-48 is by far the best piece of equipment for our purposes.*

The second statement might be correct but is still only an opinion until supported by verifiable facts. To be strictly honest, the writer should identify it as an opinion unless evidence is presented to support it as fact. In short, make sure that (1) your facts are correct when you write them down and (2) your opinions are presented as such until adequate evidence is provided to verify them.

PRESENT YOUR MATERIAL LOGICALLY

Not only should it be easy to access your document's essential message, but all your information should be in the right place. This means you must organize your material so that each idea, point, and section is clearly and logically laid out within an appropriate overall pattern. If you are following document specifications provided by someone else, you have little choice but to follow those specs, but even within a prescribed plan of organization you may have some leeway to present material the way you feel is most effective.

As always, think before writing, and keep your readers firmly in mind. If they want to know what progress you have made on a project, what you did on a trip, or how to carry out a procedure, obviously they will expect your material to be in *chronological* order. If they are expecting a description of a piece of equipment or of the layout of some facilities, they should be provided with a description that logically moves from *one physical point to another*.

On the other hand, if you have a number of points to make, such as five ways to reduce costs or six reasons why a project must be canceled, present those points from the *most to the least important*, or vice versa. Perhaps your material needs to be presented in order of *familiarity or difficulty*, as when you are writing a tutorial or textbook. Or you may want to move from the *general to the specific*, as when you write a memo first stating that more stringent safety regulations are needed at your plant and then provide concrete examples of current unsafe practices. Note that Table 9-2, Organize Your Material, in Chapter 9 of this book really applies to written material just as much as it does to oral presentations.

MAKE YOUR IDEAS ACCESSIBLE

Without even reading a word, we can look at the pages of a document and get a good idea of how efficiently the material is presented. This impression comes from the structure of the material—specifically, how well the material is laid out in visually accessible "chunks" for the reader. The two most important factors here are (1) the subdivision of material into sections and subsections with hierarchical headings, and (2) paragraph length.

HIERARCHICAL HEADINGS

Even in short engineering documents, a system of headings is essential to keep your material clearly organized and to let readers know what is in each section of the document. Headings and subheadings are also signposts that help a reader get through a report without getting lost. Moreover, they reveal the hierarchical relationships of your material, enabling readers to understand the various levels of detail or importance in your work. Clear and informative headings also give your document good "browsability," that is, they help readers quickly find the parts of your report that interest them most.

Although practice differs among engineers and organizations, a common format for the first three levels of headings is as follows:

FIRST-LEVEL HEADING

Write first-level headings in capital letters and put them flush with the left margin of the page. Use boldface to make the heading stand out and separate it from the written material above and below it by at least one space, as in this illustration.

Second-Level Heading

Also place second-level headings flush with the left margin with at least one space separating them from any text. Capitalize only the first letter of each main word, and make these headings boldface. If you don't like boldface type, you can underline your headings, although underlining does clutter the text. In any case, don't use both boldface and underlining for headings.

Third-level headings. Place third-level headings on the same line as the text they precede. They are capitalized as a sentence would be and can be in boldface or italics.

Note Each level of heading after the first can be indented two or three spaces for visual effect if you wish. The accompanying text would then also be indented with the heading. For an example, see Figure 11-4 in Chapter 11 of this book.

Numbered Headings. Sometimes you may be required to add a numbered, or decimal, system to your headings, and in fact many companies and suppliers require such numbering. A number system gives readers easier reference to parts of a very long report. Note that these different levels of headings can also be successively indented, although many companies don't follow this practice.

FIRST-LEVEL	**1. 0 QUALITY ASSURANCE PROVISIONS**
Second-level	**1.1 Contractor's Responsibility**
Third-level	**1.1.1 Component and material inspection**
Fourth-level	**1.1.1.1 Laminated material certification**

When you use this system, make sure it doesn't get out of control. If your material is so complicated or detailed that you are getting down to levels such as 2.11.3.4.6.23, as some manuals do, then maybe it's time to inspect your document closely to see where you can break it up into smaller, more manageable sections or short chapters, each with its own verbal heading and independent hierarchies within it.

These structural elements of a document (and again, it doesn't matter whether it's a two-page memo or a 500-page manual) can be planned ahead of time. Writing skills aren't needed so much for this as planning and outlining skills, plus an awareness that the headings, divisions, and subdivisions in your document play a vital part in making your information clear and easily available to your readers. So spend some time thinking about how you're going to arrange and format your document before you even *begin* to write, in order to avoid noise at the structural level. You might, of course, want to further improve the structure and organization of your paper after completing the first draft. Word processing software now makes this easy and even enjoyable.

PARAGRAPH LENGTH

No one, especially in technical fields, wants to read a solid page of wall-to-wall text of difficult material. A busy manager, for example, will want to absorb your information in as easily digestible pieces as possible.

Dense text on a page creates noise simply because it's so discouraging. When your readers are trying to follow demanding technical information, they are already challenged, and presenting it to them in solid page-long chunks is at least going to give them mental indigestion. Later, if they want to quickly find a point you made or a piece of data you presented, they are going to have trouble locating it if they have to wade through a ponderous paragraph to get to it.

A guideline in technical writing states that paragraphs should not be much over 12 lines long, but it's better if they are even fewer in general. Occasionally, you will have to go over the 12-line rule, but try not to do so too often. When editing your work, look for any overly long paragraphs and try splitting them into two; when you do, remember that you may have to add a transitional word or phrase.

As an illustration of what we're getting at we have reformatted most of this section on Paragraph Length, including the final two paragraphs of the section, with no breaks and a minimum of white space:

No one, especially in technical fields, wants to read a solid page of wall-to-wall text of difficult material. A busy manager, for example, will want to absorb your information in as easily digestible pieces as possible. Dense text on a page creates noise simply because it's so discouraging. When your readers are trying to follow demanding technical information, they are already challenged, and presenting it to them in solid page-long chunks is at least going to give them mental indigestion. Later, if they want to quickly find a point you made or a piece of data you presented, they are going to have trouble locating it if they have to wade through a ponderous paragraph to get to it. A guideline in technical writing states that paragraphs should not be much over 12 lines long, but it's better if they are even fewer in general. Occasionally, you will have to go over the 12-line rule, but try not to do so too often. When editing your work, look for any overly long paragraphs and try splitting them into two; when you do, remember that you may have to add a transitional word or phrase. Some of your paragraphs will be much shorter than 12 lines, of course, especially if they are transitional paragraphs or convey particularly complex material. If you are writing a manual or set of procedures, most "paragraphs" will probably be one-sentence directives such as *Move the pointer to the next slide and click again.* One last caution on paragraphs: Try to avoid "orphan lines" in your document — paragraphs for which the first sentence begins on the last line of a page, or the last sentence appears at the top of a page.

Looking at the above word mass, you can appreciate the need to let your text breathe. That is, make your information accessible by presenting it in fairly short chunks of information with plenty of white space around them. Some of your paragraphs will be much shorter than 12 lines, of course, especially if they are transitional paragraphs or convey particularly complex material. If you are writing a manual or set of procedures, most "paragraphs" will probably be one-sentence directives such as *Move the pointer to the next slide and click again.*

One last caution on paragraphs: Try to avoid "orphan lines" in your document—paragraphs for which the first sentence begins on the last line of a page, or the last sentence appears at the top of a page.

USE LISTS FOR SOME INFORMATION

A well-organized list is sometimes the most efficient way to communicate information. If you have to present steps in a procedure, materials to be purchased, items to be considered, reasons for a decision, a list might well be the best way to go because readers retrieve some kinds of information from a list more easily than from a passage of prose. Look at the following:

> First of all, set the dual power supply to +12 V and −12 V. Next, set up the op-amp as shown in Figure 1. Use a 1 Vpp sine wave at 1 kHz and then plot the output waveform on the HP digital scope. Then obtain a Bode plot for the gain from 200 Hz to 20 KHz.

You could present this information more efficiently in list form:

> 1. Set the dual power supply to +12 V and −12 V.
> 2. Set up the op-amp as shown in Figure 1.
> 3. Use a 1Vpp sine wave at 1 kHz.
> 4. Plot the output waveform on the HP digital scope.
> 5. Obtain a Bode plot for the gain from 200 Hz to 20 kHz.

There are three main types of efficient lists: *numbered* lists (as above), *checklists*, and *bulleted* lists. You can combine these in various ways to get sublists if you wish. Use a numbered list to indicate when a set of data follows a certain order, as in the example above. Numbered lists can also be used to indicate an order of importance in your data, such as a list of priorities or needed equipment.

Sometimes lists are formed using upper- or lowercase letters in alphabetical order. Numbers are usually best for the main entries in your list, however, since most people are more familiar with moving through steps 1 to 10 than steps (a) through (k). You can always consider using letters for sublists.

Checklists can be used to indicate that all the items on your list must be tended to, usually in the order presented:

> ☐ Connect the monitor to the computer through the monitor port.
> ☐ Connect the keyboard and mouse to the computer through the ASF port.
> ☐ Connect the power supply to the computer.
> ☐ Connect the printer to the printer port.
> ☐ Connect the modem to the modem port.

These instructions could also be presented thus:

1. Connect the monitor to the computer through the monitor port. ☐
2. Connect the keyboard and mouse to the computer through the ASF port. ☐
3. Connect the power supply to the computer. ☐
4. Connect the printer to the printer port. ☐
5. Connect the modem to the modem port. ☐

When checklists get longer than ten boxes, try to break them down into smaller, more manageable sections and give each section its own subheading.

Bulleted lists are commonly used when items in the list are in no specific order:

Some of the main concerns of environmental engineering are

- Air pollution control
- Public water supply
- Wastewater
- Solid waste disposal
- Industrial hygiene
- Hazardous wastes

Word-processing software allows you to create bullets easily and to substitute arrows, tick marks, or other graphics if you wish. Lengthy bulleted lists—over seven items—are hard for readers to refer to, so use numbers for longer lists even if no order of priority is intended. Also, if possible, space between each item in a list, as we have done in the last example above.

Punctuation and Parallelism in Lists.   If the lead-in to your list ends with a verb, no colon is necessary. *The five priorities we established are* would not require a colon after *are* since the list is needed to logically and grammatically complete the statement. (Also see the bulleted list above.) A lead-in like *We have established the following five priorities* would be followed by a colon, however, since the statement is grammatically complete.

If the items in your list are complete sentences and contain internal punctuation, you should conclude each one with a period. Otherwise, a period at the end of list items is optional. Capitalizing the first listed item is up to you, unless each entry is a complete sentence. Whichever style of punctuation and capitalization you use, be consistent.

Another concern when writing lists is to maintain *grammatical parallelism* between entries. This simply means if some entries begin with a verb, all entries should do so; if all begin with a noun, all should. This makes for smoother reading and logical neatness. Note how the following list is bumpy due to problems with parallelism:

Last week we accomplished the following for WW3-a:

- Completed BIU, ICACHE, and ABUS logic design.
- All instruction buffer blocks have had final simulations.
- Written and debugged 75 percent of test patterns.
- Scheduling of first silicon reticules for WW4-a with Vern Whittington in Fab 16.

Making the items in the list parallel cuts out some psychological noise:

Last week we accomplished the following for WW3-a:

- Completed BIU, ICACHE, and ABUS logic design.
- Ran final simulations on all instruction buffer blocks.
- Wrote and debugged 75 percent of test patterns
- Scheduled first silicon reticules for WW4-a with Vern Whittington in Fab 16.

FORMAT YOUR PAGES CAREFULLY

In addition to how you divide information up and how long you make your paragraphs, other factors can also have a positive or negative effect on your reader. People prefer print that is visually accessible and pleasing. You can create psychological noise if you fail to meet these preferences, but you can easily prevent it by keeping the following pointers in mind.

MARGINS

Leave ample margins around your text to help prevent your pages from appearing overloaded. Standard margins are 1 inch all around your page, but you can go a little above or below this if you have to. Make sure the margins are consistent on all pages. If permitted, let your lines of text wrap around with a "ragged" right-hand margin rather than aligning them on the right, since this makes for easier reading. If your report is important enough to be bound like a book, you will need a wider-than-usual left margin to accommodate the binding and ensure that the first word or so of each line is still readable.

TYPOGRAPHY

Typeface is the style of individual letters and characters. *Serif* and *sans (without) serif* are the two general type styles, with serif fonts having small strokes or stems on the edges of each letter. Books, magazines, and newspapers generally use serif fonts for their text, so this is what people are most used to seeing. Sans serif fonts can be effective for titles and headings, but serif fonts make larger quantities of text more readable since the little stems bind the letters and help guide the reader's eye from letter to letter.

Sans serif:	The electric car prototype has regenerative braking, which recharges the supply while decelerating the vehicle.
Serif:	The electric car prototype has regenerative braking, which recharges the supply while decelerating the vehicle.

Standard type size is 10 to 12 point. You should use larger or smaller sizes only for special effect in titles, captions, warnings, and such. Generally avoid sentences with all capital letters—known as "shouting"—because in a long sequence of uppercase letters you have the same visual contours, making such a sentence slower and somewhat more difficult to read:

THE GOVERNMENT PLANS TO ESTABLISH A HIGH-LEVEL ADMINISTRATIVE COUNCIL TO COORDINATE SCIENCE AND TECHNOLOGY.

Capitalized words *should* be used to emphasize a heading or directive, however:

DANGER: A 7000 V potential exists across the transformer output terminals.

WHITE SPACE

White space refers to areas of a page not filled with text or graphics. When reading, we tend to take white space for granted, but it plays an important part in a document by creating a path for a reader's eyes, isolating and emphasizing important data, and providing "breathing room" between blocks of information. Thus, it can have a positive effect by making difficult technical material appear more accessible and less threatening.

You will have enough white space on your pages if you do the following:

- Provide adequate and consistent margins
- Leave a space between all paragraphs
- Leave spaces before and after every heading and subheading
- Leave one or two spaces between text and graphics or lists
- Leave a space before and after each equation in the text
- Indent subheadings or text where appropriate
- Use a ragged (unjustified) right margin

Much of what we advise to make sure your documents are well formatted, and thus visually accessible to your readers, is illustrated in Figure 11-4 in Chapter 11. Shown as an example of effective documentation, the page is also well formatted. Notice how sections there are clearly organized and labeled, using both numeric and verbal hierarchical headings and subheadings. The page is not cluttered or dense and the prudent use of spacing creates a page that is not daunting to a busy reader.

EXPRESS YOURSELF CLEARLY

Engineering is considered a precise discipline (although in reality, as most engineers will admit, it's not always as precise as we would like it to be). Machine parts, for example, may be allowed a certain degree of variation or tolerance within a specified zone and still be interchangeable. Similarly, you have some choices in how you express yourself in engineering writing. In English, you can often say the same thing three or four different ways, but your overriding concern should always be to state what you have to say clearly and to the point. Don't force your readers to work harder than necessary to grasp what you have written; your sentences must convey a single meaning with no room for interpretation or misunderstanding. If your readers yearn for uncertainty and suspense, they can read a romantic novel or detective story, and if they enjoy different connotations and levels of meaning, they can read poetry. So, here are some pitfalls to avoid.

AMBIGUITY

The word *ambiguous* comes from a Latin word meaning to be undecided. Ambiguity primarily results from permitting words like *they* and *it* to point to more than one possible referent in a sentence, or from using short descriptive phrases that could refer to two or more parts of a sentence. In either case, your reader becomes confused—undecided—and may interpret your sentence differently than you intended, as illustrated in the following examples.

Ambiguous: *Before accepting materials from the new subcontractors, we should make sure they meet our requirements.*

(Who are *they*, the materials or subcontractors?)

Clear: Before we accept them, we should make sure the materials from the new subcontractors meet our requirements.

Ambiguous: *The microprocessor interfaced directly with the 7055 RAM chip. It runs at 5 MHz.*

(What does *it* refer to?)

Clear: The microprocessor interfaced directly with the 7055 RAM chip. The 7055 runs at 5 MHz.

Ambiguous: *Our records now include all development reports for B-44 engines received from JPL.*

(What was received from JPL — the reports or the engines?)

Clear: Our records now include all B-44 engine development reports received from JPL.

Ambiguous: *After testing out at the specified high temperatures, the company accepted the new chip.*

(Did the company or the chip test out at the high temperatures?)

Clear: The company accepted the new chip after it tested out at the specified high temperatures.

VAGUENESS

If ambiguity causes readers to see more than one meaning in your writing, vagueness causes them to see no useful meaning at all. What would you think if your doctor told you to "take a few of these pills every so often"? You would want him or her to provide some facts and figures. Explanations or directions lacking specific detail sound fuzzy and unfocused, more like personal opinion than useful data.

Abstract words are not inherently wrong, but they fail to provide the precision effective technical writing needs. Try to avoid abstract words and phrases like *pretty soon, substantial amount,* and *corrective action,* or the unspecific *etc.,* and replace them with terms that have exact meaning such as *in three days, $8,436.00, replace the altimeter.* Here are two more examples of vague writing and ways they can be remedied:

Vague: *The Robotics group is several weeks behind schedule.*

Useful: The Robotics group is six weeks behind schedule.

Vague: *The CF553 runs faster than the RG562 but is much more expensive.*

Useful: The CF553 runs 84% faster than the RG562 but costs $2840 more than the CF553.

As you can see in the second example, vague writing might require fewer words, but it's rarely wise to be concise at the expense of precision. This is especially true when writing instructions and specifications. On the other hand, vagueness can be an asset to people who don't want to reveal too much—or who have nothing to reveal because they've done nothing. The following satirical "Progress Report for All Occasions" has been going around industry for some years now, and is a monument to vague writing:

During the report period that encompasses the organized phase, considerable progress has been made in certain necessary preliminary work directed toward the establishment of initial activities. Important background information has been carefully explored and the functional structure of components of the cognizant organization has been clarified.

The usual difficulty was encountered in the selection and procurement of optimum materials, available data, experimental data, and statistical analysis, but these problems are being attacked vigorously, and we expect that the development phase will continue to proceed at a satisfactory rate.

You might write something like this—if in reality you have no progress to report!

COHERENCE

The root of the word *coherence* is *cohere*, meaning to stick together, and as you know, a cohesive does just that. Coherence in writing refers to how well paragraphs and even complete documents "stick together"—that is, stay focused on their true subject. In a coherent paragraph, all the sentences clearly belong where they are because they address only the topic of the paragraph and are logically connected to one another. You might say the sentences stick to the point and stay there. Coherence in a complete report also means how well the report is designed to take the reader through its paragraphs and sections by means of clear transitions such as headings and subheadings, and how all the sections focus on and support the subject of the report. (Our chapters on report writing will show you how to achieve coherence in longer documents.)

You can achieve coherence in your paragraphs by making sure each sentence clearly relates to the one before it and after it. This means opening with your main point or topic sentence, repeating key words where needed, and using transitional words (see Chapter 2) and pronouns to link sentences as they build up the paragraph. Note how the following paragraph lacks coherence and how it is improved by the devices in boldface in the revised version.

Poor Coherence

A significant disadvantage of the 125-H CRT is its high power consumption. The tube requires substantial power to produce the high voltages and currents that are necessary to drive and deflect the electron beam. The 125-H is inefficient — only about 10% to 20% of the power used by the tube is converted into visible light at the surface of the screen. The 125-H is poorly suited for portable display devices that run on batteries, where lower power consumption is necessary. We should consider other options before committing to purchase the 125-H.

Effective Coherence

A significant disadvantage of the 125-H CRT is its high power consumption. *This* tube requires substantial power to produce the high voltages and currents that are necessary to drive and deflect the electron beam. *In addition*, the 125-H is inefficient — only about 10% to 20% of the power used by the tube is converted into visible light at the surface of the screen. *Thus*, the 125-H is poorly suited for portable display devices that run on batteries, where lower power consumption is necessary. *Because of this drawback*, we should consider other options before committing to purchase the 125-H.

DIRECTNESS

Being as direct as possible in your writing lets your reader grasp your point quickly. Suspense might be thrilling, but a busy technical reader wants access to your information quickly and easily. The most important part of your message should come at the beginning of your sentences and paragraphs. Here are some examples of what this means at the sentence level:

Indirect:	After a long and difficult development cycle due to factory renovation, the Infrared controller will be ready for production in the very near future.
Direct:	The Infrared controller will be ready for production March 4. Its development cycle was slowed by the factory renovation.
Indirect:	Fred has been busily working on this project. This past week he also reworked the logic diagrams, rewired the controller arm, and redesigned all of the RIST circuitry.
Direct:	Fred redesigned the RIST circuitry on Thursday. He also reworked the logic diagrams and rewired the controller arm last week.

USE EFFICIENT WORDING

Opinions vary on how much it costs a company for an employee to produce one written page of technical information, but as stated in Chapter 1, it can be anywhere up to and beyond $200. When you think of all the people writing letters, memos, reports, manuals, proposals, and countless other documents for industry, you see how the costs mount up. Add to this the fact that most of us have little training in producing concise prose, and you can appreciate how sharpening your writing and editing skills can mean not only saving time, but money. Moreover, since we all tend to be wordy, carefully editing our work can often reduce or eliminate a lot of time-consuming work for our readers.

WORDINESS

Using an unnecessarily pompous word instead of a straightforward one can cause your readers to slow down. Choose the simplest and plainest word whenever you can. Your readers can be distracted or even confused by words that call attention to themselves without contributing to meaning. This pitfall becomes even more likely if some of your readers are not native speakers of English, as is often the case in engineering fields today. Write to communicate rather than to impress, or as the saying goes, "Never utilize *utilize* when you can use *use*." A few of the more ostentatious—oops, make that showy—words found in engineering writing are listed here, with some plain, equally efficient counterparts:

commence	*start*	fabricate	*make*	proceed	*go*
compel	*force*	finalize	*end*	procure	*get*
comprises	*is*	initiate	*begin*	rendezvous	*meet*
employ	*use*	optimal	*best*	terminate	*end*
endeavor	*try*	prioritize	*rank*	visitation	*visit*

Wordiness can also result from using far more words than you need to express an idea. Unkind editors sometimes refer to this as verbiage (by analogy to garbage?). Few of us appreciate hearing

> I regret to say that at this point in time I basically do not have access to that specific information. . . .

when a simple "I don't know" is enough. Similarly, your reader is unlikely to thank you for having to plow through

> It is our considered recommendation that a new computer should be purchased. . . .

when you could have simply said you recommend buying a new computer.

You can eliminate a lot of wordiness in your writing by training yourself to edit carefully and to make every word count. Look at the following three pairs; you will see which sentences are more efficient and noise-free.

> *It is essential that the lens be cleaned at frequent intervals on a regular basis as is delineated in Ops Procedure 132-c.*
>
> Clean the lens frequently and regularly (see Ops Procedure 132-c).
>
> *The location of the experimental robotics laboratory is in room 212A.*
>
> The experimental robotics lab is in 212A.
>
> *There are several EC countries that are now trying to upgrade the communication skills of their engineers.*
>
> Several EC countries are trying to upgrade the communication skills of their engineers.

You can also reduce wordiness by avoiding certain pretentious phrases that have unfortunately become common. A good stylebook will give numerous examples, but here are a few that crop up frequently in engineering writing:

Verbiage	Efficient
a large number of	many
at this point in time	now
come in contact with	contact
exhibits the ability to	can
in the event of	if
in some cases	sometimes
in the field of	in
in the majority of instances	usually
in the neighborhood of	about
in view of the fact that	because
in view of the foregoing	therefore
serves the function of being	is
subsequent to	after
the reason why is that	because
within the realm of possibility	possible

Check your writing for such unnecessary phrases and for unneeded words in general—as we do in the next sentence. You may ~~often~~ find ~~that there are a number of~~ words ~~contained in your writing~~ that can be ~~safely~~ eliminated without any ~~kind of~~ danger to your meaning ~~whatsoever~~.

Note If you let your writing "cool off" for a while and come back to edit it later, chances are you will discover more wordiness than if you try to edit immediately after writing.

REDUNDANCY

One category of verbiage is redundancy. This means using words that say the same thing, like *basic fundamentals*, or phrases that duplicate what has already been said, as in *They decided to reconstruct a hypothetical test situation that does not exist*. In fact, if you master the art of redundancy, you can make everything you write almost twice as long as need be. A few common redundant pairs are identified below, but the list is far from exhaustive.

Redundant	Efficient
alternative choices	alternatives
actual experience	experience
completely eliminate	eliminate
component part	component (or part)
connected together	connected
collaborate together	collaborate
diametrically oppose	oppose
exactly identical	identical
integral part	part
just exactly	exactly
permeate throughout	permeate
prove conclusively	prove
rectangular in shape	rectangular
12 noon	noon
very best	best

Again, we all can be wordy at times, so it's a good idea to edit your writing once simply looking for redundancy and wasted words. Grammar-checking software can help, but you still need human editing to remove this kind of noise from your writing.

TURNING VERBS INTO NOUNS

Replacing a perfectly good verb (action) with a noun (the name of an action) is unfortunately common in much engineering writing. This is often the result of wanting to write in the passive rather than active voice. Look at these three pairs of sentences:

> *An analysis of the data will be made when all the results are in.*
>
> We will analyze the data when all the results are in.
>
> *An investigation of all possible sources of noise was undertaken.*
>
> All possible noise sources were investigated.
>
> *Acknowledgment of all incoming messages is performed by the protocol handler.*
>
> The protocol handler acknowledges all incoming messages.

It's easy to see which sentences are shorter and more natural. If you take the verb that really matters in a sentence (such as *analyze, investigate*, and *acknowledge* in the examples above) and make a noun of it, you are forced to add another, generally weaker, verb to convey your meaning.

Thus you will write *made a selection of* instead of *selected*, or *procurement of services can be accomplished by*, instead of *services can be procured by*. Note that many such verbs when changed into nouns need to be followed by *of*. Grammar checkers use this as a cue to warn you of the problem, but again, there is no better tool than your own editing skills—or those of a competent and honest colleague—to free your writing of verbiage.

MANAGE YOUR TIME EFFICIENTLY

Few engineers feel they have enough time to do the writing required of them. Often a memo is hastily churned out, or a report is rapidly thrown together and tacked on the tail end of a project. As with anything done in a hurry, the results are usually not the best. As the pressure to get a piece of writing out increases, sloppiness—that is, noise—also increases. Rather than leaving your writing to the last minute, it is far better to consider it just as much a part of your professional activities as designing, building, and testing.

FINDING AND USING TIME

There are a number of ways to find time to spend on careful writing and editing, but most are not too attractive. You can get to work an hour earlier, or take work home at night (plenty of successful engineers do). You can use your breaks to get away from distractions and concentrate on your writing tasks. You might designate a specific time each day as your writing period—if your colleagues and other duties permit this. You can write on your laptop computer at airports, in flight, on trains, in hotels, or in waiting rooms.

However, as stated above, it's much more practical to make your written work an organic part of your daily schedule. In this way you can assign brief time periods to write short memos and letters or small sections of a report. Larger chunks of time can be designated to concentrate on longer writing tasks.

OUTLINES, DEADLINES, AND TIME LINES

When you have to write anything over two pages long, it's useful to first spend some time making a rough outline. This outline does not have to be set in concrete—that is, you don't have to slavishly follow it once you've written it, and it can be altered at any time—but it will give you some indication of what is involved in producing the finished paper. It will also help you divide your task into smaller sections that can then be written separately at different times, and not necessarily in any order. Less demanding sections, for example, can be relegated to short periods of available time or to times when more distractions surround you.

Even if a deadline for completing a document hasn't been imposed on you, it's a good idea to establish one for yourself. Estimate how long you expect the job to take and schedule back from there. You might even draft a timeline for yourself, showing each date by which you should have completed specific parts of the paper. (See Figure 3-2.) Always allow yourself enough time at the end to review and edit the entire document.

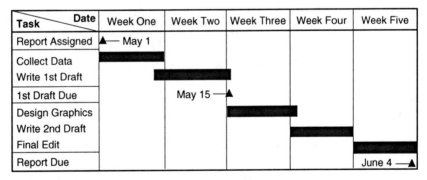

Figure 3-2   The timeline you make for your writing project can be as simple or as detailed as you wish. Make sure you have all your important tasks and due dates down, however, and then do everything you can to keep to them.

EDIT AT DIFFERENT LEVELS

Rather than glance over their finished document once or twice in hopes of randomly finding anything in need of improvement, many writers like to take a more methodical approach to editing. You might want to try this. First, check your document for *technical accuracy*. Then decide what "writing levels" to approach your editing on, and go through your document at least once on each level.

LEVEL 1

The first level, Level 1, is the nitty-gritty one of mechanics, spelling, punctuation, typos—all the basics we were supposed to master in elementary and high school. Again, a good word processing program will provide you with suggestions on spelling and grammar; however, *you* must make the final choices on many of these options. You might also call upon the services of a friend or colleague who is well grounded in these basics.

LEVEL 2

Level 2 involves looking at such things as paragraph and sentence length and structure, possible verbiage, and precise word choice. Is the tone of your document appropriate? Have you used the active voice where possible? How about transitions, parallelism, and emphasis where called for?

LEVEL 3

The final level, Level 3, is the more global level of the document, where you check the overall format, organization, readability appearance, and accuracy of content. Is the work arranged the way it should be? Are specifications (if any) followed? Is it the right length? Have you used the best font size, margins, and spacing? Are headings, subheadings, lists, and graphics used effectively and consistently? Is the title page attractive? How about the "packaging" of the document, such as the quality of paper used, the binding, and the covers?

SHARE THE LOAD: WRITE AS A TEAM

Not many engineers write lengthy reports by themselves. Technical people work together as teams for research, design, development, and testing, and often find they must team up to write proposals, manuals, completion reports, and a lot of other technical documents. Team writing is not always easy, especially when people with different degrees of writing ability or ego investment are involved, or when team members are torn between team responsibilities and other duties. If your group plans the team project carefully, however, it can turn out to be relatively painless and very rewarding, since as a team you will be tapping into far more knowledge, skill, and creativity than you can bring to a project alone.

FIVE POSITIVE APPROACHES

A team is a group of two or more people who interact and coordinate their work in order to accomplish a specific goal. When you work on a team project or help put together a long written document with others, you should be prepared to do the following:

- Communicate
- Coordinate
- Collaborate
- Cooperate
- Compromise

This list might seem obvious, but many teams fail to reach their potential because some members have difficulty in following it. Some people even see *collaboration* and *compromise* in their more negative connotations rather than as the positive attitudes they are meant to be in the context of team activities. Let's look at each one briefly.

Communicate.   Obviously, very little teamwork is possible without frank and open communication. This means that members of the team create an atmosphere that enables free discussion at all times. It also means that the channels of communication, i.e., email, telephone numbers, mail addresses, and meeting times and places, are all common knowledge to each member.

Coordinate.   Since team members are often scattered when not physically working together, it's very important that everyone knows what the others are doing, who is responsible for what, when the next deadline or meeting is, as well as other tactical details. Often one member of a team is appointed as the coordinator, and if that person does the job well, there will be a minimum of frustration, repetition, redundancy, or uncertainty among the team.

Collaborate.   The Latin root of this word means "to willingly labor." In a team setting it means just that—to willingly assist one another. In the spirit of collaboration you will, for instance, assist a partner on some work if necessary, or work at understanding what another team member is doing. You will also freely share your own work with the other team members and work at creating a final document that is unified and seamless.

Cooperate.   An attitude of cooperation is essential to the smooth working of any team project. If the project has a designated leader, you will do all you can to cooperate with that person and to accept his or her decisions, deadlines, changes, or reassignments. Such executive actions on the part of the leader hopefully will be the result of open discussion with all team members, but there may be times when this is not possible, and you may have to cheerfully accept a decision you have no control over.

Compromise.   This word has two meanings, and only one of them is somewhat derogatory. The other meaning refers to making mutual concessions in order to reach a goal. In practical teamwork, this means you may sometimes have to give a bit on an attitude, opinion, approach, method, or course of action, because by doing so you help the team reach its overall objective. Compromise should as much as possible be the outcome of open and friendly communication.

PRACTICAL TEAM WRITING

Besides maintaining the attitudes just described, there are three practical ways you can employ teamwork to produce a written document. Some work for some groups, others work for other groups. We rarely work in ideal circumstances, and you may have to be flexible when working with others on a writing task. The three methods, from the least preferable (but the most commonly used) to the most effective, are as follows:

1. Divide the length of the assignment by the number of people involved and get each to write his or her share. Individuals will do any research needed for their own section and should write and edit it. Then the document can be "glued together."

Unfortunately, this method may not result in a very efficient or effective product. Individuals bring their personal writing style, vocabulary, quirks, and weaknesses to their part, and their material may overlap with other parts of the report or fail to provide important transitions between sections. You will still need a strong writer as "overseer" and final editor who can take the completed draft and mold it into a coherent and useful document.

2. Have one person organize the material, write the entire draft, edit it, and pass the finished product on to the next member of the team. This person will add, delete, rearrange, and re-edit as he or she sees fit. The third member of the team will do the same, and so on down the line. The assumption here is that when all team members have had their say, the document will be as complete and close to perfect as can be.

With a closely knit and cooperative team this method *might* result in an effective report rather than a total mess, but you will still need a strong document manager/editor to monitor each step in the process. You might even find this system bringing friendships to an abrupt end. Moreover, if team members want to see what others have done to *their* version of the draft, and are inclined to debate and dispute each amendment, you could be a long way past deadline before everything is set right and everyone is satisfied.

3. By far the best way to produce a team document is to assign each member to different tasks according to that member's strengths and interests:

   a. Designate one person as project manager to organize and assign tasks, check that the project is on schedule, and even referee disputes if necessary.

b. Have another team member get together the needed information for the document, write notes, and put together a very rudimentary draft.

c. Get the next member, the designated "strong" writer, to generate a working draft of the paper. Ideally, this person is good at writing, enjoys writing—and has read this book.

d. If possible, get yet another team member with editing skills to act as quality control officer, reading, checking, editing, and in general perfecting the document while working closely with the previous writer.

Using this method, everyone on the team can bring particular strengths to the task and play a significant part in producing the document. Each person has direct access to the document manager, knows what the others' responsibilities are, and has the satisfaction of being uniquely involved in the job. This is the ideal situation. However, even with this method you may have to compromise sometimes, double up on tasks, or mix this method with elements of the first two described. Whatever the situation, though, carefully planning and assigning collaborative writing tasks to team members *before* the writing project begins will result in a more efficiently produced document that is both coherent and useful.

EXERCISES

1. Think of some significant communication events you have experienced in the past several months at work or in class. What kinds of audiences were involved? Did a lack of clearly defined audience and purpose cause noise in the communication process? How would a more complete analysis of the audiences have enabled technical information to be transferred more efficiently?

2. Look inside the back cover of an IEEE or other technical journal where you will find a page of advice for authors who wish to publish in that journal. To what extent does the information provide specifications for the articles to be published? Are specifications given for such details as abstracts, length, headings, margins, columns, graphics, size of print, references, and so on? If you still have questions about how a paper for that journal should be written and formatted, how would you get in touch with the editor?

3. Find a government or industry report on a subject that interests you. Who is the assumed audience? Does the report get to the point right away or does it keep you guessing until the end? How useful are the headings and subheadings? Is it easy to outline the plan of organization the author has used? How do divisions and paragraph length add to the accessibility of the information? Could any of the information be better presented in list form? Select three or four random paragraphs and closely analyze them for ambiguity, wordiness, unnecessary technical jargon, and nouns that could be turned into verbs. Then rewrite those passages.

4. Keep a log of the time you spend writing a document. How long did it take you? Were you working under a deadline? How much time each day was spent planning, writing, and editing? Did others have a part in writing the document, and if so, how were tasks or sections delegated? Were you satisfied with the completed document? Was whoever

assigned you the task satisfied with your work? What factors would have enabled you to do an even better job?

5. Take some examples of your own recent writing and analyze them in light of each of the guidelines in this chapter.

BIBLIOGRAPHY

Alred, Gerald J., Charles T. Brusaw, and Walter E. Oliu. *The Technical Writer's Companion.* New York: Bedford/St. Martin's, 2002.

Larson, Kevin. "The Technology of Text," *IEEE Spectrum,* May 2007, pp. 26–31.

Microsoft Manual of Style for Technical Publications, 3rd ed. New York: Microsoft Corporation, 2004.

Nadziejka, David E. "The Levels of Editing Are Upside Down." *Proceedings of the International Professional Communication Conference,* pp. 89–93, September 28-October 1, 1984.

Reep, Diana C. *Technical Writing: Principles, Strategies, and Readings,* 7th ed. New York: Pearson Longman, 2009.